Skeletal Muscle Structure and Function

Implications for Rehabilitation and Sports Medicine

Skeletal Muscle Structure and Function

Implications for Rehabilitation and Sports Medicine

RICHARD L. LIEBER, Ph.D.

Associate Professor

University of California

San Diego, California

Editor: John P. Butler
Associate Editor: Linda Napora
Copy Editor: Rebecca Marnhout
Designer: Karen S. Klinedinst
Illustration Planner: Wayne Hubbel
Production Coordinator: Charles E. Zeller

428 East Preston Street
Baltimore, Maryland 21202, USA

Printed in the United States of America

Library of Congress Cataloging-in-Publication Data

Lieber, Richard L.
Skeletal muscle structure and function / Richard L. Lieber.
p. cm.
Includes index.
ISBN 0-683-05026-5
1. Striated muscle. I. Title.
[DNLM: 1. Muscles—anatomy & histology. 2. Muscles—physiology. WE 500 L716s]
QP321.L62 1992
612.7′4—dc20
DNLM/DLC
for Library of Congress 91-30387
CIP

4 5 6 7 8 9 10

Preface

Understanding such universal phenomena as movement and strength requires an understanding of the neuromuscular system. Since muscle represents about three-fourths of the body mass, a healthy muscular system is usually associated with healthy cardiovascular, pulmonary, and endocrine systems. Conversely, disorders of the neuromuscular system have dramatic effects on daily activities and independence. Unfortunately, there is currently no text available that presents muscle basic science and plasticity in a simple, unified manner. Thus the purpose of this book is to provide useful information for understanding strength and movement by describing the structure, function, and plasticity of the neuromuscular system.

This book is conceptually divided into two parts: Part 1 (Chapters 1–3) represents foundations of basic science; Part 2 (Chapters 4–6) represents illustrations of the adaptive capacity of muscle.

- Chapter 1 begins with an integrated presentation of muscle development and anatomy. The significance of these anatomic features is discussed in later book chapters.
- Chapter 2 (the longest chapter) describes both the mechanical and physiologic properties of skeletal muscle. The elegance of structure-function relationships in the neuromuscular system is emphasized. Again, the significance of these properties is foreshadowed.
- Chapter 3 describes the way in which muscles, tendons, and joints interact to produce movement, using Chapters 1 and 2 as a conceptual base. The key to this chapter is that the whole muscle-tendon-joint system is not merely the sum of its parts.
- Chapter 4 presents various ways of increasing muscle use along with their effects: chronic electric stimulation, passive stretch, voluntary exercise, surgical transfer, *etc.* By looking at the detailed muscle response to these interventions, we will understand the response of muscle to almost any type of increased use and can strategize regarding optimal methods for strengthening muscle.
- Chapter 5 continues our discussion of plasticity for decreased-use models: immobilization, spinal cord injury, denervation, and weightlessness. Again,

we will learn the way in which a muscle adapts when the level of use is decreased. We will also learn why some muscles are not vulnerable to the atrophy experienced by others.

- Chapter 6 ends the book with a description of muscle's cellular and physiologic response to injury as well as some exciting experimental treatments of muscle disease. Since injury can arise in the form of trauma, intense exercise, laceration, or injection of local anesthetic, a rationale for avoiding such injuries is also presented.

Based on this broad presentation, rehabilitation professionals can focus their attention on the scientific basis for muscle treatment. Throughout the text, intervention strategies and traditional thinking are critically evaluated in light of the data presented. Many of the world's most outstanding experiments are presented and interpreted in a way that provides unifying principles to the understanding of muscle structure, function, and plasticity.

Acknowledgments

Special thanks to my colleagues who read various chapters and provided helpful criticism: Drs. Sue Bodine, Frank Booth, Bruce Carlson, V. Reggie Edgerton, Brenda Russell Eisenberg, Bob Gregor, Roland Roy, and Tricia Silva. Of course, no scientist is an island, and thus I acknowledge the support and leadership of my scientific collaborators: Reid Abrams, Bill Adams, David Amiel, Wayne Akeson, Ron Baskin, Mike Botte, Sue Bodine, Dale Daniel, Larry Danzig, V. Reggie Edgerton, David Gershuni, Bob Gregor, Alan Hargens, Jan Fridén, Don McNeal, Roland Roy, and Yin Yeh. In the same way, I have benefited from the interaction with current and former laboratory students and staff: Marcia Beckman, Field Blevins, Jenny Boakes, Cindy Brown, Clark Campbell, Brendan Donoghue, Babak Fazeli, Tom Ferro, Mark Jacobson, Jeanne Kelly, Ann Merrill, Michelle Lau, Margot Leonard, Michael Mai, Dev Mishra, Lori Pfeifer, Rajnik Raab, Tony Sanzone, Mary Schmitz, Scott Shoemaker, Dean Smith, Christy Trestik, Thalia Woodburn, Nancy Wudek, and Abbe Zaro. I am indebted to medical artist Rebecca Chamberlain, who enthusiastically and cheerfully created most of the beautiful illustrations contained herein, learning more about muscle than I'm sure she ever wanted to know! Cindy Brown masterfully replotted graphs from the original literature data, and Tiffaney Whiteside provided skillful typing and proofreading.

Thanks also to the editorial staff at Williams & Wilkins—John Butler and Linda Napora provided needed perspective and guidance at just the right time.

Finally, a most grateful and heartfelt appreciation to my family, who literally endured the long hours of reading and writing. Katie and Kristi, I feel blessed to be your Dad, to know you and to enjoy your love. Debbie, your love, encouragement, and humor are food for my soul. "Unless the Lord build the house, they labor in vain who build it" (Psalm 127:1).

Contents

Chapter

1

Skeletal Muscle Anatomy

OVERVIEW

This chapter about skeletal muscle development and structure forms the foundation for much of the information to be presented later. This initial glimpse of the amazing world of skeletal muscle presents a sample of the incredible complexity and synchrony that occur within a single cell. We will journey through time (as we watch the cell develop) and space (as we explore the components of the mature mammalian muscle cell). This will enable us to understand the way in which these cells form the building blocks of all muscles within the body.

INTRODUCTION

Skeletal muscle represents the classic example of a structure-function relationship. At both the macro- and microscopic levels, skeletal muscle is exquisitely tailored for force generation and movement. Because of this structure-function relationship, studies of muscle function are intimately tied to studies of muscle structure. We begin with a discussion of the skeletal muscle cell. A description of the process of skeletal muscle development will lead to a discussion of muscle cell structure. Our structural discussion begins at the microscopic level and builds to the macroscopic level. We then introduce the way in which the various structures contribute to muscle function. Muscle research is exciting and alive and full of unanswered questions. Enjoy your adventure in the study of one of the most well-understood biologic structures.

MUSCLE DEVELOPMENT

It is difficult to appreciate the beautifully orchestrated developmental process apart from an understanding of mature muscle structure and function. The developmental sequence may make more sense after you understand muscle cell structure presented later in the chapter. We will see the intricate and specific arrangement of the neuromuscular system: Motor nerves arise from a variety of spinal cord levels, combine into a common fascicle along with nerves from other levels, and proceed peripherally to innervate specific muscles—even specific fiber types within specific muscles! How do muscles and nerves achieve this intricate arrangement? What are some of the factors that guide the process? Are nerves told exactly where to meet their muscle, or do they search into the periphery and then recognize specific muscle fibers? Do muscles attract nerves by sending specific signals? Nerve-muscle recognition is a rapidly changing and growing field, so that definitive answers to these questions must await further discovery. However, a number of exciting discoveries made over the last 2 decades provided meaningful insights into the developmental process (Kelly, 1983).

Overview of Muscle Development

We divide the process of muscle development into four phases (Figure 1.1). Many of these phases occur simultaneously, coordinated in time. The four processes are (*a*) axonal outgrowth (the process by which axons traverse into the periphery to contract muscles), (*b*) myogenesis (muscle cell formation), (*c*) synaptogenesis (formation of neuromuscular junctions between the motor nerve and muscle fiber), and (*d*) synapse elimination (the process that eliminates "extra" neuromuscular connections). Studies of each step in the developmental process have used various experimental models. These include *in vitro* cell culture preparations (to study muscle cell differentiation) and denervation and reinnervation of mature muscle (to study nerve-muscle interaction). Taken together, these studies provide a framework on which to build our understanding of development. Let us look in more detail at these processes.

Axonal Outgrowth—Nerve Meets Muscle

It is important to state at the onset that connections between muscles and nerves are very specific. Motor nerves that exit the ventral root of the spinal cord are long processes extending from motoneuron cell bodies located along the lateral motor column of the ventral horn (Figure 1.2). There is a rough (but not perfect) correspondence between the location of the various motoneuron pools along the spinal cord and the muscles that they innervate (Table 1.1). This correspondence is extremely important in clinical diagnosis of the location of trauma or disease based on clinical observation of the specific muscles that are affected. For example, note in Table 1.1 that if weakness or paresis of the brachialis muscles is observed, the nerve root exiting at the C6 level is implicated.

The most detailed information regarding motor axon outgrowth is available from studies of the chick hindlimb. Upon inspection of the complex and numerous muscle-nerve connections, one might ask, "How do all of these

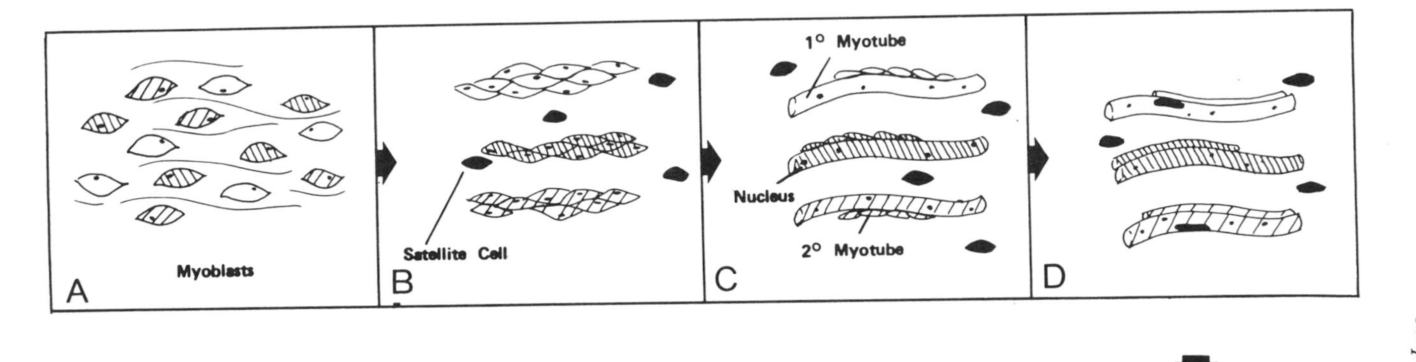

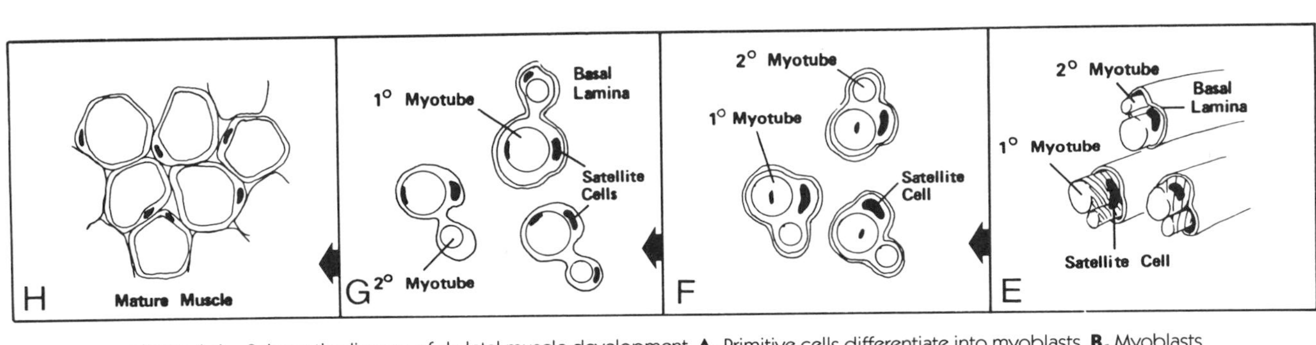

Figure 1.1 Schematic diagram of skeletal muscle development. **A,** Primitive cells differentiate into myoblasts, **B,** Myoblasts fuse together to form primary myotubes. **C,D,E,F,** Later, secondary myotubes arise beneath the basal lamina of the primary myotubes. Fusion of myoblasts radially and longitudinally results in formation of the muscle fiber beneath basal lamina (shown in cross-section in **E** through **H**). In addition, some unfused myoblasts remain as satellite cells, which are maintained in the mature cell. **G,** As the muscle matures, primary and secondary myotubes separate, each with myonuclei and satellite cells, to become a mature fiber. **H,** Finally, as the muscle fibers grow, they become arranged as tightly packed polygonal cells, characteristic of adult muscle.

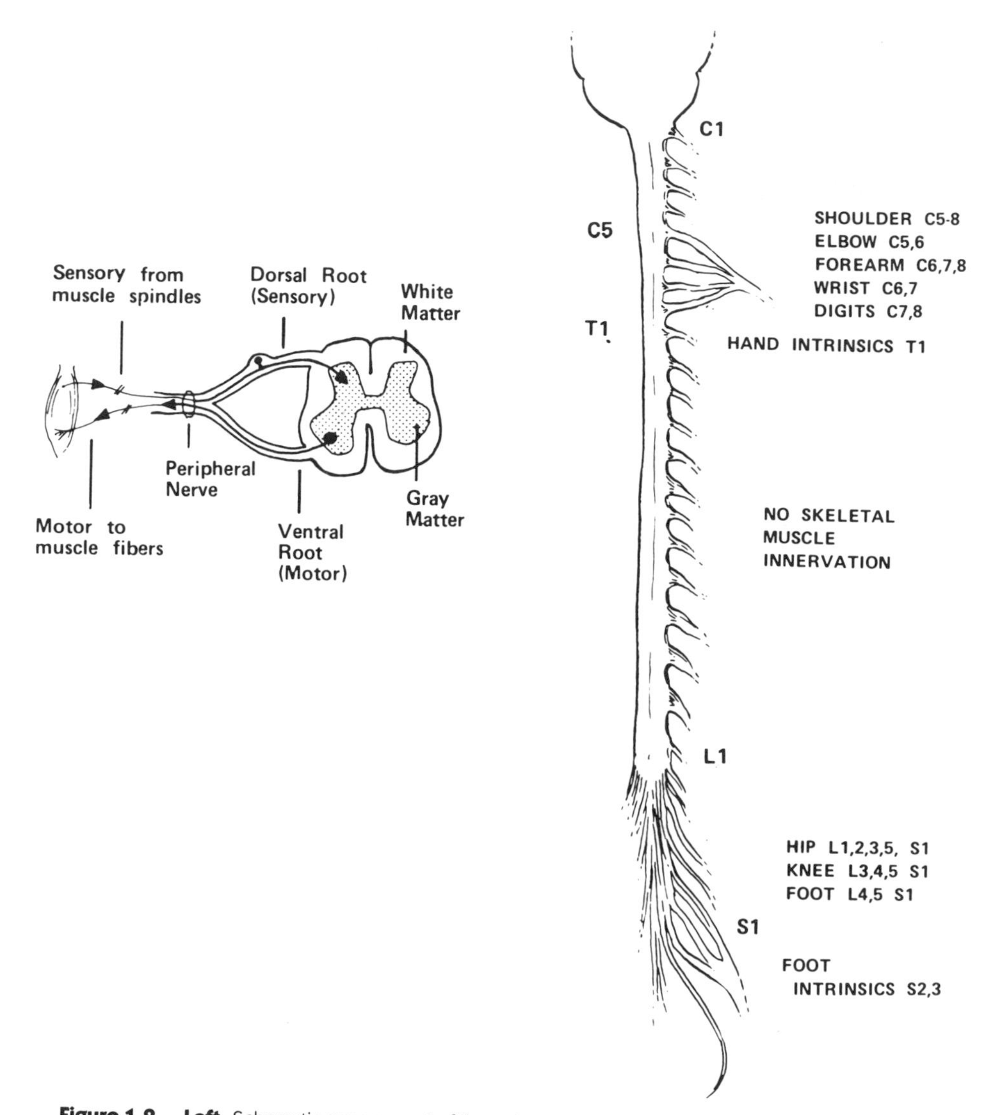

Figure 1.2. **Left,** Schematic arrangement of the spinal cord. Motoneurons have their cell bodies in the grey matter of the spinal cord and exit through the ventral root, via the peripheral nerve, to innervate skeletal muscles. Sensory nerves from muscle spindles and Golgi tendon organs have their cell bodies in the dorsal root. Sensory fibers from the muscle traverse the peripheral nerve to synapse in the spinal cord. **Right,** Cranial-caudal organization of the spinal nerve roots. Spinal nerve roots descend from the brain via specific tracts and exit the spinal cord at various locations to innervate muscle in a topographical manner (see Table 1.1). Note that nerve root level roughly corresponds with anatomic level. This is useful information clinically in that a particular muscle weakness can be associated with nerve root dysfunction at a particular spinal segment level.

Table 1.1.
Anatomy of Human Limb Muscles

Muscle	Origin	Insertion	Innervation	Nerve Root	No. Joints Crossed
Extensor hallucis longus	Medial fibular surface	Dorsal surface of distal phalanx	Deep peroneal	L5, S1	>3
Extensor digitorum longus	Anterior fibular surface	Dorsal surface of toes 2–5	Deep peroneal	L5, S1	>3
Flexor hallucis longus	Posterior fibular surface	Plantar surface of distal phalanx	Tibial	S1, S2	>3
Flexor digitorum longus	Posterior tibial surface	Plantar surface of distal phalanx of toes	Tibial	S1, S2, S3	>3
Tibialis anterior	Lateral tibial surface	Medial side of medial cuneiform	Tibial	L4, L5	2
Tibialis posterior	Posterior surfaces of tibia, fibula, and interosseus membrane	Medial navicular surface	Tibial	L4, L5	2
Peroneus tertius	Lower fibular anterior surface	Dorsal surface of base of 5th metatarsal	Peroneal	L5, S1	2
Peroneus brevis	Lateral fibular surface	Dorsal surface of base of 5th metatarsal	Peroneal	L5, S1	2
Peroneus longus	Lateral fibular surface	Plantar surface of medial cuneiform	Superficial peroneal	L5, S1, S2	2
Soleus	Posterior proximal surface of fibula	Posterior surface of calcaneus	Tibial	S1, S2	1
Gastrocnemius	Posterior/superior surfaces of lateral and medial femoral condyles	Posterior surface of calcaneus	Tibial	L5, S1	3
Plantaris	Lateral supracondylar line of femur	Posterior surface of calcaneus	Tibial	S1, S2	3
Popliteus	Lateral femoral epicondyle	Soleal line	Tibial	L5, S1, S2	1
Vastus lateralis	Lateral lip of linea aspera	Lateral patellar border	Femoral	L3, L4	1

Table 1.1.
Anatomy of Human Limb Muscles *(Continued)*

Muscle	Origin	Insertion	Innervation	Nerve Root	No. Joints Crossed
Vastus medialis	Medial lip of linea aspera	Medial patellar border	Femoral	L3, L4	1
Vastus intermedius	Anterior femoral surface	Superior patellar border	Femoral	L3, L4	1
Rectus femoris	Anterior inferior iliac spine	Superior patellar border	Femoral	L3, L4	2
Tensor fasciae lateae	Lateral surface of anterior superior iliac spine	Lateral tibial condyle	Superior gluteal	L4, L5, S1	2
Gracilis	Body of pubis	Medial tibial condyle	Obturator	L2, L3	2
Sartorius	Anterior superior iliac spine	Medial tibial condyle	Femoral	L1, L2, L3	2
Semimembranosus	Ischial tuberosity	Medial tibial condyle	Sciatic	L4, L5, S1	2
Semitendinosus	Ischial tuberosity	Medial tibial condyle	Sciatic	L4, L5, S1, S2	2
Biceps femoris	*Long head*—ischial tuberosity *Short head*—linea aspera	Fibular head	Sciatic	L5, S1, S2	2
Piriformis Obturator internus	Anterior surface of sacrum	Greater trochanter		L1, L2	1
Obturator externus	Medial pelvic surface	Greater trochanter		L1, L2	1
Gemellus superior	Dorsal surface of ischial spine	Greater trochanter		L1, L2	1
Gemellus inferior	Ischial tuberosity	Greater trochanter		L1, L2	1
Quadratus femoris	Ischial tuberosity	Greater trochanter		L1, L2	1
Gluteus maximus	Posterior gluteal line, dorsal sacral surface	Gluteal tuberosity of femur and iliotibial tract	Inferior gluteal	L5, S1, S2	2

Table 1.1.
Anatomy of Human Limb Muscles *(Continued)*

Muscle	Origin	Insertion	Innervation	Nerve Root	No. Joints Crossed
Gluteus medius	Between posterior and anterior gluteal lines of ilium	Lateral surface of greater trochanter	Superior gluteal	L4, L5, S1	2
Gluteus minimus	Between anterior and inferior gluteal lines of ilium	Anterior surface of greater trochanter	Superior gluteal	L4, L5, S1	1
Psoas major	Transverse processes of vertebral bodies	Lesser femoral trochanter	Lumbar	L1, L2, L3	3
Iliacus	Iliac fossa	Lesser femoral trochanter	Femoral	L1, L2	1
Adductor longus	Pubic body	Medial lip of linea aspera	Obturator	L2, L3, L4	1
Adductor brevis	Inferior ramus of pubis	Upper half of linea aspera	Obturator	L2, L3, L4	1
Adductor magnus	Ischial tuberosity and ischial and pubic ramus	Entire linea aspera and adductor tubercle of femur	Obturator sciatic	L2, L3, L4	1
Pectineus	Superior ramus of pubis	Beginning of linea aspera	Femoral	L2, L3	1
Deltoid	*Anterior part*—lateral half of clavicle *Middle part*—inferior border of scapular spine	Deltoid tuberosity of humerus	Axillary	C5	1
Pectoralis major	*Clavicular part*—medial half of clavicle *Sternal part*—anterior surface of sternum and costal cartilage of ribs 1–6	Lateral lip of bicipital groove	Thoracic	C8	2
Coracobrachialis	Tip of coracoid process	Medial surface of humerus	Musculocutaneous	C8	1

Table 1.1.
Anatomy of Human Limb Muscles *(Continued)*

Muscle	Origin	Insertion	Innervation	Nerve Root	No. Joints Crossed
Latissimus dorsi	Lower 6 thoracic spines and posterior layer of lumbar fascia	Floor of bicipital groove	Thoracodorsal	C7	2
Teres major	Dorsal surface of the inferior angle of the scapula	Medial lip of the bicipital groove	Lower subscapular	C6	1
Supraspinatus	Supraspinous fossa	Greater tubercle of humerus	Suprascapular	C5	1
Infraspinatus	Infraspinous fossa	Greater tubercle of humerus	Suprascapular	C5	1
Teres minor	Lateral scapular border	Posterior aspect of greater tubercle	Axillary	C5	1
Subscapularis	Subscapular fossa	Lesser tubercle of humerus	Subscapular	C5	1
Biceps brachii	*Short head*—tip of coracoid process *Long head*—supra glenoid tubercle	Radial tuberosity	Musculocutaneous	C5	3
Triceps brachii	*Long head*—infraglenoid tubercle *Lateral head*—upper posterior surface of humerus *Medial head*—posterior surface of lower humerus	Common tendon: superior aspect of olecranon process	Radial	C7	1
Brachialis	Anterior surface of humerus	Coronoid process of ulna and ulnar tuberosity	Musculocutaneous	C6	1

Table 1.1.
Anatomy of Human Limb Muscles *(Continued)*

Muscle	Origin	Insertion	Innervation	Nerve Root	No. Joints Crossed
Brachioradialis	Lateral supracondylar ridge of humerus	Lateral surface of styloid process	Radial	C6	2
Anconeus	Lateral humeral epicondyle	Olecranon process and posterior surface of ulna	Radial	C6	1
Supinator	*Humeral head*—lateral epicondyle *Ulnar head*—supinator crest	Lateral surface of the radius	Radial	C8	1
Pronator teres	Medial epicondyle	Lateral surface of midshaft of radius	Median	C7	1
Pronator quadratus	Anterior surface of distal ulna	Anterior surface of radius	Median	C8	1
Flexor carpi radialis	Medial humeral epicondyle	Base of 2nd and 3rd metacarpal	Median	C7	1
Flexor carpi ulnaris	Medial epicondyle	Pisiform, hamate, and base of 5th metatarsal	Ulnar	C8	1
Palmaris longus	Medial humeral epicondyle	Palmar fascia	Median	C7	1
Extensor carpi radialis longus	Supracondylar humeral ridge	Base of 2nd metacarpal	Radial	C7	1
Extensor carpi radialis brevis	Lateral humeral epicondyle	Posterior surface of base of 3rd metacarpal	Radial	C7	1
Extensor carpi ulnaris	*Humeral head*—lateral epicondyle	Posterior surface of base of 5th metacarpal	Radial	C8	1

Table 1.1.
Anatomy of Human Limb Muscles *(Continued)*

Muscle	Origin	Insertion	Innervation	Nerve Root	No. Joints Crossed
	Ulnar head—posterior surface of ulna				
Extensor digitorum communis	Lateral humeral epicondyle	*Central*—posterior aspect of base of middle phalanx 2 *Collaterals*—posterior aspect of distal phalanx	Radial	C7	> 3
Extensor indicis proprius	Lower dorsal surface of ulna	Unites with index finger tendon of ext. dig. comm. on middle phalanx 2	Radial	C8	> 3
Extensor digiti minimi	Lateral humeral epicondyle	Posterior aspect of middle phalanx 5	Radial	C7	> 3
Flexor digitorum superficialis	*Humeral head*—medial epicondyle *Radial head*—anterior shaft of ulna	Sides of the middle phalanx 2–5	Median	C8	> 3
Flexor digitorum profundus	Anterior surface of ulna	Anterior base of distal phalanx	Median and ulnar	C8, T1	> 3
Abductor pollicis longus	Posterior surface of radius, ulna, and interosseus membrane	Lateral surface of base of 1st metacarpal	Radial	C8	2
Extensor pollicis brevis	Posterior surface of radius and interosseus membrane	Posterior surface of base of proximal phalanx of thumb	Radial	C8	3

Table 1.1.
Anatomy of Human Limb Muscles *(Continued)*

Muscle	Origin	Insertion	Innervation	Nerve Root	No. Joints Crossed
Extensor pollicis longus	Posterior surface of ulna and interosseus membrane	Posterior surface base of distal phalanx of thumb	Radial	C8	>3
Flexor pollicis longus	Anterior radial surface	Base of distal phalanx	Median	C8	>3

axons find one another and then find the appropriate muscle? After finding the muscle, how do they find the correct muscle fiber?" Before reviewing what is known, think about the potential answers to these questions. Maybe nerves don't know where they are supposed to go and instead go *everywhere*. Maybe the nerves that do find the appropriate destination (*i.e.*, the appropriate target muscle) survive, and all other nerves degenerate. Maybe the process of neuronal cell death (which is known to occur in CNS) creates a very specific innervation pattern out of one that starts randomly.

Lynn Landmesser and her colleagues (Landmesser, 1980) demonstrated that very early in development, nerves that project into muscle masses are composed almost entirely of the appropriate spinal nerves. This neural outgrowth occurs before the primitive muscle masses contain any muscle fibers. Thus specificity begins at the beginning—when nerves first grow out from the ventral horn. The nerve performs this remarkable bit of navigation virtually without error; all nerves find the correct muscles.

An interesting experiment by Landmesser and colleagues emphasized the interaction between muscle and nerve (Lance-Jones and Landmesser, 1978). They surgically switched the topographic location to two muscle masses before nerve outgrowth to see if outgrowing axons would simply innervate whichever muscle mass was found at the end location, or, instead, would search for the correct muscle mass. In fact, the outgrowing axons detoured and innervated the correct muscles. Thus some cueing exists between nerves and muscles. Current ideas suggest that some of the specificity results from the guidance of axons in response to specific extracellular molecules.

Myogenesis—Birth of the Muscle Fiber

Skeletal muscles derive from the somites of the embryo (somites are the masses of primordial muscle tissue that occur at intervals along the embryo length). As development proceeds within the somites, single-celled mononucleated muscle fiber precursors appear—the myoblasts (Figure 1.1 **A**; you will see the prefix "myo" in many locations throughout this book. *Myo* is the Latin prefix for "muscle.") Myoblasts are a primitive cell type, much like the more generic fibroblast. One of the first steps in myogenesis is the aggregation of myoblasts into clusters (Figure 1.1**B**). Clusters of myoblasts begin to fuse to form small, multinucleated cells known as myotubes, which are about 100–300 μm in length (0.1–0.3 mm). Further length increase (which is necessary since mature fibers may extend several centimeters in length) takes place by fusion of more myoblasts onto the myotube ends. The earliest myotubes formed are known as primary myotubes. With further development, primary myotubes, along with some associated, less differentiated cells, separate from the other primary myotubes into clusters that are surrounded by a sheath known as the basement membrane or basal lamina (Figure 1.1**C**). At this time, more myoblasts begin to aggregate beneath the primary myotube's basal lamina, using the primary myotube as a structural scaffold. These subsequent myoblasts also fuse into myotubes, known as secondary myotubes. Now, the cell cluster is composed most noticeably of the primary myotube, its associated secondary myotube, and unfused myoblasts known as satellite cells. With further development, the primary myotube begins to look more like a mature muscle fiber, with the nuclei being forced from the fiber center to the periphery because of the progressive "filling" of the fiber with contractile proteins (Figure 1.1**D,E**). If a muscle fiber is stained at this point for the contractile protein myosin, the most abundant muscle protein, the fiber appears as a doughnut shape since the contractile proteins first fill the periphery, leaving the central cell core devoid of contractile proteins and, therefore, unstained. Secondary myotubes also begin to mature into fibers at this stage, so that the satellite cell and the primary and secondary myotubes are contained within a single basement membrane (Figure 1.1**F**). At this time, the motor nerve is ready to form a connection with the muscle—a neuromuscular junction.

Synaptogenesis—Birth of the Neuromuscular Junction

The neuromuscular junction (or motor endplate) is a true synapse in the sense that a chemical neurotransmitter (acetylcholine, abbreviated ACh) is used to convey information from the nerve (presynaptically) to the muscle (postsynaptically) across a synaptic cleft. Thus the neuromuscular junction (NMJ) is a specific example of a cholinergic synapse (a synapse where the neurotransmitter is ACh). We will see later in our discussion of muscle contraction (Chapter 2, page 51) that ACh is intimately involved in the contractile process. Early in development, ACh receptors diffusely cover the entire muscle fiber (Figure 1.3). These receptors are specific proteins that integrate themselves into the muscle membrane (sarcolemma). As the nerve contacts the muscle fiber, ACh receptors begin to cluster around the site of nerve contact and become "trapped" by the nerve (Figure 1.3). The mechanism of this trapping process is not known, but two interesting observations have been made that may provide insights into the trapping mechanism. First, Mu-Ming Poo demonstrated that ACh receptors migrate in a specific way when exposed to external electrical fields (Poo, 1982). He suggested that the nerve excitation process itself, which generates small electrical fields at the endplate (the so-called endplate potential) may "attract" and trap ACh receptors in a positive feedback fashion: As ACh receptors are trapped, larger endplate potentials are generated, trapping more receptors, causing larger endplate potentials, *etc.* Another interesting observation was that, following nerve-muscle contact, the nuclei *beneath the ACh receptor* (recall that muscle cells are multinucleated and, therefore, have literally hundreds of nuclei along their length) began to direct the synthesis of ACh

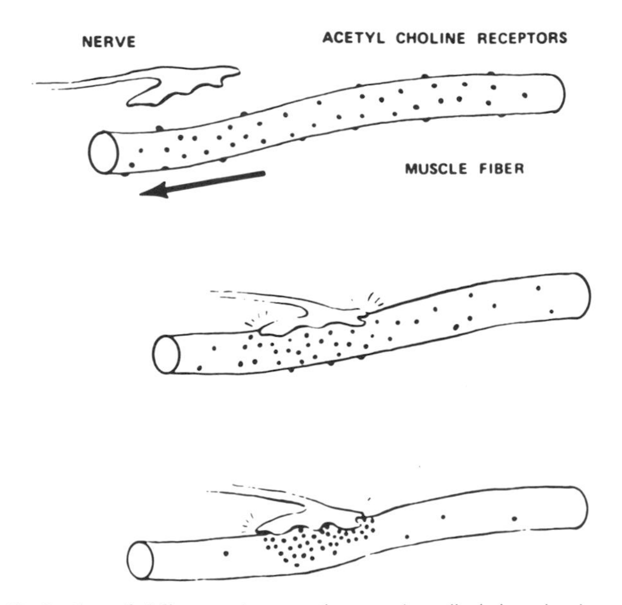

Figure 1.3. Distribution of ACh receptors on the muscle cell during development. Prior to innervation ACh receptors are distributed across the muscle fiber surface (*top*). As the nerve contacts the muscle fiber and electric activity ensues (denoted by lines around the neuromuscular junction), receptors aggregate in the region of the neuromuscular junction (*middle*). As maturation progresses, extrajunctional receptors decrease significantly while junctional receptors increase in number and density (*bottom*).

receptor mRNA and the ACh receptor! The overall effect of nerve contact with the fiber, therefore, is to decrease the number of extrajunctional ACh receptors and to increase the number of junctional ACh receptors (Figure 1.3), perhaps by these two mechanisms.

In the adult neuromuscular system, denervation results in a reversal of this process—an increase in extrajunctional ACh receptors. This is one reason that the mature muscle denervation-reinnervation model has been used to study some aspects of development. It is also extremely interesting that the increase in extrajunctional ACh receptors that occurs following denervation can be prevented by direct muscle fiber electrical stimulation! Does this mean that electrical stimulation of denervated muscles may prevent or inhibit reinnervation? This possibility is considered in Chapter 5 (page 244).

After the NMJ forms, the muscle fiber loses its receptivity to formation of a NMJ at other sites. Thus as long as a muscle is normally innervated, it is not receptive to innervation by other axons. The activity of the muscle fiber itself seems to be a crucial factor in this lack of receptivity because if a normally innervated muscle is chemically paralyzed, an ectopic synapse (synapses at locations other than the NMJ) is easily formed (Bennett, 1983).

After initial NMJ formation, several other motoneurons form synapses at this point. The actual number of motoneurons per synapse ranges from two to six, depending on the muscle, and begins to decline just after birth. It declines so that, at maturity, all mammalian fibers have only a single motoneuron per fiber. This decrease is very specific and is accomplished by the process of synapse elimination (Jansen and Fladby, 1990).

Synapse Elimination—The Finishing Touches

The reason muscle fibers are initially hyperinnervated is not clear. Early thinking suggested that hyperinnervation followed by synapse elimination served as a mechanism to correct for hardwiring errors that occurred during haphazard axonal outgrowth. However, we saw that axonal outgrowth is a well-defined process in which few mistakes are made. (It was known that synapse elimination does occur elsewhere in the CNS to correct for "wiring errors," and this was thus a plausible hypothesis.) Thus the precise teleological explanation for synapse elimination is still not clear. What are the factors that determine which motoneuron will win "the battle of the synapse"? Clearly, competition between the hyperinnervating motoneurons exists, for when competing motoneurons to a particular muscle are surgically removed, the motoneurons remaining innervate a larger number of fibers than usual. However, these surviving motoneurons do not simply innervate all available fibers. There is thus some other factor (perhaps the metabolic capacity of the cell body) that limits the final innervation ratio (number of muscle fibers per axon).

In addition to the internal "programmed" changes that are scheduled to occur during development, it is clear that environmental cues also play a role. One important environmental cue is the tension experienced by the fusing and developing muscle. Herman Vandenburg developed a method for mechanically stretching cultured muscle cells *in vitro* to observe the effects of stretch timing and magnitude on the developmental process (Vandenburg, 1982). He found that cells subjected to stretch increased their rates of protein synthesis, causing an increase in cell size, and changed their orientation to become parallel to the stretch axis (Chapter 6, page 268). While it was not clear exactly how this mechanical event was transduced by the cell, it was clear that the mechanical events strongly influenced the cellular machinery and its developmental fate. This sets up another positive feedback system in that movement enhances differentiation which increases movement which increases differentiation, *etc.*

DEVELOPMENT OF SPECIFIC MUSCLE FIBER TYPES

Most mammalian muscles contain a mixture of fiber types and motor units (see Chapter 2). Generally, whole muscles contain a heterogeneous mixture of fast-contracting and slow-contracting fibers, as well as fibers with high endurance and low endurance. The processes of myogenesis, axonal outgrowth, and synapse elimination thus usually produce a muscle of mixed fiber type. At what stage is a fiber type determined? What is the influence, if any, of the nerve in determining fiber type? Buller and colleagues clearly demonstrated the influence of motor nerves on mature muscle fiber types by surgically reattaching a motor nerve from a fast-contracting muscle to a slow-contracting muscle and *vice versa.* They found that, following cross-reinnervation, the fast-contracting muscle became slow and the slow-contracting muscle became fast (Buller *et al.,* 1960a and 1960b). This experiment formed the basis for the idea that nerves strongly influence the fate of muscle fibers. However, using molecular biology techniques, Frank Stockdale demonstrated that even at the myoblast stage, myoblasts exist as different types (Miller and Stockdale, 1986 and 1987). Using histochemical and immunohistochemical fiber typing techniques in avian muscle, Stockdale demonstrated that even primary and secondary myotubes existed as "fast" and "slow" types. He showed that the myoblasts exist in three distinct clonal types. By isolating a single type of myoblast, allowing it to proliferate, and then causing it to form myotubes, he confirmed this idea in that the resulting myotubes were all of a single type. Innervation was clearly not required in this avian preparation for the three embryonic myoblast populations to arise, proliferate, and form distinct myotube populations, although further differentiation did require innervation. This suggested that during development, the arriving nerve encounters a population of muscle fibers that are already heterogeneous.

Since we know that in mature mammals all fibers within a motor unit are of a particular type (Chapter 2, page 93), axons must either initially innervate the correct fibers, innervate the wrong muscle fiber type and then change the fiber type, or innervate randomly and use synapse elimination to correct the faulty connections. Again, it appears that the initial innervation is fairly specific.

Wes Thompson experimentally tested this question (Thompson *et al.*, 1984). Fiber type composition of developing rat soleus motor units was investigated at two different postnatal ages—8 and 16 days. At 8 days postnatally, every fiber of the rat soleus is innervated by at least two motoneurons. Over the next 8 days, synapse elimination occurs and results in single innervation of each fiber. If the initial fiber innervation had occurred at randomly, Thompson expected that muscle fibers belonging to a single motor unit aged 8 days would be nearly 50% fast and 50% slow fibers, since an equal number of fast and slow fibers were present at this time. Then, following synapse elimination, motor units might become predominantly fast or slow as the "wrong" connections were deleted. However, this was not the case. Thompson *et al.* found that the fiber type composition of motor units at both ages was largely of one type. In fact, although the innervation ratio fell from about 250–100 during the period of synapse elimination, the preponderance of a particular fiber type in a given motor unit did not change. This implied selective initial innervation of muscle fibers by specific axons. The precise mechanism for such a process is not yet clear, and many details of neuromuscular specificity await the results of future studies.

THE SKELETAL MUSCLE CELL

Skeletal muscle fibers are cells that, in many ways, are like any other bodily cell. However, because muscle cell function is highly specialized to produce force and movement, the cellular components are also highly specialized (Figure 1.4). Muscle cells (fibers) are cylindrical, with a diameter ranging from about 10 μm to about 100 μm—less than the diameter of a human hair! As we shall see, muscle fiber diameter is of profound importance for at least two reasons: First, a muscle fiber's diameter determines its strength, and second, when altered fiber diameters are observed in mature muscle, this suggests that the level of muscle use has changed. Muscle fiber length is highly

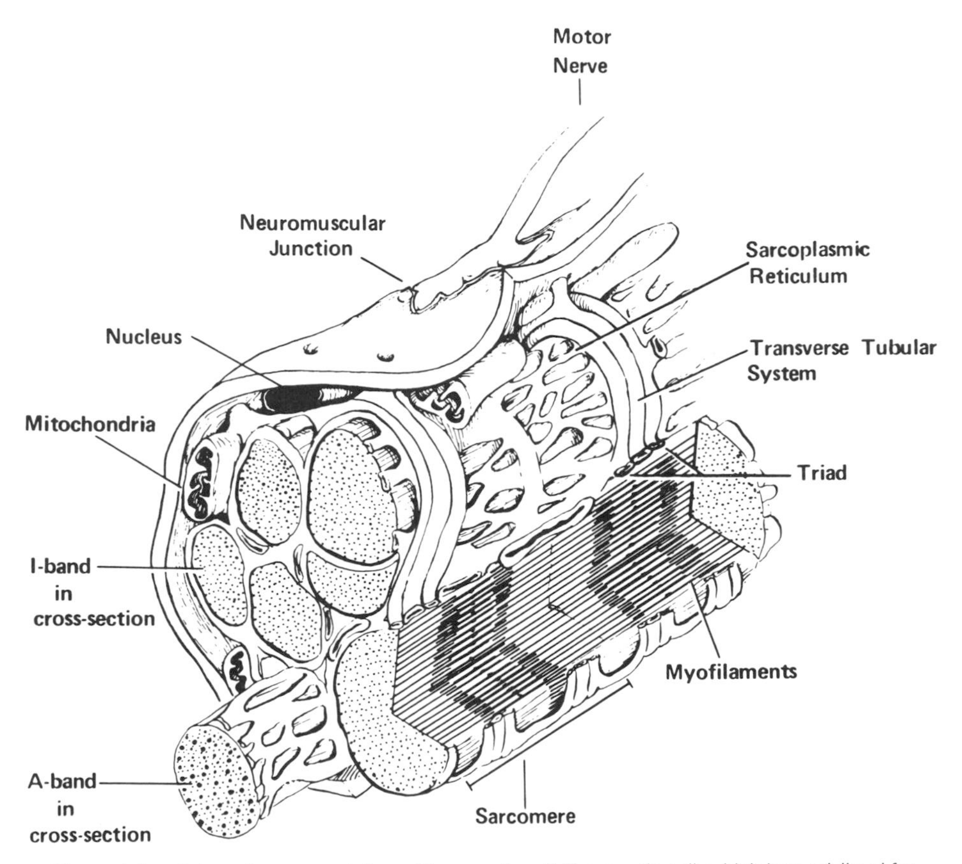

Figure 1.4. Schematic representation of the muscle cell. The muscle cell, which is specialized for the production of force and movement, contains an array of filamentous proteins as well as other subcellular organelles such as mitochrondria, nuclei, satellite cells, sarcoplasmic reticulum, and transverse tubular system. Note the formation of "triads," which represent the T-tubules flanked by the terminal cisternae of the sarcoplasmic reticulum. Also note that when the myofilaments are sectioned longitudinally, the stereotypic straited appearance is seen. When myofilaments are sectioned transversely at the level of the A- or I-bands, the hexagonal array of the appropriate filaments is seen.

variable, depending on the muscle architecture (see below). As we shall see, fiber length has a profound influence on fiber contraction velocity and the distance over which the fiber can shorten (*i.e.*, the excursion).

The scaffolding that surrounds the muscle cell is known as the basal lamina. The normal physiologic role of the basal lamina is poorly understood. However, it is clear the basal lamina plays an important role in muscle fiber recovery from injury. In injuries where the basal lamina remains intact (*i.e.,* crush injury), recovery is relatively complete. Conversely, in cases where the basal lamina is destroyed (*i.e.,* traumatic cut), fiber regeneration first requires the laying down of a new scaffold. In fact, if a muscle fiber is destroyed and denervated, but the basal lamina remains intact, when a motor nerve reinnervates the fiber, it will do so at the original site dictated by the basal lamina, even though no muscle fiber is present! Thus the basal lamina, which demarcated functional myotubes during development, retains a good deal of identity later in cellular life (see discussion in Chapter 6).

A meshlike sheath of collagenous tissue, called the endomysium, surrounds the muscle fiber. The endomysium may play some role in the passive mechanical properties of the fiber, but presently this has not been clearly determined. Bundles of fibers, each surrounded by endomysial tissue, are organized into muscle fascicles, each surrounded by a more stout perimysial tissue. Finally, bundles of fascicles are organized into muscles, surrounded by epimysial connective tissue. Recent studies suggest that muscle fibers are intimately associated with this complex connective tissue matrix and that muscle fibers themselves do not simply extend from one tendon to the other (Loeb *et al.,* 1987; Ounjian *et al.,* 1991). Thus this complex matrix may play a *central* role in muscle fiber to tendon tension transmission and not merely a supportive role.

Cellular Organelles—Discrete Components with Specific Jobs

NUCLEI—COMMAND CENTRAL

As with any other eukaryotic cell, muscle cells contain collections of organelles that are responsible for meeting the day-to-day needs of the cell. However, unlike other eukaryotic cells, muscle cells are multinucleated. Nuclei are located at intervals along the cell length. The scheduling and type of cellular material that must be distributed throughout the cell is controlled by the cell's nucleus. As a result of the many nuclei present within the cell, there is probably some form of internuclear communication to ensure that cellular properties are compatible along the cell length. Interestingly, as a muscle cell adapts (for example, to electrical stimulation), changes that occur along the cell length occur at slightly different rates. Each nucleus must, therefore, retain some degree of autonomy. For example, we already noted that, during development, nuclei in the region of the NMJ specified the synthesis of ACh receptor mRNA while adjacent nuclei did not. We thus have a picture of a community of subcellular structures in communication with one another to meet the needs of the cell.

SYNTHETIC MACHINERY

Associated with the cell nucleus is the endoplasmic reticulum (ER), which, as with any other cell, is responsible for the transportation of cellular material outside the cell. Since the muscle cell primarily manufactures components for local use (as opposed to the adrenal glands, for example, which synthesize hormones for general body use), this system is not very well developed. Ribosomes are associated with the poorly developed ER, on which the numerous muscle proteins are synthesized. Proteins synthesis that is coordinated by the nucleus is performed on ribosomes and is tremendously important to the cell. The amount and type of muscle proteins present confer on the cell its strength, speed, and endurance properties. The regulation of protein synthesis is thus of paramount importance to the cell in surviving the many different environments in which it finds itself, whether it be one of increased use (exercise, electrical stimulation, stretch) or decreased use (denervation, immobilization, tenotomy). Specific studies that highlight the impressive upward and downward regulation of the synthetic machinery will be detailed in Chapters 4 and 5, respectively.

MITOCHONDRIA—SOURCES OF AEROBIC ENERGY

Muscle mitochondria are responsible for generation of the main cellular energy molecule, adenosine triphosphate (ATP). Mitochondria contain all the enzymes that are responsible for oxidation of the high-energy precursor (*i.e.,* nicotinamide adenine dinucleotide [NADH]) into molecular oxygen and water. During this process, much chemical energy is trapped as ATP that is used throughout the cell for muscle contraction, ion transport, protein synthesis, cellular repair, *etc.* (In Chapter 2 we will describe metabolism in more detail and discuss its importance to muscle function.) The location and distribution of mitochondria are dependent on the particular muscle fiber type (see below). There is also evidence that muscle mitochondria are not actually discrete organelles but rather interconnected by a reticular network (Kirkwood *et al.,* 1986; Kayar *et al.,* 1988). This network might provide a means by which mitochondria in one portion of the cell communicate intracellularly with others. Because mitochondria generate ATP aerobically, their presence in the cell permits it to function continuously in the presence of oxygen. It is not surprising, therefore, that mitochondrial density is extremely plastic in the cell, responding rapidly and adequately to exercise training. It is also not surprising that mitochondrial density is important in determining the muscle fiber's endurance properties. The actual amount of mitochondria in the highly oxidative fibers may exceed 20% of the total cell volume (Eisenberg, 1983)!

Cytoplasmic Components

The cytoplasm of the muscle cell is rich in soluble proteins, filamentous cytoskeletal components, and other substances. Some of the more important of these substances are the muscle fiber energy sources: glycogen and lipids. The machinery that is used to generate ATP anaerobically from glucose is also present in the cytoplasm. Glycogen is a polymer of linked glucose molecules used by muscle cells as the immediate source of glucose, which can be metabolized with or without oxygen. Large glycogen granules are located in two main locations within the cell: near the terminal cisternae of the sarcoplasmic reticulum (SR) (see below) and in between the myofibrils. Soluble enzymes within the cytoplasm perform glycolysis. Because oxygen is required to metabolize lipids (see discussion of metabolism in Chapter 2), lipid droplets are most prevalent in cells with high mitochondrial densities. Finally, numerous free-floating ribosomes synthesize proteins at locations throughout the cell.

Filamentous Components—What Makes a Muscle Cell Look Like a Muscle Cell

MYOFIBRILS

Perhaps the most distinctive feature of the muscle cell is the ordered array of contractile filaments that are arranged throughout the cell (Figure 1.5; Table 1.2). There is a well-defined hierarchy of filament organization that proceeds from a large scale (on the order of microns, 10^{-6} m, abbreviated μm) to a small scale (on the order of angstroms, 10^{-10} m, abbreviated Å). The largest functional unit of contractile filaments is the myofibril (literally, "muscle thread"). Myofibrils are simply a string of sarcomeres arranged in series. Myofibrillar diameter is about 1 μm, which means that thousands of myofibrils can be packed into a single muscle fiber. One way in which a muscle fiber grows is to increase the number of myofibrils that it contains. Myofibrils are arranged in parallel (side by side) to make up the muscle fiber. However, their arrangements might not simply be like a bundle of spaghetti; there is some evidence that myofibrils within the fiber are arranged similar to the weave in a rope (Peachey and Eisenberg, 1978). (The functional consequence of this arrangement is that various myofibrils may not act completely independently during normal contraction.) Groups of muscle fibers are surrounded by a connective tissue sheath known as perimysium (literally, "around muscle") and arranged in bundles called fascicles. These fascicles are also bundled together, surrounded by more connective tissue (epimysium, literally, "on top of muscle") to form the whole muscle, which we can inspect visually.

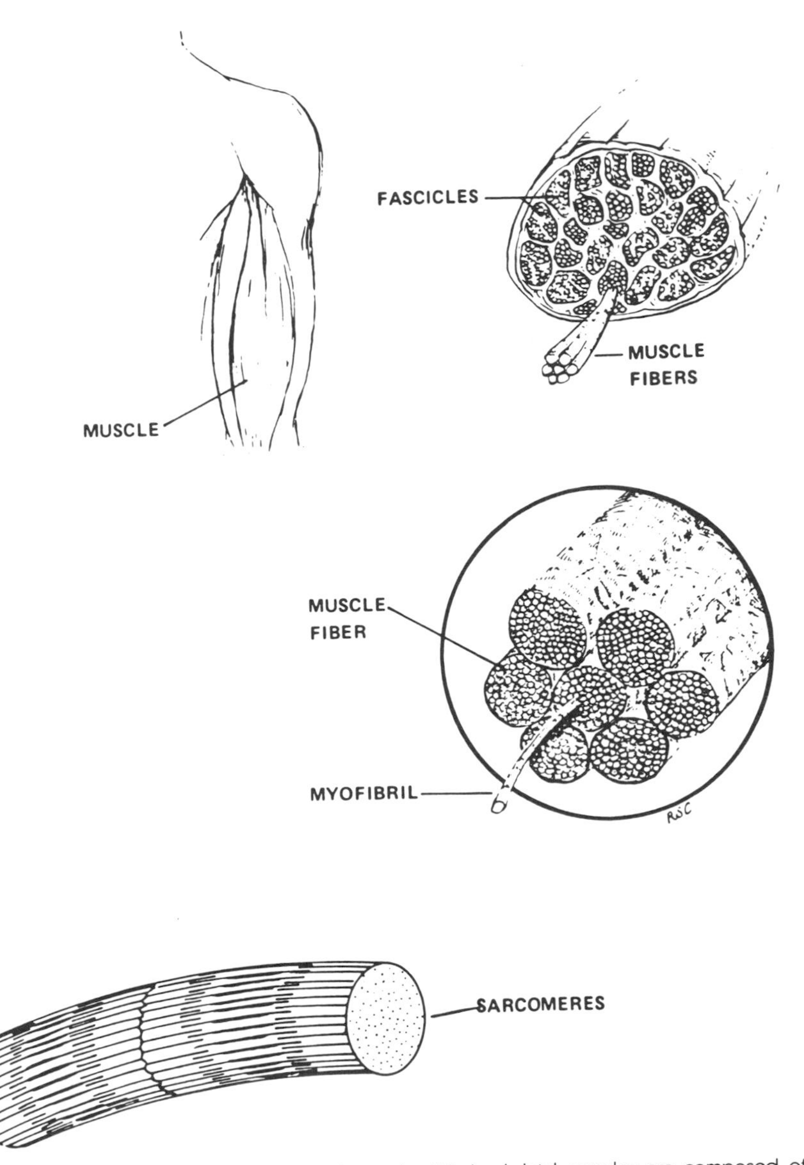

Figure 1.5. Structural hierarchy of skeletal muscle. Whole skeletal muscles are composed of numerous fascicles of muscle fibers. Muscle fibers are composed of myofibrils arranged in parallel. Myofibrils are composed of sarcomeres arranged in series. Sarcomeres are composed of interdigitating actin and myosin filaments.

Table 1.2.
Hierarchy of Skeletal Muscle Organization

Whole skeletal muscles
Muscle fascicles
Muscle fibers
Myofibrils
Sarcomeres
Myofilaments

SARCOMERES—FUNCTIONAL UNIT OF CONTRACTION

Myofibrils can also be subdivided into their component units known as sarcomeres (Figure 1.5), the functional unit of muscle contraction. A myofibril is therefore a number of sarcomeres (literally, "muscle segment") arranged in series (end-to-end). The total number of sarcomeres within a fiber depends on the muscle fiber length and diameter. Because of the series arrangements of sarcomeres within a myofibril, the total distance of myofibrillar shortening is equal to the sum of the individual shortening distances of the individual sarcomeres. This is why a whole muscle may shorten several centimeters even though each sarcomere can only shorten about 1 μm! It should also be stated that the number of sarcomeres in a mature muscle can change given the appropriate stimulus (Chapter 4, page 166). This means that muscle fibers have a great capacity for adaptation, which is the subject of the last half of this book.

Sarcomeres are composed of contractile filaments termed "myofilaments." Two major sets of contractile filaments exist in the sarcomere: One set is relatively thick, and the other set is relatively thin (Figure 1.6). These thick and thin filaments represent large polymers of the proteins myosin and actin, respectively. The myosin-containing filaments (thick filaments) and the actin-containing filaments (thin filaments) interdigitate to form a hexagonal lattice (Figure 1.6). It is the active interdigitation of these microscopic filaments that produces muscle shortening (Chapter 2, page 55). It is also this interdigitated pattern that gives the muscle its striated or striped appearance that is observable microscopically (Figure 1.7). In fact, another term for skeletal muscle is striated muscle, by virtue of the dark and light banding pattern.

Various regions of the sarcomere are named so that reference can be made to them (Figure 1.6). For example, the sarcomere region containing the myosin filaments is known as the A-band (A stands for "anisotropic," which is an optical term describing what this band does to incoming light). The region containing the actin filament is known as the I-band (I stands for "isotropic"). The region of the A-band where there is no actin-myosin overlap is called the

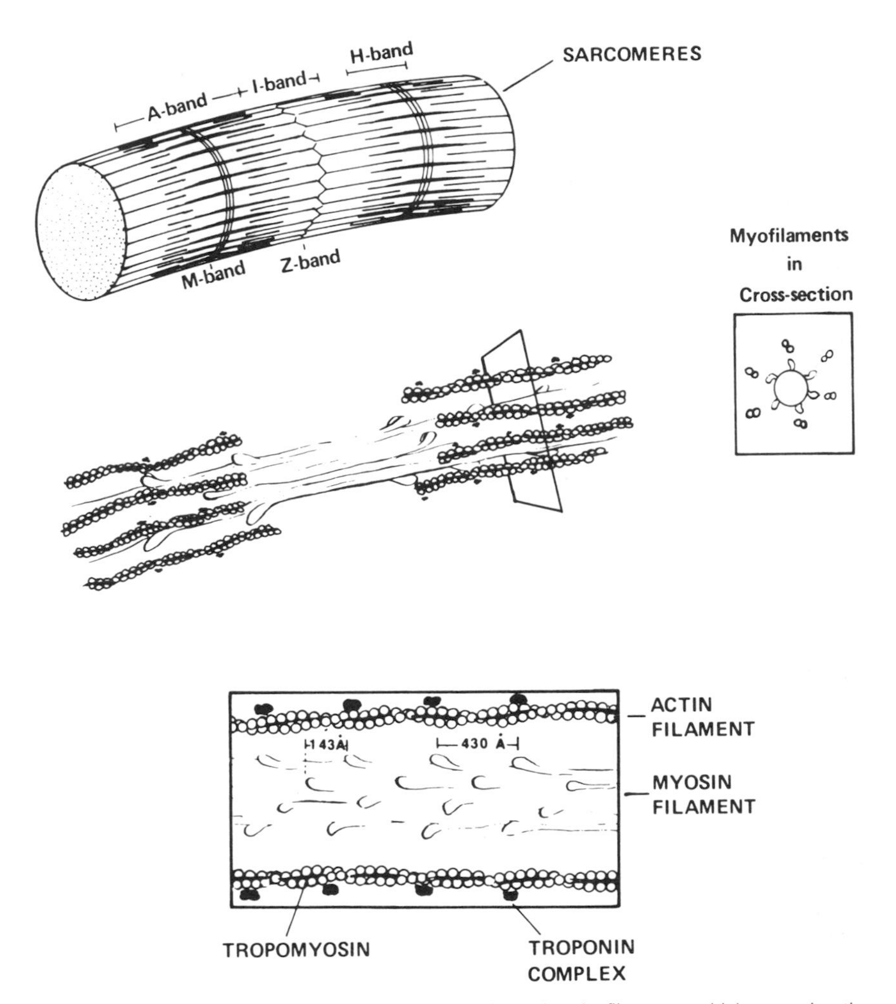

Figure 1.6. Hexagonal array of interdigitating myosin and actin filaments, which comprise the sarcomere. The myosin filament is composed of myosin molecules, and the actin filament is composed of actin monomers. Arranged at intervals along the actin filament are the regulatory proteins troponin and tropomyosin. See text for details.

H-zone (H stands for *helle,* which is German for "light"). The dark narrow line that bisects the I-band is the Z-band (Z stands for *zwitter,* which is German for "between"). Finally, the relatively dense structures noted in the center of the A-band are known as the M-band. The distance from one Z-band to the next is defined as the sarcomere length, which is an important variable relative to force generation (Chapter 2, page 55).

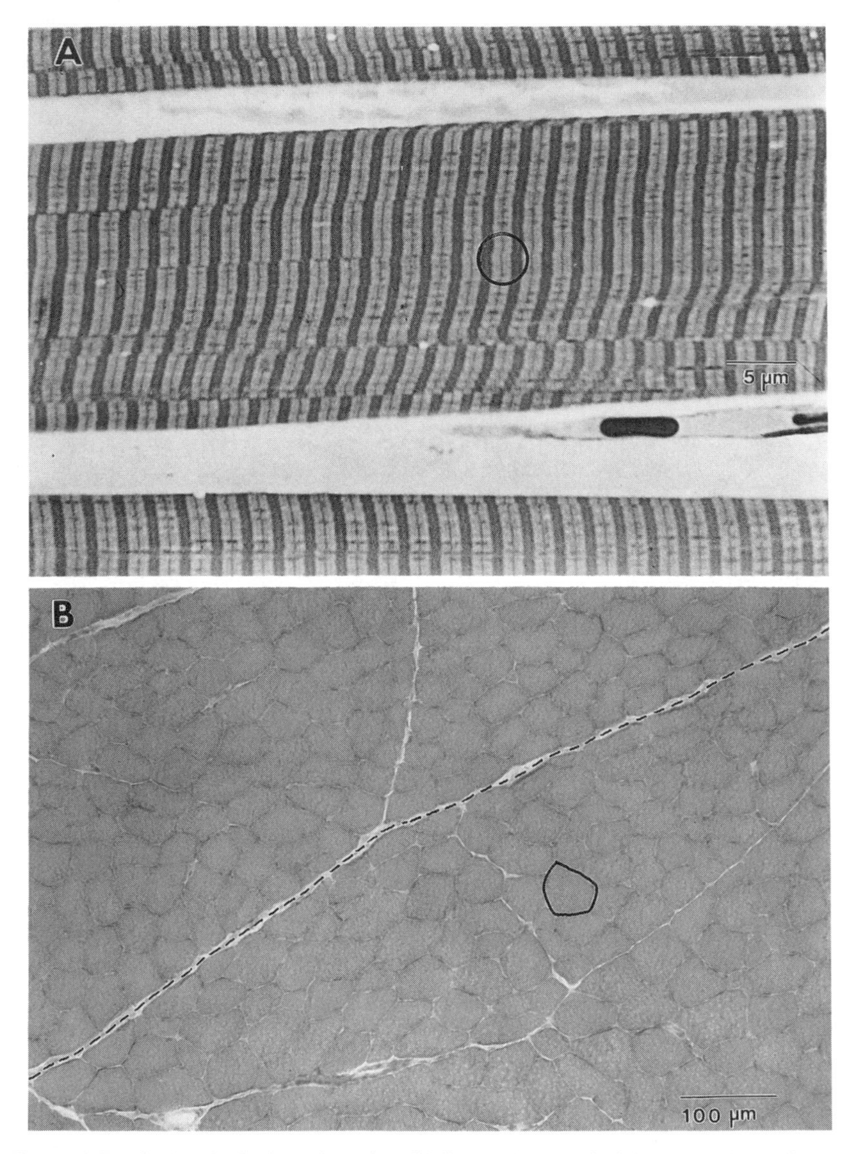

Figure 1.7. A, Longitudinal section of a tibialis anterior muscle biopsy specimen that was chemically fixed, embedded in plastic, sectioned at 1 μm thickness, and stained with toluidine blue. The alternating light and dark regions correspond to the sarcomere A- and I-bands. Several fibers are shown in this section. Circled region enlarged in Figure 1.9. **B,** Cross-section of a vastus lateralis muscle biopsy specimen that was frozen, sectioned at 8 μm thickness, and stained with hematoxylin and eosin for inspection of normal fiber morphology. Note that, in cross-section, muscle appears to be a collection of densely packed polygonal fibers. Each muscle fiber is surrounded by a sheath of endomysial connective tissue (shown for one fiber as a *solid line*), and collections of muscle fibers (fascicles) are surrounded by more dense perimysial connective tissue (shown for one region as a *dashed line*).

MYOSIN—HOME OF THE CROSS-BRIDGE

The myosin-containing filament is composed of myosin molecules that are relatively large proteins (about 470 kDaltons) known as the myosin heavy chain. This protein is one of the most widely studied proteins in all of biology since it is found in cells other than muscle cells (Squire, 1981). Myosin proteins are arranged in a so-called antiparallel fashion to made up the myosin filament (Figure 1.8). As individual myosin molecules pack together to form the thick filament, one molecule rotates about 60° relative to the molecules on either side (this arrangement varies slightly between vertebrate muscles). The myosin molecule has a long backbone and a region that extends from the backbone. Due to the antiparallel (tail-to-tail) arrangement of these molecules, as the filament is formed, it takes on a characteristic feathered appearance, with projections coming out at either end of the filament, but with the middle portion of the filament void of these projections. Because of the systematic packing of the myosin molecules into a filament, every 430 Å, the myosin molecules make a complete revolution (Figure 1.8), which means that myosin heads are spaced approximately 143 Å along the length of the myosin filament. The antiparallel arrangement of myosin actually gives the sarcomere symmetry down the middle so that each half-sarcomere is functionally identical and has a mirror image on the opposite side of the sarcomere.

Myosin was one of the first proteins to be isolated from muscle. It was found that when this molecule is placed in a mild protease (an enzyme that digests proteins, *e.g.,* chymopapain, the same enzyme that makes up meat tenderizer), the myosin molecule "falls apart" into two discrete components (Figure 1.8): the so-called light meromyosin (LMM) and heavy meromyosin (HMM), terms based on the observation that LMM has a formula weight of only 135 kDaltons while HMM weighs about 335 kDaltons. Further incubation of HMM in protease results in the production of two subfragments: subfragment 1 (S-1; molecular weight 115 kDaltons) and subfragment 2 (S-2; molecular weight 60 kDaltons). The reason that this sequential digestion of the myosin molecule furthered our understanding of muscle contraction was that each "piece" of the molecule could be tested individually to determine how myosin itself causes muscle contraction to occur. For example, it was determined that the protein's backbone is the LMM while the portion of the molecule projecting from the backbone is HMM. S-2 was shown to be the portion of HMM that projects out from the LMM backbone, while S-1 is the globular "head" of the projection. It is the combination of S-1 and S-2 that forms the well-known "cross-bridge," so central to the theory of force generation in muscle (see below). Also associated with the myosin heavy chain are two light chains (not shown in Figure 1.8). The precise function of these chains is not known. It is believed that they may provide structural support for the S-1 head. It is known,

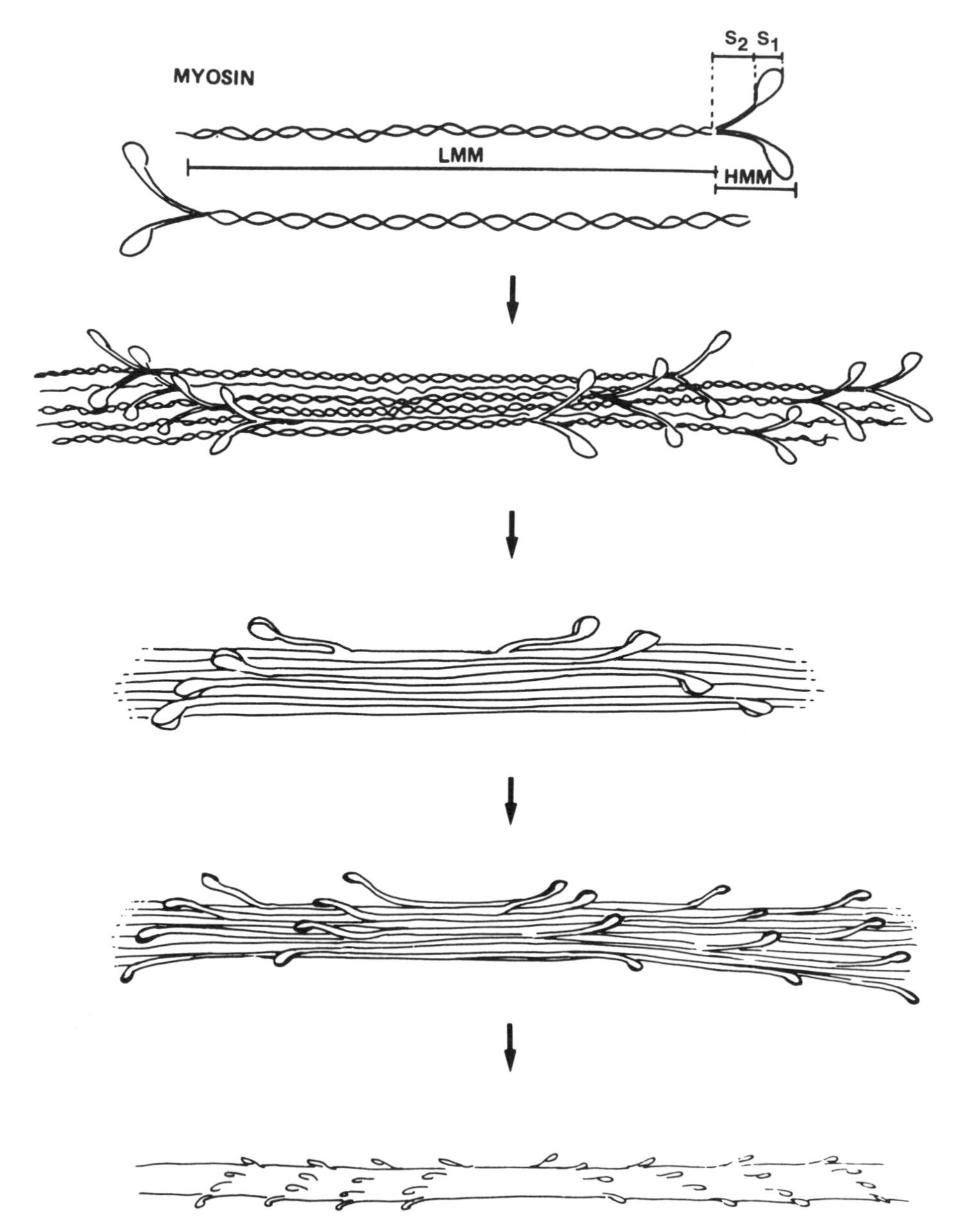

Figure 1.8. Schematic composition of the myosin filament. The myosin molecule can be digested enzymatically into its fragments. Subfragments one and two (S-1 and S-2) are composed of the globular head and neck regions of the cross-bridge, respectively. Together, S-1 and S-2 comprise the heavy meromyosin (HMM), while the remainder of the molecules comprise the light meromyosin (LMM). Myosin monomers polymerize into myosin filaments.

however, that various forms of the same light chain, so-called isoforms, exist in different muscles fibers. We will see in Chapter 4 that analysis of the myosin light chain isoform distribution provides insights into the nature of muscle adaptation (Chapter 4, page 163). Let's take a moment to state that, actually, isoforms of most of the sarcomere proteins exist. Myosin isoforms, however, have been most widely studied.

ACTIN—REGULATION OF CONTRACTION

The structure of the actin-containing filament is equally as elegant as that of myosin. While the myosin-containing filament *generates* tension during muscle contraction, the actin-containing filament *regulates* tension generation. (Not all muscle systems are regulated by the actin-containing filament. In several invertebrate systems, force regulation and generation are both performed on the myosin-containing filament [Ebashi *et al.,* 1980]. These so-called thick-filament-regulated systems are actually quite common. However, mammals use only thin-filament-based regulation.) The actin filament is composed of a long α-helical arrangement of actin monomers. In contrast to the myosin-containing filament, there is no directional symmetry to this helix (*i.e.,* the filament does not have a distinct middle section as does the myosin filament).

Actin is a ubiquitous protein found in virtually all cells as part of the intermediate filament network—part of the cell's cytoskeleton. (Interestingly, the discovery of actin filaments in cells other than muscle was first confirmed by "decorating" the actin-containing intermediate filaments with HMM! This was done by adding HMM to cytoskeletal cellular components and observing the feathered appearance of the HMM heads at regular intervals along the actin filaments.) Actin monomers are relatively small compared to myosin (only about 40 kDaltons) and are roughly spherical in shape. Thus as a result of their helical arrangement, a long groove is created along the filament's length (Figure 1.6). The regulatory protein, tropomyosin, fits nicely into this groove along the filament length. At intervals along the filament (approximately every seven actin monomers), the protein troponin is located. Troponin is the protein that is actually responsible for turning on contraction. Troponin (abbreviated Tn) is composed of three subunits: Tn-I,Tn-C, and Tn-T. The functions of these subunits will become more clear following discussion of the cross-bridge cycle (Chapter 2, page 52). Suffice it to say that Tn-T binds troponin to tropomyosin (hence "T"), Tn-C binds calcium during contraction (hence "C"), and Tn-I exerts an inhibitory influence on tropomyosin when calcium is not present (hence the "I").

In summary, contractile proteins within the muscle cells are arranged according to an elegant hierarchy. From the myosin molecule to the myofibril, structural hierarchy and organization are the rule, not the exception.

Membrane Components—A Vehicle for Excitation

In addition to the well-defined arrangement of force-generating components present in muscle cells, there exists an intricate system for activating these force generators. Recall that the skeletal muscle fiber is a highly differentiated cell that is specialized for producing force and movement. The membrane system present is actually a specially designed version of the membrane systems within normal cells. The two main components of this system are the transverse tubular system (T-system) and the SR.

TRANSVERSE TUBULAR SYSTEM AND SARCOPLASMIC RETICULUM

The T-system begins as invaginations of the surface membrane and is therefore physically contiguous with the sarcolemma. These invaginations extend transversely across the long axis of the muscle fiber (hence their name). The function of the T-system is to convey an activation signal to the myofibrils, which are themselves not in direct contact with the motoneuron (Chapter 2, page 51). The T-system thus acts as an electrical conduit for the nervous signal that reaches deep into the fiber. This provides a means for myofibrillar activation that is much faster than, say, diffusion of molecules from the cell surface. It is thought that the T-system is an electrically excitable membrane much like the sarcolemma and the membranes surrounding neurons.

The SR is a much more complex membrane system whose main function is to release and take up calcium during contraction and relaxation, respectively. As such, the SR envelops each myofibril to permit intimate contact between the activation and force-generation systems (Figure 1.4). The SR is also in contact with the T-system and therefore acts as the "middleman" in skeletal muscle activation and relaxation. Physical structures that link the T-system to the SR have been identified and termed the "junctional feet" of the SR (Peachey and Franzini-Armstrong, 1983).

The relative spatial arrangement of the T-system and SR gives rise to a characteristic pattern that is observable in a high magnification longitudinal section (a section parallel to the long axis) of a muscle cell (this is usually seen by chemically fixing the tissue, embedding it in plastic, and thin sectioning, staining, and viewing it under the electron microscope [*e.g.,* Figure 1.9]. Obviously, the true arrangement of the T-system and SR is a quite complex three-dimensional matrix. However, when one takes a slice through this matrix, a stereotypic arrangement of a T-tubule surrounded by two SR tubules is seen (Figure 1.4). This arrangement is so common that it has been named the "triad." Extending longitudinally from either side of the "triad" are membrane systems, which appear as flattened sacs. These represent longitudi-

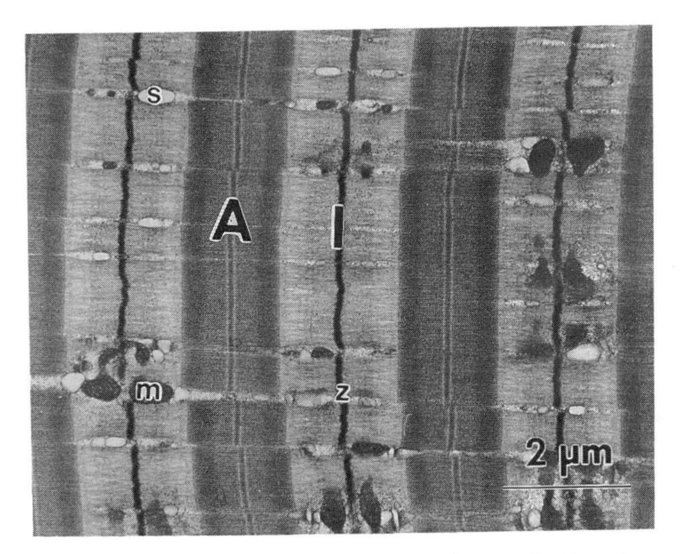

Figure 1.9. Electron micrograph of a rabbit tibialis anterior muscle. Note the much greater magnification of the A- and I-bands (compare to Figure 1.7**A**). Calibration bar corresponds to a distance of 2 μm. m, mitochondrion; z, Z-band; s, sarcoplasmic reticulum.

nal extensions of the SR that surround the myofibril. Remember, although these flattened sacs appear as a relatively minor area fraction in the longitudinal view, this is only because they have a complex three-dimensional arrangement that is not adequately represented in only two-dimensions. The various portions of the T-system and SR are so distinct that they can be physically isolated and experimentally manipulated using modern fractionation techniques now commonly used in cell biology.

STUDIES OF MUSCLE FILAMENT AND SARCOMERE STRUCTURE

The uncanny degree of order that was just described lends muscle to investigation using a variety of tools. Some tools used to study muscle structure will now be presented, along with the evidence they provide for the description given above.

Light Microscopy

Numerous microscopic methods have been used to examine skeletal muscle structure. In fact, many of the modern-day electron and light microscopic techniques were pioneered by Hugh Huxley and Andrew Huxley, respectively (no relation). Obviously, these methods have provided unique insights into the structure of many other biologic tissues.

Early light microscopes did not provide a great deal of insight into the structure of the *living* muscle fiber. This was because, unstained, the muscle fiber appeared as a translucent cylinder. However, in the early 1950s Andrew Huxley developed an interference microscope that enhanced the contrast of different muscle fiber regions. Under the interference microscope, instead of appearing translucent, the muscle fiber took on a striped or striated appearance (Figure 1.7**A**). This striation pattern, which is so well-known today, resulted from the interdigitation of the sarcomere thick and thin filaments. Huxley and Niedergerke demonstrated that as muscle length changed, the striation pattern also changed (Huxley and Niedergerke, 1954). Thus in their interference microscope they knew that they had a tool that enabled investigation of contractile mechanisms on isolated, living muscle fibers.

Huxley and Niedergerke performed a series of experiments in which a muscle was caused to contract and was observed in the interference microscope. They observed that during muscle shortening, the width of the A-band remains constant while the I-band length changes. Based on what you know of the A and I bands, can you explain this observation? Huxley and Niedergerke hypothesized that this observation might result from interdigitation of filaments that maintained a constant length. The A-band stays a constant length since it represents the myosin filaments, which remain at a constant width. The I-band decreases in width since it represents the non-overlap region of the actin filament. I haven't detailed the experiments, but the highly technical nature of the microscopy along with the clever interpretation of the data (especially in light of the thinking at that time) are truly awe-inspiring.

Electron Microscopy

During a similar period, Hugh Huxley and his coworker, Jean Hanson, were developing preparative methods for the then-new electron microscope (EM; Figure 1.9). The advantage of the EM over the light microscope was its much greater magnification power (over 50,000X compared to the light microscope's 500X capability). Unfortunately, the material to be viewed under the EM could not be living—all of the water had to be extracted because the specimen was viewed in a vacuum chamber. Thus Huxley and Hanson developed methods for dehydrating, embedding, and sectioning skeletal muscles at various lengths (Huxley and Hanson, 1954). When sectioned longitudinally, the banding pattern described above was observed. Most scientists at the time thought that this banding pattern was simply an artifact of the rather harsh dehydration and embedding procedures. However, Huxley and Hanson succeeded in demonstrating that this almost crystalline array was truly representative of muscle. They then sectioned the muscle fiber transversely and demonstrated hexagonal lattices of myofilaments (Figure 1.6). The interesting point is that if they sectioned through the H-band, only a myosin filament lattice was observed. If they sectioned through the I-band, only a thin filament lattice was observed. However, if they sectioned through the overlap region of the A-band, an interdigitating hexagonal array of actin and myosin filaments was observed.

Huxley and Hanson's findings were complementary to Huxley and Niedergerke's and were equally as startling. Both groups of investigators independently proposed (and published the results side by side in the prestigious scientific journal *Nature*) that muscle contraction occurred by the relative sliding of the thick and thin filaments past one another. This became known as the "sliding filament hypothesis" and has since been elevated to the "sliding filament theory" as a plethora of data have been acquired that provide support. (However, not all scientists are sold on the sliding filament theory; see Pollack, 1983.)

X-ray Diffraction

A third tool for structural investigation of muscle that was highly influential in developing theories of muscle contraction is the method of X-ray diffraction. The principle of X-ray diffraction is similar to that observed in the ripple pond experiment from basic physics. If an X-ray beam is projected onto a living muscle, the X-rays are scattered by the tissue. If some of the scattering elements are arranged in a regular array (which the myofilaments are), an interference pattern results that represents constructive and destructive interference of the scattered X-rays. This diffraction pattern appears as a series of lines (the so-called layer lines), which can later be analyzed. The spacings between the layer lines can then be directly related to microscopic spacings in the sarcomere itself.

As with all diffraction methods, the size of the structures "seen" by the X-rays is close to the wavelength of the X-rays themselves. Since most X-rays fall in the Å range, X-ray diffraction resolves spacing between objects with molecular dimensions. Consider what some of these dimensions might represent in muscle. In a series of experimental studies, Hugh Huxley and colleagues measured the spacing between the actin and myosin filaments, the spacing between the various myosin heads (as we just saw, the 143 Å and 430 Å distances), and even the spacing between the monomers of the actin molecule in muscle at rest (Haselgrove, 1983). Sophisticated analysis of the intensities of these layer lines during passive muscle length change and active muscle contraction was consistent with muscle contraction occurring by the relative sliding of filaments past one another and provided direct support for a portion of the myosin molecule extending out toward the actin filament during force generation! More recent studies have confirmed these earlier studies on a much faster time scale and with much better resolution (Huxley *et al.,* 1981).

Thus while X-ray diffraction may not be a technique with which you are familiar, in muscle structural studies it has provided some of the most direct, quantitative data available on normal muscle structure and muscle structural changes during contraction.

SATELLITE CELLS: RESERVES FOR INJURY AND REPAIR

An important but relatively rarely observed component of the skeletal muscle cell is the satellite cell. Maybe we shouldn't even include the satellite cells as part of the muscle cell since they are really distinct cellular entities with their own nuclei. These small cells are located beneath the fiber basal lamina (see page 3) and are approximately the same size as a muscle cell nucleus. While the satellite cell plays no known role in normal cell function, it has a central role to play in recovery of muscle fibers from injury. Satellite cells have the ability to differentiate into myoblasts and to form new muscle fibers. Clearly such an ability is central in formation of new muscle fibers following injury. This regeneration process will be discussed in Chapter 6 (page 261) as we consider muscle injury and recovery.

WHOLE SKELETAL MUSCLE STRUCTURE

Inspection of the body's numerous skeletal muscles reveals a number of common themes. Skeletal muscles attach to bones via connective tissue structures known as tendons. At times the amount of tendon is so small that the muscle fibers themselves appear to arise from bone. However, microscopic analysis of the muscle fiber end shows connective tissue such as tendon interposed between fibers and bones. A tendinous origin or insertion that is very broad and thin is termed an aponeurosis. At the gross anatomic level, each muscle has an origin (the proximal muscle end) and insertion (the distal muscle end, Table 1.1). Often the muscle origin is more broad than the insertion, in which the fibers converge onto a stout tendon (*e.g.,* m. soleus).

However, in spite of knowing a muscle's origin and insertion, it is not possible to simply describe a motion resulting from muscle contraction based only on this information. This is because muscles often cross more than one joint and therefore exert an influence at multiple location (inspection of Table 1.1 shows that most muscles cross multiple joints). For example, the rectus femoris crosses both the knee and hip joints. If the knee were fixed and the hip were free to move, rectus femoris contraction would cause hip flexion. Conversely, if the hip were fixed and the knee free to move, rectus femoris contraction would cause knee extension. Thus is the rectus femoris a hip flexor or a knee extensor or both? It is not possible to answer this question unambiguously without specifying a movement. For example, in standing from a squat, the rectus femoris is activated and generates tension, but clearly the knee and hip *both extend*. Therefore, anatomically the rectus is a hip flexor and knee extensor, but this is not always the case during the muscle's physiologic action. We must therefore resist the temptation to classify muscles in terms of anatomy. Instead, we can state that the rectus femoris always acts to generate a hip flexion moment and a knee extension moment (Zajac and Gordon, 1989). We will discuss this point further in analyzing biarticular muscle function in Chapter 3 (page 149).

Muscle Architecture—Fiber Arrangement Is Everything

Skeletal muscle is not only highly organized at the microscopic level; the *arrangement* of the muscle fibers at the macroscopic level also demonstrates a striking degree of organization. In making comparisons between various muscles, certain factors such as fiber type distribution are important, but there is no question that an important factor in determining a muscle's contractile properties is the muscle's architecture.

Skeletal muscle architecture is defined as "the arrangement of muscle fibers relative to the axis of force generation." While muscle fibers have a relatively consistent fiber diameter between muscles of different sizes, the *arrangement* of these fibers can be quite different. The various types of arrangement are as numerous as the muscles themselves, but for convenience we can discuss three types of fiber architecture.

Muscles with fibers that extend parallel to the muscle force-generating axis are termed parallel or longitudinally arranged muscles (Figure 1.10, page 38). While the fibers extend parallel to the force-generating axis, they never extend the entire muscle length (Tables 1.3 and 1.4). Muscles with fibers that are oriented at a single angle relative to the force generating axis are termed unipennate muscles (Figure 1.10, middle). The angle between the fiber and the force-generating axis generally varies from 0° to 30°. It is obvious when preparing muscle dissections that most muscles fall into the most general category, multipennate muscles—muscles composed of fibers that are oriented at several angles relative to the axis of force generation (Figure 1.10, right). As we will discuss in the next chapter, an understanding of muscle architecture is critical to understanding the various functional properties of different sized muscles.

Table 1.3.
Architectural Properties of the Human Hand, Arm, and Forearm[a,b]

Muscle	Muscle Mass (g)	Muscle Length (mm)	Fiber Length (mm)	Pennation Angle (°)	Cross-Sectional Area (cm^2)	FL/ML Ratio
			Extrinsic Muscles			
AbPL (n=9)	9.96±2.01	160.4±15.0	58.1±7.4	7.5±2.0	1.93±.59	.36±.05
BR (n=8)	16.6±2.8	175±8.3	121±8.3	2.4±.6	1.33±.22	.69±.062
EDC I (n=8)	3.05±.45	114±3.4	56.9±3.6	3.1±.5	.52±.08	.49±.024
EDC M (n=5)	6.13±1.2	112±4.7	58.8±3.5	3.2±1.0	1.02±.20	.50±.014
EDC R (n=7)	4.70±.75	125±10.7	51.2±1.8	3.2±.54	.86±.13	.42±.023
EDC S (n=6)	2.23±.32	121±8.0	52.9±5.2	2.4±.7	.40±.06	.43±.029
EDQ (n=7)	3.81±.70	152±9.2	55.3±3.7	2.6±.6	.64±.10	.36±.012
EIP (n=6)	2.86±.61	105±6.6	48.4±2.3	6.3±.8	.56±.11	.46±.023
EPB (n=9)	2.25±1.36	105.6±22.5	55.0±7.5	7.2±4.4	.47±.32	.54±.13
EPL (n=7)	4.54±.68	138±7.2	43.6±2.6	5.6±1.3	.98±.13	.31±.020
FDP I (n=9)	11.7±1.2	149±3.8	61.4±2.4	7.2±.7	1.77±.16	.41±.018
FDP M (n=9)	16.3±1.7	200±8.2	68.4±2.7	5.7±.3	2.23±.22	.34±.011
FDP R (n=9)	11.9±1.4	194±7.0	64.6±2.6	6.8±.5	1.72±.18	.33±.009
FDP S (n=9)	13.7±1.5	150±4.7	60.7±3.9	7.8±.9	2.20±.30	.40±.015
FDS I(C) (n=6)	12.4±2.1	207±10.7	67.6±2.8	5.7±.2	1.71±.28	.33±.025
FDS I(D) (n=9)	6.6±.8	119±6.1	37.9±3.0	6.7±.3	1.63±.22	.32±.013
FDS I(P) (n=6)	6.0±1.1	92.5±8.4	31.6±3.0	5.1±0.2	1.81±.83	.34±.022
FDS M (n=9)	16.3±2.2	183±11.5	60.8±3.9	6.9±.7	2.53±.34	.34±.014
FDS R (n=9)	10.2±1.1	155±7.7	60.1±2.7	4.3±.6	1.61±.18	.39±.023
FDS S (n=9)	1.8±.3	103±6.3	42.4±2.2	4.9±.7	0.40±.05	.42±.014
FPL (n=9)	10.0±1.1	168±10.0	45.1±2.1	6.9±.2	2.08±.22	.24±.010

Table 1.3.
Architectural Properties of the Human Hand, Arm, and Forearm[a,b] *(Continued)*

Muscle	Volume Muscle Mass (g)	Muscle Length (mm)	Velocity Fiber Length (mm)	Pennation Angle (°)	Cross-Sectional Area (cm^2)	FL/ML Ratio
PL (n=6)	3.78 ± .82	134 ± 11.5	52.3 ± 3.1	3.5 ± 1.2	.69 ± .17	.40 ± .032
PQ (n=8)	5.21 ± 1.0	39.3 ± 2.3	23.3 ± 2.0	9.9 ± .3	2.07 ± .33	.58 ± .021
PT (n=8)	15.9 ± 1.7	130 ± 4.7	36.4 ± 1.3	9.6 ± .8	4.13 ± .52	.28 ± .012
			Intrinsic Muscles			
AbDM (n=9)	3.32 ± 1.67	68.4 ± 6.5	46.2 ± 7.2	3.9 ± 1.3	.89 ± .49	.68 ± .10
AbPB (n=9)	2.61 ± 1.19	60.4 ± 6.6	41.6 ± 5.6	4.6 ± 1.9	.68 ± .28	.69 ± .09
AddPol (n=9)	6.78 ± 1.84	54.6 ± 8.9	34.0 ± 7.5	17.3 ± 3.4	1.94 ± .39	.63 ± .15
DI I (n=9)	4.67 ± 1.17	61.9 ± 2.5	31.7 ± 2.8	9.2 ± 2.6	1.50 ± .40	.51 ± .05
DI II (n=9)	2.65 ± 1.01	62.8 ± 8.1	25.1 ± 6.3	8.2 ± 3.1	1.34 ± .77	.41 ± .13
DI III (n=9)	2.01 ± 0.60	54.9 ± 4.6	25.8 ± 3.4	9.8 ± 2.8	.95 ± .45	.47 ± .07
DI IV (n=9)	1.90 ± 0.62	50.1 ± 5.3	25.8 ± 3.4	9.4 ± 4.2	.91 ± .38	.52 ± .11
FDM (n=9)	1.54 ± .44	59.2 ± 10.4	40.6 ± 13.7	3.6 ± 1.0	.54 ± .36	.67 ± .17
FPB (n=9)	2.58 ± .56	57.2 ± 3.7	41.5 ± 5.2	6.2 ± 4.5	.66 ± .20	.73 ± .08
Lum I (n=9)	0.57 ± .019	64.9 ± 10.0	55.4 ± 10.2	1.2 ± .9	.112 ± .03	.85 ± .03
Lum II (n=9)	0.39 ± .22	61.2 ± 17.8	55.5 ± 17.7	1.6 ± 1.3	.079 ± .04	.90 ± .05
Lum III (n=9)	0.37 ± .16	64.3 ± 8.9	56.2 ± 10.7	1.1 ± .8	.081 ± .04	.87 ± .07
Lum IV (n=9)	0.23 ± 0.11	53.8 ± 11.5	50.1 ± 8.4	0.7 ± 1.0	.063 ± .03	.90 ± .05
OpDM (n=9)	1.94 ± .98	47.2 ± 3.6	19.5 ± 4.1	7.7 ± 2.9	1.10 ± .43	.41 ± .09
OpPol (n=9)	3.51 ± .89	55.5 ± 5.0	35.5 ± 5.1	4.9 ± 2.5	1.02 ± .35	.64 ± .07
PI II (n=9)	1.56 ± .22	55.1 ± 5.0	25.0 ± 5.0	6.3 ± 2.2	.75 ± .25	.45 ± .08

Table 1.3.
Architectural Properties of the Human Hand, Arm, and Forearm[a,b] *(Continued)*

Muscle	Muscle Mass (g)	Muscle Length (mm)	Fiber Length (mm)	Pennation Angle (°)	Cross-Sectional Area (cm^2)	FL/ML Ratio
PI III (n=9)	1.28±.28	48.2±2.9	26.0±4.3	7.7±3.9	.65±.26	.54±.08
PI IV (n=9)	1.19±.33	45.3±5.8	23.6±2.6	8.2±3.5	.61±.23	.52±.10

[a]Data on extrinsic muscles from Lieber *et al.,* 1990, 1991; data on intrinsic muscles from Jacobson MD. J Hand Surg, submitted, 1992.
[b]AbDM: abductor digiti minimi; AbPB: abductor pollicis brevis; AddPol: adductor pollicis; BR: brachioradialis; DI I to DI IV: dorsal interosseous muscles; EDC I, EDC M, EDC R, and EDC S: extensor digitorum communis to the index, middle, ring, and small fingers, respectively; EDQ: extensor digiti quinti; EIP: extensor indicis proprious; EPL: extensor pollicis longus; FDM: flexor digiti minimi; FDP I, FDP M, FDP R, and FDP S: flexor digitorum profundus muscles; FDS I, FDS M, FDS R, and FDS S: flexor digitorum superficialis muscles; FDS I (P) and FDS I (D): proximal and distal bellies of the FDS I; FDS I (C): the combined properties of the two bellies as if they were a single muscle; FPB: flexor pollicis brevis; FPL: flexor pollicis longus; L I to L IV: lumbrical muscles; OpDM: opponens digiti minimi; OpPol: opponens pollicis; PI I to PI IV: palmar interosseous muscles; PQ: pronator quadratus; PS: palmaris longus; PT: pronator teres.

Table 1.4.
Architectural Properties of Human Lower Limb[a,b]

Muscle	Muscle Mass (g)	Muscle Length (mm)	Fiber Length (mm)	Pennation Angle (°)	Cross-Sectional Area (cm^2)	FL/ML Ratio
AB (n=3)	43.8±8.4	156±12	103±6.4	0.0±0.0	4.7±1.0	.663±.036
AL (n=3)	63.5±16	229±12	108±2.0	6.0±1.0	6.8±1.9	.475±.023
AM (n=3)	229±32	305±12	115±7.9	0.0±0.0	18.2±2.3	.378±.013
BF_I (n=3)	128±28	342±14	85.3±5.0	0.0±0.0	12.8±2.8	.251±.022
BF_S (n=3)		271±11	139±3.5	23±0.9		.517±.032
EDL (n=3)	35.2±3.6	355±13	80.3±8.4	8.3±1.7	5.6±0.6	.226±.024
EHL (n=3)	12.9±1.6	273±2.4	87.0±8.0	6.0±1.0	1.8±0.2	.319±.030
FDL (n=3)	16.3±2.8	260±15	27.0±0.58	6.7±1.7	5.1±0.7	.104±.004
FHL (n=3)	21.5±3.3	222±5.0	34.0±1.5	10.0±2.9	5.3±0.6	.154±.010
GR (n=3)	35.3±7.4	335±20	277±12	3.3±1.7	1.8±0.3	.828±.017
LG (n=3)		217±11	50.7±5.6	8.3±1.7		.233±.016
MG (n=3)	150±14	248±9.9	35.3±2.0	16.7±4.4	32.4±3.1	.143±.010

Table 1.4.
Architectural Properties of Human Lower Limb[a,b] *(Continued)*

Muscle	Muscle Mass (g)	Muscle Length (mm)	Fiber Length (mm)	Pennation Angle (°)	Cross-Sectional Area (cm^2)	FL/ML Ratio
PB (n=3)	17.3±2.5	230±13	39.3±3.5	5.0±0.0	5.7±1.0	.170±.006
PEC (n=3)	26.4±6.0	123±4.5	104±1.2	0.0±0.0	2.9±0.6	.851±.040
PL (n=3)	41.5±8.5	286±17	38.7±3.2	10.0±0.0	12.3±2.9	.136±.010
PLT (n=3)	5.30±1.9	85.0±15	39.3±6.7	3.3±1.7	1.2±0.4	.467±.031
POP (n=2)	20.1±2.4	108±7.0	29.0±7.0	0.0±0.0	7.9±1.4	.265±.048
RF (n=3)	84.3±14	316±5.7	66.0±1.5	5.0±0.0	12.7±1.9	.209±.002
SAR (n=3)	61.7±14	503±27	455±19	0.0±0.0	1.7±0.3	.906±.017
SM (n=3)	108±13	262±1.5	62.7±4.7	15±2.9	16.9±1.5	.239±.017
SOL (n=2)	215 (n=1)	310±1.5	19.5±0.5	25±5.0	58.0 (n=1)	.063±.002
ST (n=2)	76.9±7.7	317±4	158±2.0	5.0±0.0	5.4±1.0	.498±0.0
TA (n=3)	65.7±10	298±12	77.3±7.8	5.0±0.0	9.9±1.5	.258±.015
TP (n=3)	53.5±7.3	254±26	24.0±4.0	11.7±1.7	20.8±3	.095±.015
VI (n=3)	160±59	329±15	68.3±4.8	3.3±1.7	22.3±8.7	.208±.007
VL (n=3)	220±56	324±14	65.7±0.88	5.0±0.0	30.6±6.5	.203±.007
VM (n=3)	175±41	335±15	70.3±3.3	5.0±0.0	21.1±4.3	.210±.005

[a]Data from Wickiewicz *et al.,* (1982).
[b]AB, adductor brevis; AL, adductor longus; AM, adductor magnus; BF_l, biceps femoris, long head; BF_S, biceps femoris, short head; EDL, extensor digitorum longus; EHL, extensor hallucis longus; FDL, flexor digitorum longus; GR, gracilis; FHL, flexor hallucis longus; LG, lateral gastrocnemius; MG, medial gastrocnemius; PEC, pectineus; PB, peroneus brevis; PL, peroneus longus; PLT, plantaris; POP, popliteus; RF, rectus femoris; SAR, sartorius; SM, semimembranosus; SOL, soleus; ST, semitendinosus; TA, tibialis anterior; TP, tibialis posterior; VI, vastus intermedius; VL, vastus lateralis; VM, vastus medialis.

Experimental Determination of Skeletal Muscle Architecture

Early studies of muscle architecture were pioneered by the anatomist Carl Gans (Gans, 1982). Gans and his colleagues developed precise methods for muscle architecture determination based on microdissection of whole muscles. These muscles were chemically fixed in order to maintain their

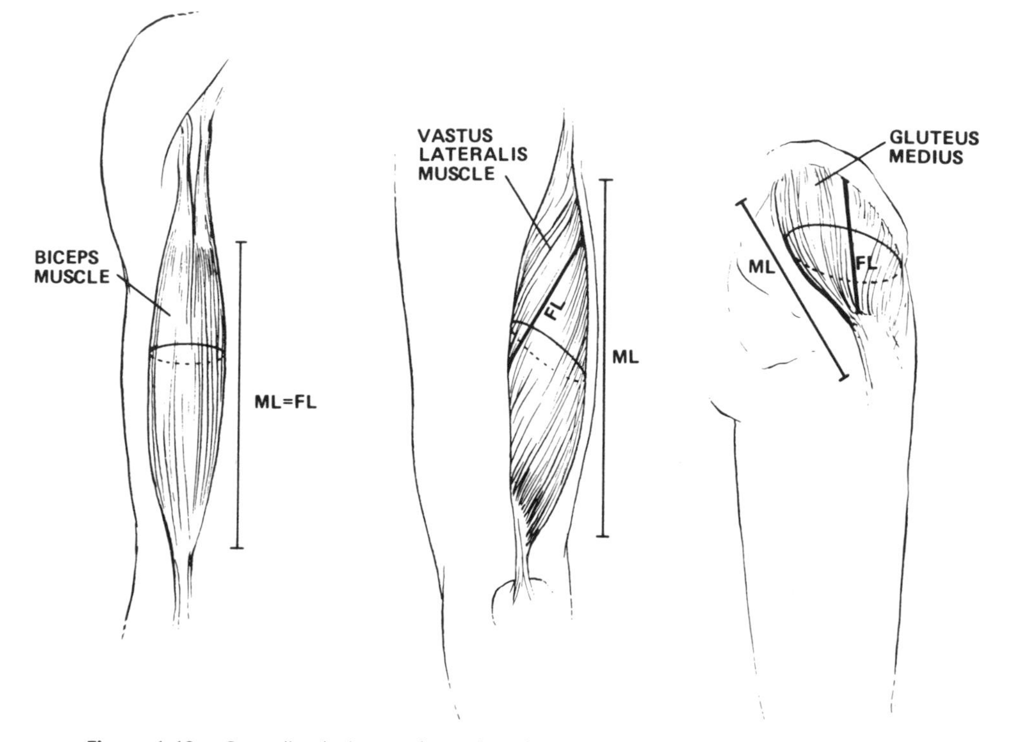

Figure 1.10. Generalized picture of muscle architectural types. Skeletal muscle fibers may be oriented along the muscle's force-generating axis (*left*), at a fixed angle relative to the force-generating axis (*middle*), or at multiple angles relative to the force-generating axis (*right*). Each of these represents an idealized view of muscle architecture and probably does not adequately describe any single muscle. ML, muscle length; FL, fiber length.

integrity during dissection. Ideally, the muscle was fixed while attached to the skeleton to roughly preserve its physiologic length. Following fixation, muscles were dissected from the skeleton, their mass determined, and their pennation angle (*i.e.,* the fiber angle relative to the force-generating axis) and muscle length were measured.

Currently, pennation angle is measured by determining the average pennation angle of fibers on the superficial muscle surface. While more sophisticated methods could be devised, it is doubtful they would provide a great deal more information. In measuring muscle length, it is important to note that muscle length is defined as the distance from the origin of the most proximal muscle fibers to the insertion of the most distal fibers. As mentioned above, muscle fiber length is *never* the same as the whole muscle length.

Muscle fiber length can only be determined by microdissection of individuals fibers from the fixed tissues. In general, unless investigators are explicit, when they refer to muscle fiber length, they are actually referring to muscle fiber *bundle* length. It is extremely difficult to isolate intact fibers, which run from origin to insertion, especially in mammalian tissue (Sacks and Roy, 1982; Loeb *et al.,* 1987; Ounjian *et al.,* 1991).

The final experimental step in performing architectural analysis is to determine the sarcomere length within the isolated bundles. This is necessary in order to compensate for differences in muscle lengths that occur during fixation. In other words, if we conclude that a muscle is "long," we must be sure that it is truly "long" and that is it was not simply fixed in a stretched position. Similarly, muscles measured to be "short" must be further investigated to ensure that they were not simply fixed at a short sarcomere length. In order to permit such conclusions, architectural measurements are always normalized or "adjusted" to a constant sarcomere length.

Having obtained muscle mass, fiber length, sarcomere length, muscle length, and pennation angle, a number of parameters can be calculated that summarize the muscle architecture. These parameters have a direct relation to the whole muscle's contractile properties.

Physiologic Cross-Sectional Area

Following measurement of the typical architectural parameters, the so-called physiologic cross-sectional area (PCSA) can be calculated. What is the significance of such a calculation? In short, the PCSA is directly proportional to the maximum tetanic tension of the muscle. This value is almost never the cross-sectional area of the muscle in any of the traditional anatomic planes, as would be obtained, for example, using a noninvasive imaging method such as magnetic resonance imaging (MRI) or computerized tomography (CT). Theoretically, PCSA represents the sum of the cross-sectional areas of all the muscle fibers within the muscle. It is calculated using Equation 1.1, which was pioneered by Carl Gans and verified experimentally by Roland Roy and Reggie Edgerton. In Equation 1.1, ρ represents muscle density (1.056 g/cm^3 for mammalian muscle) and θ represents surface pennation angle.

$$\text{PCSA (cm}^2) = \frac{\text{Muscle Mass (g)} \cdot \text{cosine } \theta}{\rho\ (\text{g/cm}^3) \cdot \text{Fiber Length (cm)}} \tag{1.1}$$

If we partition the equation into its components, the rationale for this expression becomes more clear. First, note that the muscle mass divided by density equals muscle volume (Equation 1.2). If the muscle were roughly cylindrical in shape, dividing volume by length (fiber length) would represent the cylinder cross-sectional area (Equation 1.3). Since, in our example, fiber length does not equal cylinder length, the area is not an actual area; rather, it is a theoretical area that would be occupied by a cylinder with a length equal to that of the fibers. Now, since the fibers may be oriented at some angle relative to the axis of force generation, the cosine term must be included (Equation 1.4). For those of you with a background in physics, the basis for this term is obvious. Because this is a concept that will be useful to us at a later stage, consider the situation shown in Figure 1.10.

$$\text{Volume (cm}^3) = \frac{\text{Muscle Mass (g)}}{\rho\ (\text{g/cm}^3)} \tag{1.2}$$

$$\text{CSA (cm}^2) = \frac{\text{Volume (cm}^3)}{\text{Fiber Length (cm)}} \tag{1.3}$$

Suppose a muscle fiber pulls with x units of force at an angle θ relative to the muscle axis of force generation. Clearly, some of the force of the fiber will not be transmitted along the axis but will be lost. Thus only a *component* of muscle fiber force will actually be transmitted along the muscle axis. Noting the right triangle in Figure 1.11**A,** it can be seen that the component of muscle force transmitted will be x-cosθ, since $\cos\theta = \frac{F}{x}$, which will *always* be less than x since cosθ is always less than 1. In other words, pennation itself results in a loss of muscle force relative to a muscle with the same mass and fiber length but with zero pennation angle. Why would the system be designed this way, such that force was lost? Consider the alternative. If the pennation angle were zero, the absolute size of the muscle would prohibit placing it in many bodily locations due to the large number of fibers that would have a PCSA equal to an anatomic CSA in, say, the transverse plane (Figure 1.11**B**). Thus it appears that pennation is a space-saving strategy even though it costs a bit in force generation. Pennation angle (θ) does not appear to have a large detrimental influence on PCSA in spite of this argument. This is because, as you know, the cosine of 0° is 1 and the cosine of 30° (which would be a very large pennation angle and is rarely encountered) is 0.87 which represents only a 13% force loss for a huge increase in fiber packing ability.

$$\text{PCSA (cm}^2) = \text{CSA (cm}^2) \cdot \cos\theta \tag{1.4}$$

The usefulness of this equation was recently highlighted by Roy and Edgerton in an experimental comparison between the *estimated* maximum muscle tetanic tension (based on PCSA calculations) and *measured* maximum tetanic tension (measured using traditional physiologic testing techniques). These investigators found that the estimations and predictions agreed within experimental error (Powell *et al.,* 1984). The only exception to that conclu-

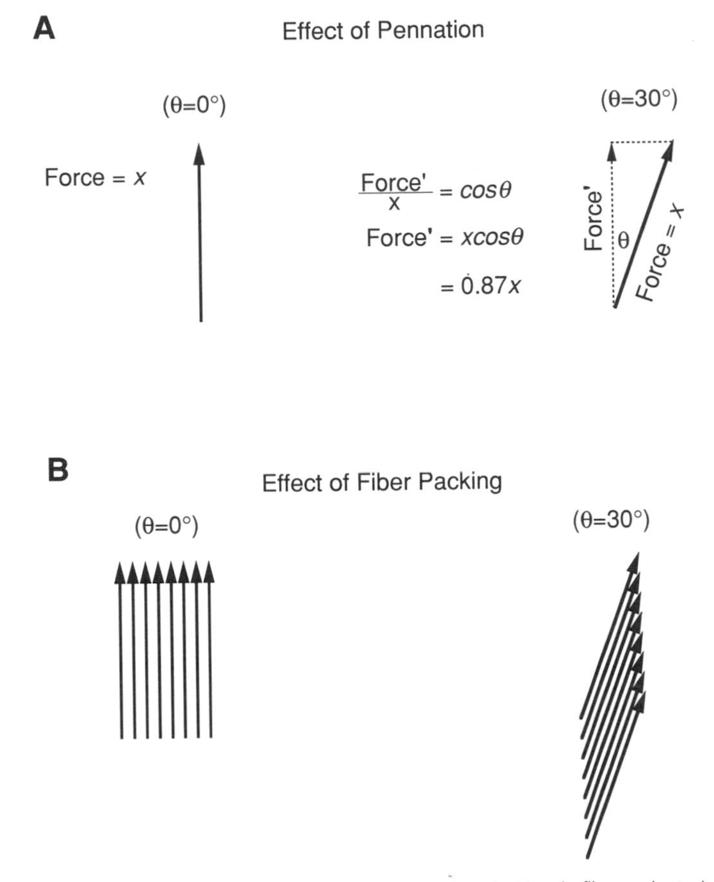

Figure 1.11. Schematic illustration of the effect of pennation. **A,** Muscle fibers oriented parallel to the axis of force generation transmit all of their force to the tendon. However, muscle fibers oriented at a 30° angle relative to the force-generating axis transmit only a portion of their force (cosine[30°] = 0.87, or 87%) to the tendon. **B,** While only about 90% of the muscle fiber force is transmitted to the tendon due to pennation, pennation itself permits packing of a large number of fibers into a smaller cross-sectional area.

sion was that the soleus muscle (a monoarticular plantar flexor of the deep calf) did not seem to agree. Interestingly, this was the only muscle tested that contained a large proportion of slow muscle fibers (see Chapter 2). These data may suggest that slow fibers generate less tension than fast fibers, but the jury is clearly not in in this regard.

It is important to highlight the observation that PCSA (and therefore maximum muscle tension) is *not* simply proportional to muscle mass (as is clear from the equation). In other words, given information on muscle mass or on muscle mass change (say, due to immobilization or spinal cord injury), we can make *no* statement with respect to muscle force. This is another way of saying that while mass is proportional to the amount of contractile material in the muscle, the *arrangement* of that material is of critical importance. We should also state that in some of these pathologic conditions, mass may change due to noncontractile proteins (*e.g.,* increased connective tissue or inflammatory cells). In such cases, PCSA will not accurately predict tetanic tension.

ARCHITECTURE OF HUMAN SKELETAL MUSCLES

Experimental measurement of human muscle architecture has a great deal of importance not only in understanding normal muscle function but also in understanding muscle adaptation (as we shall see in Chapters 4–6). In fact, if you as a therapist know nothing else about a muscle, knowing its architecture may be *the* most important.

Several architectural investigations have been performed in human upper and lower limbs. Tables of the relevant architectural features are presented in Tables 1.3 and 1.4. These tables look intimidating at first but will serve you primarily as a reference. Take some time to look at the normal range of parameters seen. For example, notice that pennation angles normally range from about 0° to 30°. Thus as mentioned above, pennation probably has a relatively small influence on muscle PCSA (and function). Note also that the ratio of muscle fiber length to muscle length (FL/ML ratio) typically ranges from about 0.2 to about 0.6. In other words, even the most longitudinally oriented muscles have fibers that extend only about 60% of the muscle length. Finally, note that there is a very poor correlation between muscle mass and muscle PCSA. Again, mass gives *little* information that is relevant to function.

Muscles of the Lower Limb

While each muscle is unique in terms of its architecture, taken as functional groups (*e.g.,* hamstrings, quadriceps, dorsiflexors, plantar flexors), a number of generalizations can be made (Figure 1.12). In terms of architecture, the typical properties of the various groups can be articulated (compare properties shown in Table 1.4; Wickiewicz *et al.,* 1983). The quadriceps are characterized by their relatively high pennation angles, large PCSAs, and short fibers. In terms of design, these muscles appear suited for the generation of large forces. The hamstrings, on the other hand, by virtue of their relatively long fibers and intermediate PCSAs, appear to be designed for large excursions (because excursions are proportional to fiber length). The same

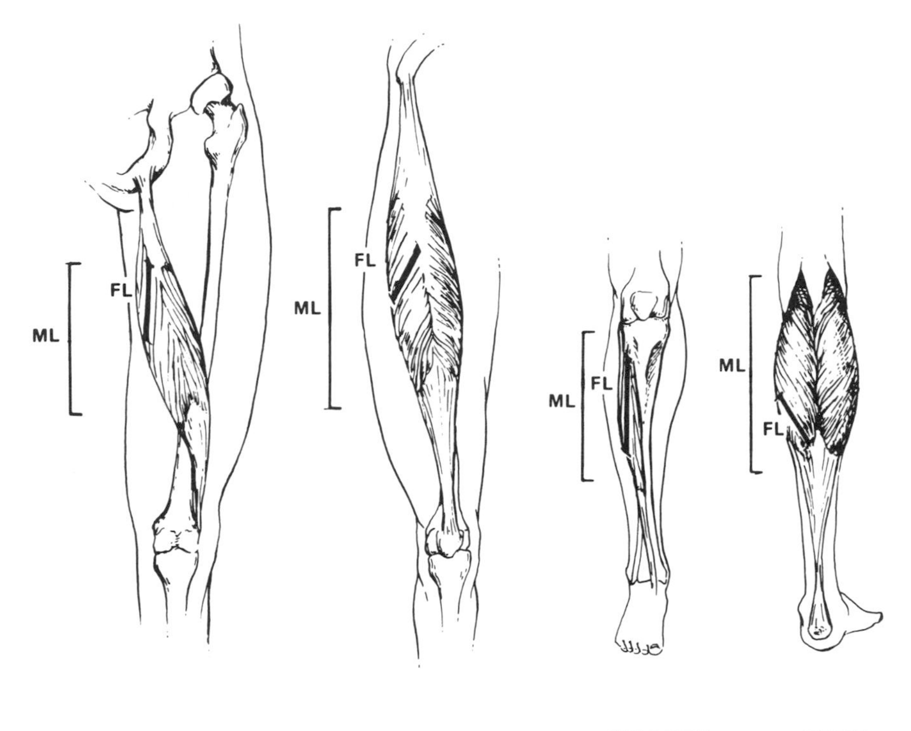

Figure 1.12. Schematic illustration of muscle architectural properties in the lower limb. Functionally, quadriceps and plantar flexors are designed for force production due to their low fiber length/muscle length ratios and large cross-sectional areas. Conversely, in general, hamstrings and dorsiflexors are designed for high excursions and velocity by nature of their high fiber length/muscle length ratios and long muscle fibers.

appears to be true of the plantar flexors and dorsiflexors, respectively. A very general (and a bit dangerous) conclusion might be that the antigravity extensors are more designed toward force production, while the flexors are more designed for high excursions. This generalization will break down upon close scrutiny but might provide a useful memory tool. You should be careful when considering architecture alone when trying to deduce function since normally muscles act via a moment arm to produce joint torque. Moment arms and strength will be discussed in Chapter 3 (page 121).

Probably the two most important muscle architecture parameters are muscle PCSA (which is proportion to maximum muscle force) and muscle fiber length (which is proportional to maximum muscle excursion). These

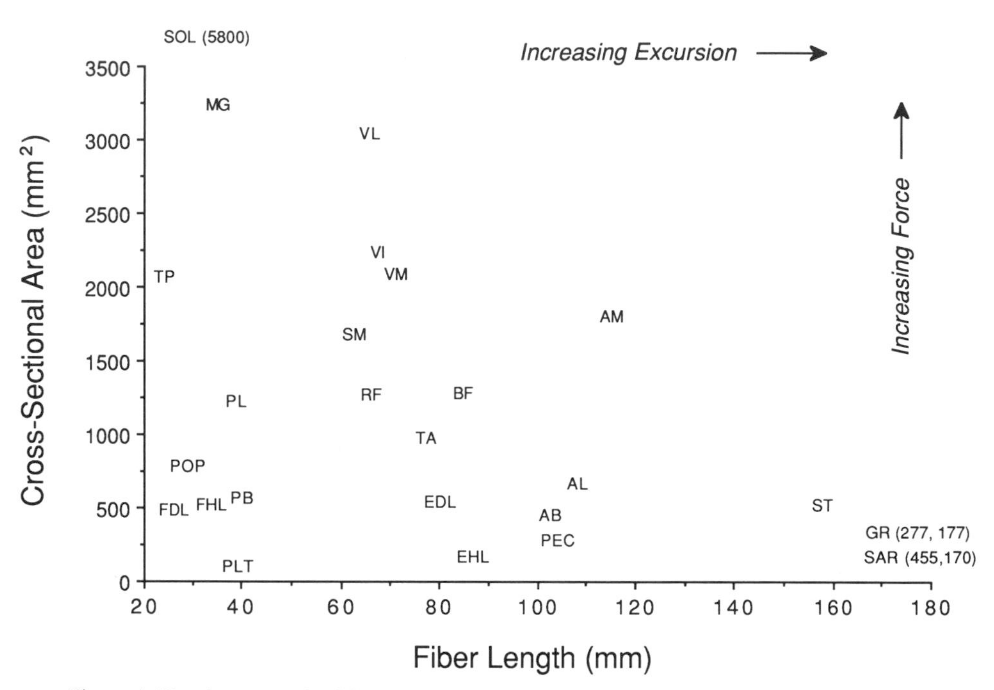

Figure 1.13. Scatter graph of fiber length and cross-sectional areas of muscles in the human lower limb (data from Wickiewicz *et al.,* 1983). Fiber length is proportional to muscle excursion, and cross-sectional area is proportional to maximum muscle force. Thus this graph can be used to compare the relative forces and excursions of leg muscles. See Table 1.4 for definitions of abbreviations.

two parameters are shown in graphical form for each muscle (Figures 1.13 and 1.14) and can be used to make general comparisons between muscles in terms of design. For example, note that the sartorius, semitendinosus, and gracilis muscles have extremely long fiber lengths and low PCSAs, which permit long excursions at low forces (Figure 1.13). At the other end of the spectrum is the soleus muscle, with its high PCSA and short fiber length, suitable for generating high forces with small excursions. Based on our understanding of the normal use of each of these muscles in the gait cycle (Chapter 3, page 143), these designs appear to be reasonable.

Muscles of the Upper Limb

In light of the specialization observed in the lower limb, the author and his colleagues were interested in understanding the architectural features of muscles in the human arm and forearm. While no such clear-cut generaliza-

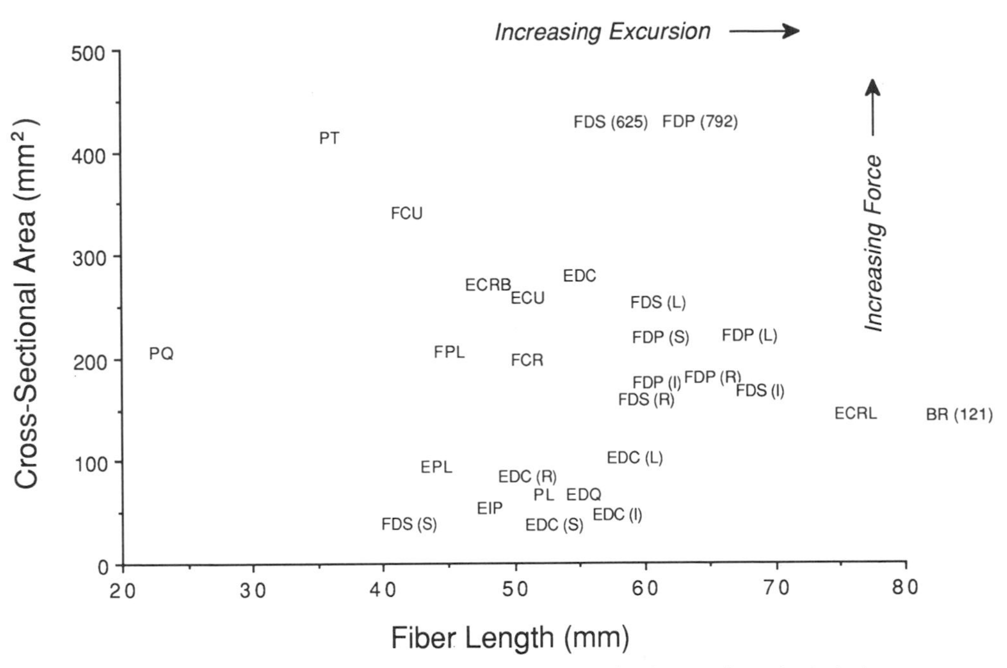

Figure 1.14. Scatter graph of the fiber length and cross-sectional areas of muscles in the human arm (data from Lieber *et al.,* 1990 and 1991). Fiber length is proportional to muscle excursion, and cross-sectional area is proportional to maximum muscle force. Thus this graph can be used to compare the relative forces and excursions of arm and forearm muscles. See Table 1.3 for definitions of abbreviations.

tions could be made (as were made for the lower limb), it was possible to demonstrate the extreme degree of specialization present in many upper limb muscles (Lieber *et al.,* 1990 and 1991). The details of these results are presented in Table 1.3. Again, note the high degree of specialization "built into" each of these muscles by virtue of their design. For example, the superficial and deep digital flexors to each digit are very similar to one another but very different from the digital extensors (Figure 1.14). Again, this type of scatter plot can be used to compare functional properties between muscles of the forearm. Clearly, such differences could be considered in surgical and rehabilitative procedures involving the upper limb. One might expect that when a muscle is surgically transferred to perform the function of another muscle whose function has been lost, matching of architectural properties may prove beneficial. We will return to this topic in Chapter 4 (page 172).

Significance of Muscle Architecture

After this relatively lengthy discussion of architecture, one might ask, "So what? What do we know now that we didn't know before?" My answer to that question is two-pronged. First, the fact that muscles that are composed of identical building blocks (sarcomeres) can have such dramatically different force-generating properties highlights the clever design of the human body. The body uses identical components and arranges them in different ways to construct different "motors," which produce movement. Second, by virtue of architectural specialization, it is clear that the neuromuscular system does not simply modify muscular force and excursion by changing the nervous input to the muscles. Muscles are *designed* for a specific function—large excursion, for example. The nervous system provides the signal for the muscle to "do its thing" but does not necessarily specify the details of that action. It is as if the nervous system acts as the central control while the muscle interprets the control signal into an external action by virtue of its intrinsic design. This elegant design is but one of many that we will encounter in our voyage through the neuromusculoskeletal system.

SUMMARY

Skeletal muscles arise by a unique developmental process that includes axonal outgrowth, myogenesis, neuromuscular junction formation, and synapse elimination. The fully differentiated muscle cell is uniquely suited to perform force generation and movement. A sterotypical view of the muscle fiber can be presented in which force-generating, force-regulating, and force-activating roles are assigned to various structures. The structural muscle hierarchy proceeds from the whole muscle all the way to the myofilaments. The arrangement of muscle fibers within whole muscles is known as muscle architecture and is an important factor in determining whole muscle properties. We will soon see that skeletal muscle fibers can be differentiated into various types, the details of which will be presented in the following chapter.

REFERENCES

Bennett, M.R. (1983). Development of neuromuscular synapses. Physiol. Rev. 63:915–1048.

Buller, A.J., Eccles, J.C., and Eccles, R.M. (1960a). Differentiation of fast and slow muscles in the cat hindlimb. J. Physiol. 150:399–416.

Buller, A.J., Eccles, J.C., and Eccles, R.M. (1960b). Interactions between motorneurons and muscles in respect to the characteristic speeds of their responses. J. Physiol. 150:417–439.

Ebashi, S, Maruyama, K., and Endo, M. (1980). Muscle Contraction. Its Regulatory Mechanisms. New York: Springer-Verlag.

Eisenberg, B.R. (1983). Quantitative ultrastructure of mammalian skeletal muscle. In: Peachey, L.D., ed. Handbood of Physiology. Bethesda, MD: American Physiological Society, 73–112.

Gans, C. (1982). Fiber architecture and muscle function. Exerc. Sports Sc. Rev. 10:160–207.

Haselgrove, J.C. (1983). Structure of vertebrate striated muscle as determined by x-ray-diffraction studies. In: Peachey, L.D., Adrian, R.H., Geiger, S.R., eds. Handbook of Physiology. Bethesda, MD: American Physiological Society, 143–171.

Huxley, A.F., and Niedergerke, R. (1954). Structural changes in muscle during contraction. Interference microscopy of living muscle fibers. Nature. 173:971–973.

Huxley, H.E., and Hanson, J. (1954). Changes in the cross-striations of muscle during contraction and stretch, and their structural interpretation. Nature. 173:973–976.

Huxley, H.E., Simmons, R.M., Faruqi, A.R, Kress, M., and Koch, M.H.J. (1981). Millisecond time-resolved changes in x-ray reflections from contracting muscle during rapid mechanical transients, recorded using synchrotron radiation. Proc. Natl. Acad. Sci. 78:2297–2301.

Jansen, J.K.S, and Fladby, T. (1990). The perinatal reorganization of the innervation of skeletal muscle in mammals. Prog. Neurobiol. 34:39–90.

Kayar, S.R., Hoppeler, H., Mermod, L., and Weibel, E.R. (1988). Mitochondrial size and shape in equine skeletal muscle: a three-dimensional reconstruction study. Anat. Rec. 222:333–339.

Kelly, A.M. (1983). Emergence of specialization in skeletal muscle. In: Peachey, L.D., ed. Handbook of Physiology. Bethesda, MD: American Physiological Society, 417–486.

Kirkwood, S.P., Munn, E.A., and Brooks, G.A. (1986). Mitochondrial reticulum in limb skeletal muscle. Am. J. Physiol. 251:C395–C402.

Lance-Jones, C., and Landmesser, L. (1978). Effect of spinal cord deletions and reversals on motoneuron projection patterns in the embryonic chick hindlimb. Soc. Neurosci. 4:118(Abstr.).

Landmesser, L.T. (1980). The generation of neuromuscular specificity. Ann. Rev. Neurosci. 3:279–302.

Lieber, R.L., Fazeli, B.M, and Botte, M.J. (1990). Architecture of selected wrist flexor and extensor muscles. J. Hand Surg. 15:244–250.

Lieber, R.L., Jacobson, M.D., Fazeli, B.M., Abrams, R.A., and Botte, M.J. (1992). Architecture of selected muscles of the arm and forearm: anatomy and implications for tendon transfer. J. Hand Surg. In press.

Loeb, G.E., Pratt, C.A., Chanaud, C.M., and Richmond, F.J.R. (1987). Distribution and innervation of short, interdigitated muscle fibers in parallel-fibered muscles of the cat hindlimb. J. Morphol. 191:1–15.

Miller, J.B., and Stockdale, F.E. (1986). Developmental origins of skeletal muscle fibers; clonal analysis of myogenic cell lineages based on expression of fast and slow myosin heavy chains. Proc. Natl. Acad. Sci. 83:3860–3864.

Miller, J.B., and Stockdale, F.E. (1987). What muscle cells know that nerves don't tell them. Trends Neurosci. pp. 10–12.

Ounjian, M., Roy, R.R., Eldred, E., Garfinkel, A., Payne, J.R., Armstrong, A., Toga, A.W., and Edgerton, V.R. (1991). Physiological and developmental implications of motor unit anatomy. J. Neurobiol. 22:547–559.

Peachey, L.D., and Franzini-Armstrong, C. (1983). Structure and function of membrane systems of skeletal muscle cells. In: Peachey, L.D., ed. Handbook of Physiology. Bethesda, MD: American Physiological Society, 23–73.

Peachey, L.D., and Eisenberg, B.R. (1978). Helicoids in the T system and striations of frog skeletal muscle fibres seen by high voltage electron microscopy. Biophys. J. 22:145–154.

Pollack, G.H. (1983). The cross-bridge theory. Physiol. Rev. 63:1049–1113.

Poo, M.M. (1982). Rapid lateral diffusion of functional ACh receptors in embryonic muscle cell membrane. Nature. 295:333–334.

Powell, P.L., Roy, R.R., Kanim, P., Bello, M.A., and Edgerton, V.R. (1984). Predictability of skeletal muscle tension from architectural determinations in guinea pig hindlimbs. J. Appl. Physiol. 57:1715–1721.

Sacks, R.D., and Roy, R.R. (1982). Architecture of the hind limb muscles of cats: functional significance. J. Morphol. 173:185–195.

Squire, J. (1981). The structural basis of muscular contraction. New York: Plenum Press.

Thompson, W.J., Sutton, L.A., and Riley, D.A. (1984). Fibre type composition of single motor units during synapse elmination in neonatal rat soleus muscle. Nature. 309:709–711.

Vandenberg, H.H. (1982). Dynamic mechanical orientation of skeletal myofibers *in vitro*. Dev. Biol. 93:438–443.

Wickiewicz, T.L., Roy, R.R., Powell, P.L., and Edgerton, V.R. (1983). Muscle architecture of the human lower limb. Clin. Orthop. Rel. Res. 179:275–283.

Zajac, F.E., and Gordon, M.E. (1989). Determining muscle's force and action in multi-articular movement. Exerc. Sport Sci. Rev. 17:187–230.

Chapter

2

Skeletal Muscle Physiology

OVERVIEW

This chapter describes the way that muscle structures produce the desired function. Under the general classification of "physiology," the chapter has been divided into three parts. First, we discuss the activation and contraction sequence of muscle, along with some functional consequences of this scheme. Next, the two basic mechanical properties of muscle—the length-tension and force-velocity properties—are highlighted. The basis for all muscle contraction is presented as well as the details of the cross-bridge cycle. The manner in which architecture affects these mechanical properties is included. Finally, the topics of muscle fiber types and motor units are presented, which enables discussion of recruitment, locomotion, and fatigue.

INTRODUCTION

In this chapter, anatomy meets physiology. In other words, the rationale for the physical arrangement of the various muscle components will become apparent. An understanding of muscle physiology is predicated on a good understanding of muscle macro- and microanatomy. This chapter represents the payoff for having waded through the previous one. It should be noted that anatomic studies are rarely performed in isolation of physiologic studies and *vice versa*. In fact, anatomists routinely refer to physiologic data in describing the significance of their findings, and physiologists routinely refer to anatomic studies in proposing mechanisms for their observations. Thus the distinction between muscle anatomy and physiology is often one of orientation. Significant cross-referencing between muscle anatomic and physiologic studies will be required as we continue our discussion of skeletal muscle structure and function.

PART 1: FIBER ACTIVATION

EXCITATION-CONTRACTION COUPLING

Our discussion of skeletal muscle physiology begins with the process of muscle activation itself. It is well known that peripheral nerves innervate skeletal muscles and that neural activation precedes muscle contraction. The precise process by which this neural activation signal culminates in muscle contraction is known as excitation-contraction coupling, or EC coupling (Figure 2.1). EC coupling is viewed as a sequence of events, each of which is necessary for contraction to occur. If any single step of EC coupling is impaired, muscle contraction does not occur normally. This impairment might be interpreted as muscle paralysis or fatigue. However, such a general classification is not useful unless the underlying cause is known (Ebashi *et al.*, 1980).

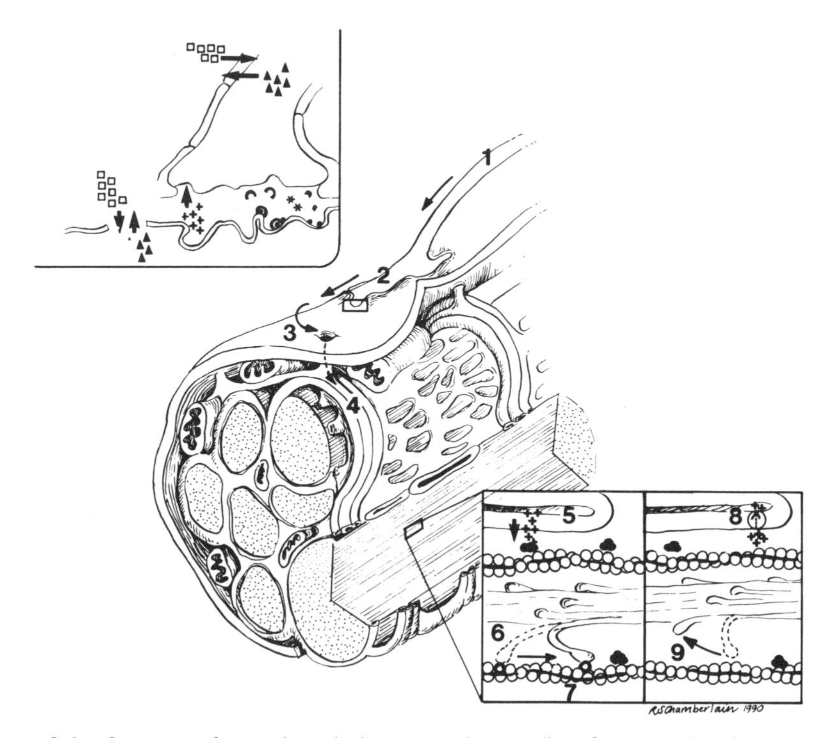

Figure 2.1. Sequence of events in excitation-contraction coupling of a nervous impulse to muscle contraction. **1,** Action potential conducted by nerve to muscle (squares represent Na^{+} ions entering nerve, and triangles represent K^{+} ions leaving nerve to conduct the action potential). **2,** Nervous impulse transmitted across neuromuscular junction to muscle fiber (crosses represent Ca^{++} ions entering nerve end, half-moons represent the neurotransmitter ACh, and asterisks represent the enzyme acetylcholinesterase degrading ACh). **3,** Action potential conducted along fiber surface. **4,** Action potential conducted deep into fiber via the T-system. **5,** Ca^{++} released from SR to activate actin filament. **6** and **7,** Cross-bridge produces force and filament sliding **8,** Ca^{++} pumped back into SR. **9,** Cross-bridge relax due to lack of Ca^{++} filament activation.

Action Potential—the First Step

The first step in the EC coupling chain is the generation of the peripheral nerve action potential. The action potential results from depolarization of the peripheral nerve axon that innervates the muscle. In addition to a signal from the CNS, the axon may be depolarized in a number of ways, including trauma to the peripheral nerve or application of an external electrical stimulating device. In any case, the resulting action potentials that propagate down the peripheral nerve are identical. The action potential arrives at the neuromuscular junction, the interface between muscle and nerve. The neuromuscular junction is itself a complex structure. The nerve ends in a small indentation on the muscle fiber surface, known as the synaptic cleft.

Acetylcholine Release—the Neurotransmitter

The end of the nerve contains packets of the neurotransmitter acetylcholine (ACh), which causes muscle fiber excitation. ACh is synthesized by the cell body of the motor nerve and transported down the axon where it is stored at nerve endings for later use. Following nerve depolarization, a quantum or unit of ACh is released into the small space between the muscle and nerve, the synaptic cleft (Figure 2.1). ACh then diffuses across the synaptic cleft and binds to the ACh receptor, which is integrated into the muscle membrane (refer to the discussion of synaptogenesis in Chapter 1, page 12). ACh binding results in depolarization of the muscle fiber sarcolemma and an action potential that propagates from the neuromuscular junction outward in all directions.

Transverse Tubular System and Sarcoplasmic Reticulum Involvement in Excitation-Contraction Coupling

At various intervals along the fiber surface, the action potential encounters invaginations of the sarcolemma that extend into the fiber—the transverse tubular system (T-system; Figure 2.1; Peachey and Franzini-Armstrong, 1983). The action potential is conducted deep into the fiber by the T-system. The interface between the "outside world" of the muscle fiber and the "inside world" of the contractile apparatus occurs at the next step where the T-system signals the sarcoplasmic reticulum (SR) to release calcium. The precise mechanism for this communication is not completely understood. However, it is believed that the SR feet, which anchor the terminal region of the SR to the T-system, are involved in some way. An important observation is that the lumen of the T-system contains *extracellular* fluid while the lumen of the SR contains *intracellular* fluid. Thus extracellular fluid is actually contained deep within the muscle fiber!

Calcium Release Results in Muscle Contraction

After the T-system signals the SR, the SR releases calcium ions in the region of the myofilaments (Figure 2.1). This release process is extremely fast. The calcium ions bind to troponin, the actin filament regulatory protein, which in turn releases the inhibition on the actin filament, permitting interaction with the myosin filament and resulting in cross-bridge cycling (see details below) *i.e.,* force generation.

Calcium Uptake Results in Muscle Relaxation

As long as neural impulses arrive at the neuromuscular junction and, therefore, calcium concentrations remain high in the region of the myofilaments, force generation continues. However, when the impulses cease, calcium is pumped back into the SR by the calcium-activated adenosine triphosphatase (ATPase) enzyme. The calcium-activated ATPase enzyme is an integral protein that is embedded in the bilayer of SR membrane. The mechanism of action of this enzyme has been thoroughly studied and is one of the best understood of the ion transport enzymes (Entman and Van Winkle, 1986). The calcium pumping process is energy dependent and requires ATP. When calcium levels in the region of the myofilaments drop below a critical level, thin filament inhibition again resumes, and actin-myosin interaction is prevented. This inhibition is manifest externally as muscle fiber relaxation.

We thus have the chain of events required for muscle contraction to occur following nerve depolarization:

1. Generation of the peripheral nerve action potential
2. Release of ACh from the nerve terminal
3. Binding of ACh to the muscle fiber ACh receptor
4. Depolarization of the sarcolemma after ACh receptor binding
5. Conduction of the action potential into the fiber by the T-system
6. Signaling of the SR by the T-system to release calcium
7. Binding of calcium to the regulatory protein troponin, permitting actin-myosin interaction
8. Force generation resulting from actin-myosin interaction
9. Pumping of calcium back into the SR when neural activation ceases, resulting in inhibition of actomyosin interaction and muscle relaxation

TEMPORAL SUMMATION

A well-known muscle contractile property follows directly from an understanding of the EC coupling sequence presented above. First, it should be obvious that the time required for activation, contraction, and then relaxation to occur is finite. That is, excitation (with accompanying calcium release) is relatively rapid (on the order of about 5 msec) while contraction and relaxation are relatively slow (on the order of about 100 msec). The mechanical consequence of the activation process (*i.e.,* the muscle twitch) lags far behind the activation process itself. For example, let us suppose that the entire EC coupling process requires 100 msec. If, after the first impulse, we deliver a second impulse before 100 msec have elapsed, the muscle will be signaled to contract before it has fully relaxed. In other words, the second impulse will be superimposed somewhat on part of the cycle initiated by the first one, resulting in summation. Because the two events have summated due to their relative temporal relationship, this process is referred to as temporal summation.

The physiologic effects of temporal summation are quite dramatic. One effect of the first stimulus is to cause the contracting sarcomeres to "stretch out" the passive structures that lie in series with them (*e.g.,* tendons or passive sarcomeres). When the second impulse "arrives at the scene," it is not required to stretch out any of these structures and causes a greater force to be generated at the ends of the muscle fiber. Thus two impulses that are delivered to a muscle fiber and separated by only about 50 msec result in more force than the same two pulses delivered to the muscle but separated by more time. If a "train" of such pulses (say, 50 pulses in a row) is delivered to the muscle, separated in time by different amounts, this results in a tetanic contraction, and the resulting force is quite different (Figure 2.2). Higher forces result when stimuli are delivered at higher frequencies since there is less time for relaxation (frequency = 1/interpulse interval; low intervals correspond to high frequencies). Notice, in Figure 2.2, that at relatively low frequencies (*e.g.,* 10 Hz), the contractile record almost completely relaxes between successive pulses. This is referred to as an *unfused contraction,* because it is still possible to distinguish individual contractile events within the force record. However, note that as stimulation frequency increases, the tetanic record becomes more fused, until at very high frequencies (*e.g.,* 100 Hz), the contractile record becomes a *fused contraction.* A fused tetanic contraction appears as such because the repeated calcium release onto the myofilaments is much faster than the rate at which the myofilaments can relax.

Rate Coding: The Physiologic Significance of Temporal Summation

The variation in force obtained by altering activation frequency is known as frequency or rate coding. Because muscle force varies as a function of activation frequency, this is one method the CNS can use to alter muscle force. If high forces are required at the periphery, the CNS can deliver high-frequency pulses. Conversely, if only low forces are required, the CNS can deliver low-frequency

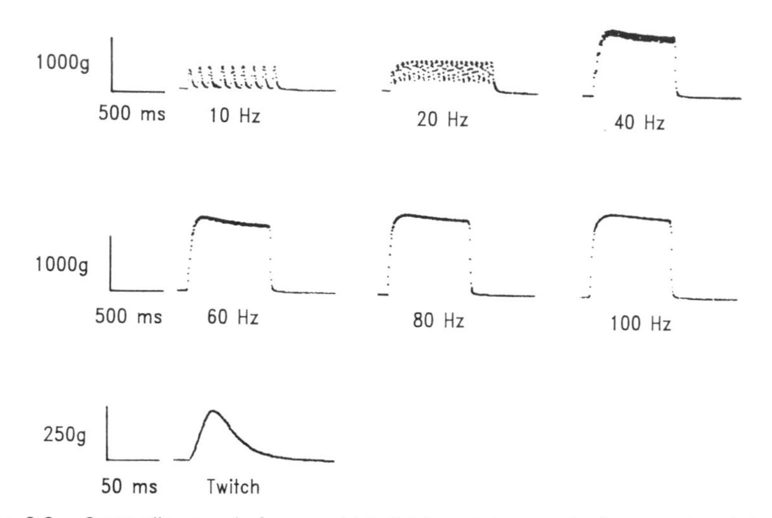

Figure 2.2. Contractile records from a rabbit tibialis anterior muscle demonstrating fusion of mechanical twitches as stimulation frequency increases (temporal summation). Note different tension calibration bars for tetani (upper panels) and twitch (lower panel). (From Lieber RL, Smith DE, Hargens AR. Real-time acquisition and data analysis of skeletal muscle contraction in a multi-user environment. Comp Prog Methods Biomed 1986;22:259–265.)

pulses. Of course, this type of effect is very difficult to demonstrate directly. In a very technical procedure involving single motor unit recording, ventral root recording, and muscle tension recording *in situ* during locomotion (as well as some fancy postexperimental data processing), Andy Hoffer and colleagues were able to demonstrate what they interpreted to be rate coding, which occurred during normal locomotion (Hoffer *et al.*, 1981 and 1987).

We shall see, however, that the control of muscle force by the CNS is very much more sophisticated than this. The area of study that includes the control of muscles by the nervous system is known as neuromotor control (Binder and Mendell, 1990), which will be discussed in the chapters to follow. Suffice it to say for the present that muscle and nerve properties are matched in a very sophisticated fashion in order to accomplish a particular task. Rate coding is only one of the methods by which this match is accomplished.

PART 2: SKELETAL MUSCLE MECHANICS

LENGTH-TENSION RELATIONSHIP: ISOMETRIC MUSCLE CONTRACTION

Since the late 1800s, it has been known that the force developed by a muscle during isometric contraction (*i.e.*, when the muscle is not allowed to shorten) varies with its starting length (see review by Podolsky and Shoenberg, 1983). The isometric length-tension curve is generated by maximally stimulating a skeletal muscle at a variety of discrete lengths and measuring the tension generated at each length. When maximum tetanic tension at each length is plotted against length, a relationship such as that shown in Figure 2.3 is usually obtained. While a general description of this relationship was established early in the history of biologic science, the precise structural basis for the length-tension relationship in skeletal muscle was not elucidated until the sophisticated mechanical experiments of the early 1960s were performed. It was these experiments that defined the precise relationship between myofilament overlap and tension generation, which we refer to today as the length-tension relationship. In its most basic form, the length-tension relationship states that tension generation in skeletal muscle is a direct function of the magnitude of overlap between the actin and myosin filaments.

Sarcomere Length-Tension Relationship

In the late 1950s and early 1960s, Andrew Huxley, Albert Gordon, and Fred Julian, working in England (Gordon *et al.*, 1966), and Paul Edman, working in

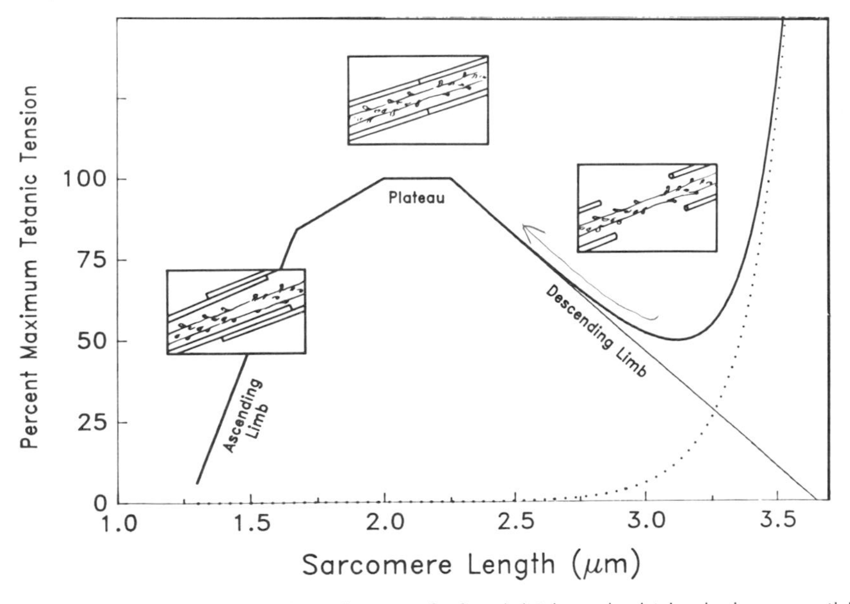

Figure 2.3. The sarcomere length-tension curve for frog skeletal muscle obtained using sequential isometric contractions in single muscle fibers. Insets show schematic arrangement of myofilaments in different regions of the length-tension curve. Dotted line represents passive muscle tension.

Sweden (Edman, 1966), defined what might be one of the most explicit structure-function relationships in all of biology. It was obvious to these investigators that in order to determine the detailed structural basis for the length-tension relationship, isolated, intact single skeletal muscle fibers would be required. The muscle used was that of the frog since a great deal was known at the time about its structure, and intact single fibers could be isolated. Picture this: The experiments I am about to describe were performed on tissue (single muscle cells) approximately 8 mm long and 75 μm in diameter! That's small!

Andrew Huxley (who by this time had already received the Nobel prize with Alan Hodgkin for determining the mechanism of the nerve action potential) invented a mechanical version of his voltage clamp apparatus that was used in the nerve studies. This apparatus was designed to keep a small segment of the fiber at a constant length (and therefore keep a region of the fiber at a constant sarcomere length). This enabled him to make a unique correlation between muscle tension and sarcomere length. It was much easier said than done because it turned out that 'isometric" force generation in the single fiber was anything but isometric! Sarcomeres in the end region of the fiber tended to stretch sarcomeres in the central region, and thus the apparatus Huxley developed was absolutely critical.

The results of the classic experiments by Gordon, Huxley, and Julian (1966) are summarized in Figure 2.3. In this figure, muscle relative tetanic tension (as a percentage of maximum) is plotted as a function of sarcomere length (in μm). This was one case where anatomy met physiology in dramatic fashion, because knowledge of the precise anatomic lengths of the myosin and actin filaments was crucial for understanding the basis of this relationship.

Descending Limb of the Length-Tension Curve

As a muscle was highly stretched by the investigators to a sarcomere length of 3.65 μm, the muscle developed no active force. Why did the muscle develop zero force at this length? The answer lay in the observation that, since the myosin filament is 1.65 μm long and the actin filament is 2.0 μm in length, at a sarcomere length of 3.65 μm, there is no overlap (interdigitation) between the actin and myosin filaments. Therefore, although the EC coupling process might *permit* actin-myosin interaction by removing the inhibition on the actin filament, because no myosin cross-bridges are in the vicinity of the actin active sites, no force generation can occur.

As the muscle was allowed to shorten, overlap between actin and myosin was possible, and the amount of force generated by the muscle increased as sarcomere length decreased. Increasing force with decreasing sarcomere length occurred until the muscle reached a sarcomere length of 2.2 μm. Why did tension slowly increase? Over the range of sarcomere lengths from 2.2–3.65 μm, as sarcomere length decreases, the number of cross-bridges between actin and myosin increases, resulting in increased force. This region of the length-tension curve is known as the descending limb.

Plateau Region of the Length-Tension Curve

As sarcomere length changed from 2.0 μm to 2.2 μm, muscle force remained constant. Again, this was a direct result of thick filament structure. Recall from Chapter 1 that the myosin filament is a polymeric arrangement of myosin molecules arranged in an antiparallel fashion. Because many myosin "backbones" (the light meromyosin portion of the myosin molecules) come together in the center of the myosin filament, there exists a bare region of the myosin molecule that is devoid of cross-bridges. You guessed it—the length of the bare region was 0.2 μm! Thus while sarcomere length shortening over the range 2.2–2.0 μm results in greater filament overlap, it does not result in increased force generation since no additional cross-bridge connections are made. The region of the length-tension curve over which length change results in no change in force is known as the plateau region. The maximum tetanic tension of the muscle in this region is abbreviated P_O. The length at which P_O is attained is known as optimal length (L_O).

Ascending Limb of the Length-Tension Curve

At a sarcomere length of 2.0 μm, notice that the actin filaments from one side of the sarcomere juxtapose the actin filaments from the opposite side of the sarcomere (Figure 2.3). It might be predicted that shortening past this point would be impossible. However, as sarcomere length decreases below the plateau region, actin filaments from one side of the sarcomere double overlap with the actin filaments on the opposite side of the sarcomere. That is, at these lengths, actin filaments overlap both with themselves and with the myosin filament. Under these double-overlap conditions, the actin filament from one side of the sarcomere interferes with cross-bridge formation on the other side of the sarcomere, and this results in decreased muscle force output. This occurs from 2.0–1.87 μm, and this region is known as the shallow ascending limb of the length-tension curve. The word "shallow" distinguishes it from the next portion of the length-tension curve, which is known as the steep ascending limb, because at these very short lengths, the myosin filament actually begins to interfere with shortening as it abuts the sarcomere Z-disk, reducing force precipitously.

An interesting observation relative to muscle force generation at short lengths was made in the late sixties by Rüdel and Taylor (1971). They observed that when an intact muscle fiber was stimulated at very short sarcomere lengths (*i.e.,* sarcomere lengths on the ascending limb), electrical failure of the EC coupling apparatus occurred. This raised the question as to whether the decreased force at short sarcomere lengths was actually due to myofilament properties or was simply an electrical failure phenomenon. To address this question, Taylor and Rüdel (1970) ensured maximal single fiber activation by bathing the fiber in caffeine (which enhances calcium release from the SR) and obtained the same relationship as Gordon, Huxley, and Julian had obtained. Rick Moss repeated the experiment on small *pieces* of single muscle fibers, which were activated chemically using a calcium buffering system (Moss, 1979). Again, the same relationship was obtained. Thus while shortening deactivation as described by Rüdel and Taylor could occur, it did not seem to detract from the elegance and truth of the sarcomere length-tension relationship itself.

To summarize, the length-tension relationship states that muscle force varies as a function of sarcomere length (myofilament overlap). This is a physiologic property of the force-generating system and should not simply be viewed as an anatomic artifact. Recent experimental studies suggest that this length-tension relationship can be advantageous to the musculoskeletal torque-generating system, as will be described in Chapter 3.

Origin of the Passive Portion Length-Tension Curve

The solid line in Figure 2.3 represents the tension generated if a muscle is stretched to various lengths without stimulation. Note that near the optimal length, passive tension is almost zero. However, as the muscle is stretched to longer lengths, passive tension increases dramatically. These relatively long lengths can be attained physiologically, and therefore, passive tension can play a role in providing resistive force even in the absence of muscle activation. What is the origin of passive tension? Obviously, the structure(s) responsible for passive tension are outside of the cross-bridge itself since muscle activation is not required. Recent studies performed by Alan Magid have shown that the origin of passive muscle tension is actually *within* the myofibrils themselves. He demonstrated this by chemically stripping the sarcolemma from a single muscle fiber and measuring passive tension (Magid and Law, 1985). Interestingly, a new structural protein has also been identified, which may be the source of this passive tension. The very large protein, creatively named "titin," connects the thick myosin filaments end to end. This very large protein is also relatively fragile and thus has probably been missed in earlier studies because the laboratory techniques destroyed the protein. In addition to passively supporting the sarcomere, titin stabilizes the myosin lattice so that high muscle forces do not disrupt the orderly hexagonal array. If titin is selectively destroyed, normal muscle contraction causes significant myofibrillar disruption (Horowits and Podolsky, 1987).

Before leaving the length-tension relationship, let me present one caution: Never try to describe a shortening muscle using the length-tension relationship. In other words, looking at Figure 2.3, one might be tempted to predict that as a muscle shortens from a long length, force increases. However, one must remember that the length-tension relationship is strictly valid only for *isometric* contractions. Thus the curve represents the artificial connection of individual data points from isometric experiments. In order to describe *motion,* we will require an understanding of the force-velocity relationship, presented below.

FORCE-VELOCITY RELATIONSHIP: ISOTONIC MUSCLE CONTRACTION

Unlike the length-tension relationship, the force-velocity relationship does not have a precise, anatomically identifiable basis. The force-velocity relationship states that the force generated by a muscle is a function of its velocity. It can also be stated in the reverse, such that the velocity of muscle contraction is dependent on the force resisting the muscle. Historically, the force-velocity relationship was used to define the kinetic properties of the cross-bridges as well as the precise force-velocity relationship itself.

Experimental elucidation of the force-velocity relationship was first presented by A. V. Hill and Bernard Katz in their classic papers (Hill, 1938; Katz, 1939), but the current description of the force-velocity relationship has been ascribed to the physiologist A. V. Hill (see summary in Hill, 1970). Hill, in his decades of important muscle studies, generated an equation for the muscle force-velocity relationship that is still in use today. Interestingly, Andrew Huxley, in 1957, developed a theory of isotonic muscle contraction based on specific cross-bridge properties, which yielded the actual force-velocity relationship and explained the amount of energy used by a muscle during contraction at different velocities (Huxley, 1957; Hill, 1964). The beauty of this theory was its ability to explain both mechanical and energetic data.

Experimentally, the force-velocity relationship, like the length-tension relationship, is a curve that actually represents the results of many experiments plotted on the same graph. Experimentally, a muscle is stimulated maximally and allowed to shorten (or lengthen) against a constant load. The muscle velocity during shortening (or lengthening) is measured and then plotted against the resistive force. The general form of this relationship is plotted in Figure 2.4. On the horizontal axis we have plotted muscle velocity relative to maximum velocity (V_{max}) while on the vertical axis we have plotted muscle force relative to maximum force (P_O).

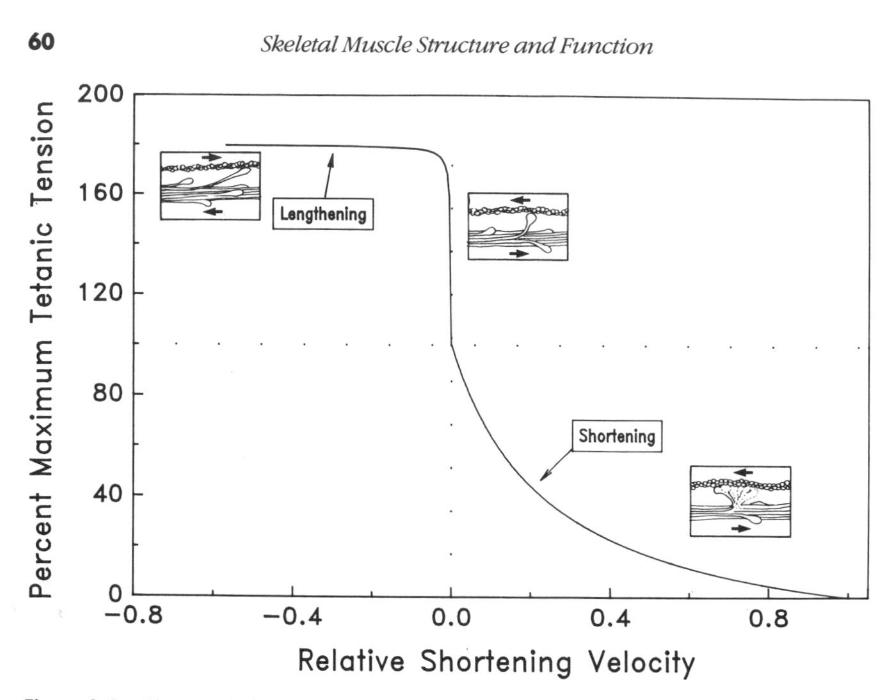

Figure 2.4. The muscle force-velocity curve for skeletal muscle obtained using sequential isotonic contractions in single fibers. *Insets* show schematic representation of cross-bridges. Note that force increases dramatically upon forced muscle lengthening.

Concentric Contractions—Muscle Actively Shortening

When a muscle is activated and required to lift a load that is less than its maximum tetanic tension, the muscle begins to shorten. Contractions that permit the muscle to shorten are known as concentric contractions. In concentric contractions, the force generated by the muscle is always less than the muscle's maximum (P_O). As the load the muscle is required to lift decreases, contraction velocity increases. This occurs until the muscle finally reaches its maximum contraction velocity, V_{max}. V_{max} is a parameter we can use to characterize muscle, which is related to both fiber type distribution and architecture. The mathematical form of the force-velocity relationship is a rectangular hyperbola and is given in Equation 2.1:

$$(P+a)v = b(P_O - P) \tag{2.1}$$

where a and b are constants derived experimentally (usually about 0.25), P is muscle force, P_O is maximum tetanic tension, and v is muscle velocity. This equation can be used to determine the relative muscle force that occurs as a muscle is allowed to shorten. Some of these values are presented below in Table 2.1. It is important to note that the force-velocity relationship is a steep rectangular hyperbola. In other words, force drops off rapidly as velocity increases. For example, in a muscle that is shortening at only 1% of its maximum contraction velocity (extremely slow), tension drops by 5% relative to maximum isometric tension. Similarly, as contraction velocity increases to only 10% maximum (easily attainable physiologically), muscle force drops by 35%! Note that even when muscle force is only 50% maximum, muscle velocity is only 17% V_{max}. The take-home lesson is that as a muscle is allowed to shorten, force drops precipitously.

What is the physiologic basis of the force-velocity relationship? It has been determined that the cross-bridges between actin and myosin attach at a certain rate and detach at a certain rate (see below). These rates are referred to as *rate constants.* At any point in time, the force generated by a muscle depends on the number of cross-bridges attached. Because it takes a certain amount of time for the cross-bridges to attach (based on the rate constant of attachment), as filaments slide past one another faster and faster (*i.e.,* as the muscle shortens with increasing velocity), force decreases due to the lower number of cross-bridges attached. Conversely, as the relative filament velocity decreases (*i.e.,* as muscle velocity decreases), more cross-bridges have time to attach and to generate force, and thus force increases. This discussion is not meant to be a definitive description of the basis for the force-velocity relationship, only to provide some insight as to how cross-bridge rate constants can affect muscle force generation as a function of velocity.

Eccentric Contractions—Muscle Actively Lengthening

As the load on the muscle increases, it reaches a point where the external load is greater than the load which the muscle itself can generate. Thus the muscle is activated, but it is forced to lengthen due to the high external load.

Table 2.1.
Relative Muscle Force at Various Muscle Velocities

Relative Force	Velocity
100% P_o	0% V_{max}
95% P_o	1% V_{max}
90% P_o	2.2% V_{max}
75% P_o	6.3% V_{max}
50% P_o	16.6% V_{max}
25% P_o	37.5% V_{max}
10% P_o	64.3% V_{max}
5% P_o	79.1% V_{max}
0% P_o	100% V_{max}

This is referred to as an eccentric contraction (please remember that contraction in this context does not necessarily imply shortening!). There are two main features to note regarding eccentric contractions. First, the absolute tensions are very high relative to the muscle's maximum tetanic tension generating capacity. Second, the absolute tension is relatively independent of lengthening velocity. This suggests that skeletal muscles are very resistant to lengthening, a property which we shall see comes in very handy for many normal movement patterns (Chapter 3).

Eccentric contractions are currently under study for three main reasons: First, much of a muscle's normal activity occurs while it is actively lengthening, so that eccentric contractions are physiologically common. Second, muscle injury and soreness are selectively associated with eccentric contraction. Finally, muscle strengthening is greatest using exercises that involve eccentric contractions. These phenomena will be elaborated upon in Chapters 4 and 6.

LENGTH-TENSION-VELOCITY RELATIONSHIP

From the preceding discussion, it is apparent that muscle force changes due to changing length and/or due to changing velocity. It should not be surprising, therefore, to suggest that when muscle length *and* muscle velocity change simultaneously, it is still possible to define the muscle force produced. It should also not be surprising that while the length-tension and force-velocity relationships are useful, such isometric and isotonic conditions are almost never encountered in daily activities. However, the length-tension experiment can be viewed simply as a series of length-force-velocity experiments performed at constant (zero) velocity. Similarly, the force-velocity relationship can be viewed as a series of length-force-velocity experiments performed at constant length (L_O). The point shared between the classic force-velocity and length-tension curves is the point of maximum isometric tension (L_O, at zero velocity, resulting in a tension of P_O). If both length and velocity simultaneously change, the result is the superposition of the two relationships.

The appearance of the length-tension-velocity relationship is shown in Figure 2.5. Don't let the three-dimensional nature of the relationship intimidate you. If the surface is viewed along one set of axes, it is simply a series of force-velocity curves at different lengths. When viewed along the other set of axes, it is simply a series of length-tension curves at different velocities. In this surface we have all possible combinations of muscle length and velocity and their resulting force. What can we conclude? For one thing, if muscle velocity is very high, force will be low no matter what the length. In other words, at high velocities, length is not very important. At low concentric

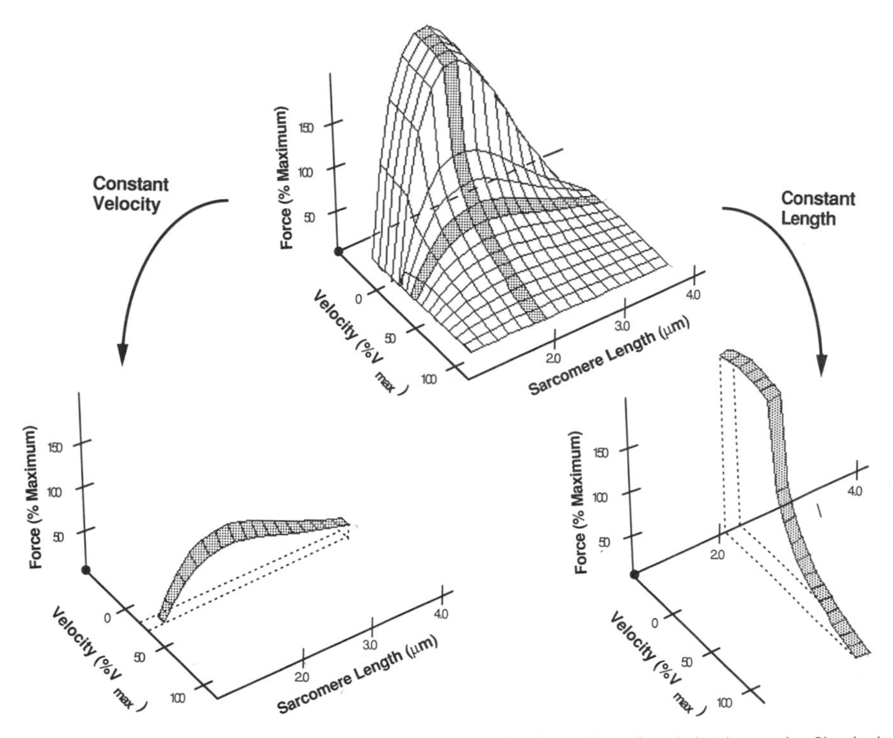

Figure 2.5. The hypothetical muscle length-force-velocity surface for skeletal muscle. Shaded regions represent a slice of the surface at either constant length or velocity. A slice of the surface at constant length is simply a force-velocity curve (compare with Figure 2.4). A slice of the surface at constant velocity is simply a length-tension curve (compare with Figure 2.3). (From Fridén J, Lieber RL. The structural and mechanical basis of exercise-induced muscle injury. Med Sci Sports Exerc, in press.

velocities, muscle length becomes an important force modulator. At eccentric velocities, again muscle velocity dominates length as the determinant of force. This relationship is of course important in neuromotor control as we attempt to understand how muscle actions can be responsible for external movements observed. We will have more to say on this topic in Chapters 3 and 4.

THE CROSS-BRIDGE CYCLE

We have alluded to cyclic interaction between actin and myosin in our structural discussion in Chapter 1 and in explaining the force-velocity curve above. How were such hypotheses generated? Much of our understanding of the mechanism of muscle contraction has come in large part from excellent biochemical studies performed from the 1950s to the mid-1970s (Webb and

Trentham, 1983). It was during this period that methods for isolating specific muscle proteins were developed as well as the methods for measuring their physicochemical and biochemical properties. For example, if a muscle was homogenized in a blender and mixed with a concentrated salt solution (ionic strength of about 600 μM, well-above that observed physiologically), the high-ionic-strength solution caused individual myosin molecules to let go of their ionic interactions with one another and become soluble. Then this soluble portion was removed from the rest of the muscle debris, the ionic strength was slowly lowered to physiologic levels (about 120 μM), and lo and behold, myosin and actin filaments reformed and, in fact, formed a solid precipitate complex (Szent-Györgyi, 1953). Simple addition of ATP rendered this precipitate perfectly clear (Maruyama and Gergely, 1962). Why? Obviously, ATP affected the relationship between actin and myosin. Biochemists believed that by performing experiments such as these, they could investigate different steps of the cyclic interaction between muscle proteins. These types of experiments are quite popular in recent muscle literature and have provided insights into muscle contraction as well as actin-myosin interaction in all eukaryotic cells.

In its simplest form, based on experiments such as those presented above, the cross-bridge cycle can be envisioned as actin (A) combining with myosin (M) and ATP to produce force, adenosine diphosphate (ADP), and inorganic phosphate, P_i. This can be represented as a chemical reaction in the form

$$A + M + ATP \rightarrow A + M + ADP + P_i + \text{Force} \quad (2.2)$$

However, we also know that upon the death of a muscle, a rigor state is entered whereby actin and myosin interact to form a very stiff connection. This can be represented as

$$A + M \rightarrow A{\cdot}M \text{ ``rigor'' complex} \quad (2.3)$$

If actin and myosin can interact by themselves, where does ATP come into the picture during contraction? As discussed in Chapter 1, the myosin molecule can be enzymatically split into its subfragments. Experiments that cleaved myosin into light meromyosin (LMM), subfragment 1 (S-1), and subfragment 2 (S-2) demonstrated that the myosin S-1 portion retained the ability to hydrolyze ATP into ADP and P_i. In other words,

$$M + ATP \rightarrow M + ADP + P_i \quad (2.4)$$

It is now clear that ATP serves at least two functions in skeletal muscle systems: First, ATP disconnects actin from myosin, and second, ATP is hydrolyzed by the S-1 portion of the myosin molecule. Can you see the competition this sets up? In contracting skeletal muscle, ATP binds to the actin-myosin complex, causing actin and myosin to dissociate. When it does,

ATP is hydrolyzed by myosin into ADP and P_i, which then allows actin and myosin to reassociate! Thus Equation 2.2 (our simple cross-bridge cycle) can be combined with Equations 2.3 and 2.4 to yield the more detailed "two step" cross-bridge cycle shown below:

$$\begin{aligned} A + M &\rightarrow A{\cdot}M \text{ "rigor" complex} \\ A{\cdot}M \text{ "rigor" complex} + ATP &\rightarrow A + M + ADP + P_i + \text{Force} \end{aligned} \qquad (2.5)$$

This basic scheme has been expanded by many excellent works over the years, but the same basic idea remains: ATP is required to dissociate actin from myosin and is hydrolyzed by the S-1 portion of the myosin head. Several interesting experiments refined this concept to provide actual rate constants for the various reactions. For example, we now know that when myosin is alone in solution, it hydrolyzes ATP *very* slowly. Thus Equation 2.4 occurs only at a rate of about 0.1/second. However, an interesting observation is that when actin is added to a solution of S-1 and ATP, the previously slow hydrolysis rate increases about 200-fold! Thus actin acts as a catalyst for ATP hydrolysis by S-1. We can modify our ATP hydrolysis mechanism to include two paths for hydrolysis—a path with and a path without actin. The path without actin is shown is Equation 2.7 and the path with actin is shown is Equation 2.6:

$$A{\cdot}M + ATP \rightarrow A{\cdot}M{\cdot}ATP \rightarrow A + M + ADP + P_i \qquad \text{(fast)} \; (2.6)$$

$$M + ATP \rightarrow M{\cdot}ATP \rightarrow M + ADP + P_i \qquad \text{(slow)} \; (2.7)$$

It is also easy to interconnect these two schemes by adding a step whereby actin can dissociate from the A·M·ATP complex, and the scheme thus becomes

$$\begin{aligned} &A{\cdot}M + ATP \rightarrow A{\cdot}M{\cdot}ATP \rightarrow A + M + ADP + P_i \qquad \text{(fast)} \\ &\qquad \text{ACTIN DISSOCIATES} \downarrow\uparrow \text{ ACTIN ASSOCIATES} \qquad (2.8) \\ &M + ATP \rightarrow M{\cdot}ATP \rightarrow M + ADP + P_i \qquad \text{(slow)} \end{aligned}$$

or, more simply,

$$\begin{aligned} A{\cdot}M + ATP \boldsymbol{\rightarrow} A{\cdot}M{\cdot}ATP &\rightarrow A + M + ADP + P_i \\ -A \boldsymbol{\downarrow}&\uparrow +A \\ M + ATP \rightarrow M{\cdot}ATP &\boldsymbol{\rightarrow} M + ADP + P_i \end{aligned} \qquad (2.9)$$

You can see that we have generated a reasonable approximation of cyclic interaction between actin, myosin, and ATP! The reactions with the fastest rate constants are shown in Equation 2.9 in boldface; they are the most likely to occur. This very simplified scheme actually explains a great deal of the experimental data and, conceptually, will allow you to understand many physiologic properties to be presented later in this chapter.

For those of you who are interested, there is an interesting caveat to the scheme of Equation 2.9. Experimental studies by Richard Lymn and Ed Taylor (1971) demonstrated that, after formation of the A·M·ATP complex, the actin quickly dissociates, so that ATP hydrolysis actually occurs with the actin and myosin filaments separated! After the hydrolysis, the actin quickly reassociates with the $M\cdot ADP\cdot P_i$ complex to cause dissociation of the hydrolysis products. The rate-limiting step of the entire sequence is the release of the reaction products from myosin, so that actin increases the ATP hydrolysis rate by speeding the release of hydrolysis products from myosin. The ATP hydrolysis scheme is therefore more accurately represented as

$$\begin{array}{ccc} \mathrm{A\cdot M + ATP} \boldsymbol{\rightarrow} \mathrm{A\cdot M\cdot ATP} & & \mathrm{A\cdot M\cdot ADP\cdot P_i} \boldsymbol{\rightarrow} \mathrm{A + M + ADP + P_i} \\ -\mathrm{A}\,\boldsymbol{\downarrow} & & \boldsymbol{\uparrow}\,+\mathrm{A} \\ \multicolumn{3}{c}{\mathrm{M + ATP} \rightarrow \mathrm{M\cdot ATP} \boldsymbol{\rightarrow} \mathrm{M\cdot ADP\cdot P_i} \rightarrow \mathrm{M + ADP + P_i}} \end{array} \qquad (2.10)$$

with the normal route of hydrolysis (that is, the fastest rate constants) shown in boldface. This sequence of biochemical steps is referred to as the Lymn-Taylor actomyosin ATPase hydrolysis mechanism (Lymn and Taylor, 1971).

The relationship between the Lymn-Taylor kinetic scheme and the mechanical cross-bridge cycle is not fully known. However, Lymn and Taylor proposed that their biochemical data could be incorporated into a four-step cross-bridge cycle that could be envisioned thus (Figure 2.6):

1. The actin-myosin bridge very rapidly dissociates due to ATP binding.
2. The free myosin bridge moves into position to attach to actin, during which ATP is hydrolyzed.
3. The free myosin bridge along with its hydrolysis products rebinds to the actin filament.
4. The cross-bridge generates force, and actin displaces the reaction products (ADP and P_i) from the myosin cross-bridge. This is the rate-limiting step of contraction. The actin-myosin cross-bridge is now ready for the ATP binding of step 1.

It might be appreciated that confirmation of this mechanism would be very difficult indeed. In fact, a recent advance in biochemistry has allowed direct testing and manipulation of this scheme. The advance involves the development of "caged" compounds—compounds which are inactive in their caged form and become active when the cage is instantaneously removed by a pulse of high-energy laser light (McCray *et al.,* 1980). Using caged ATP, single muscle fibers have been subjected to experiments such as those described above and found to behave much as predicted based on the biochemical data (Goldman, 1987). These experiments, performed by Yale Goldman and his colleagues, are truly a case where biochemistry and physiology have met head-on.

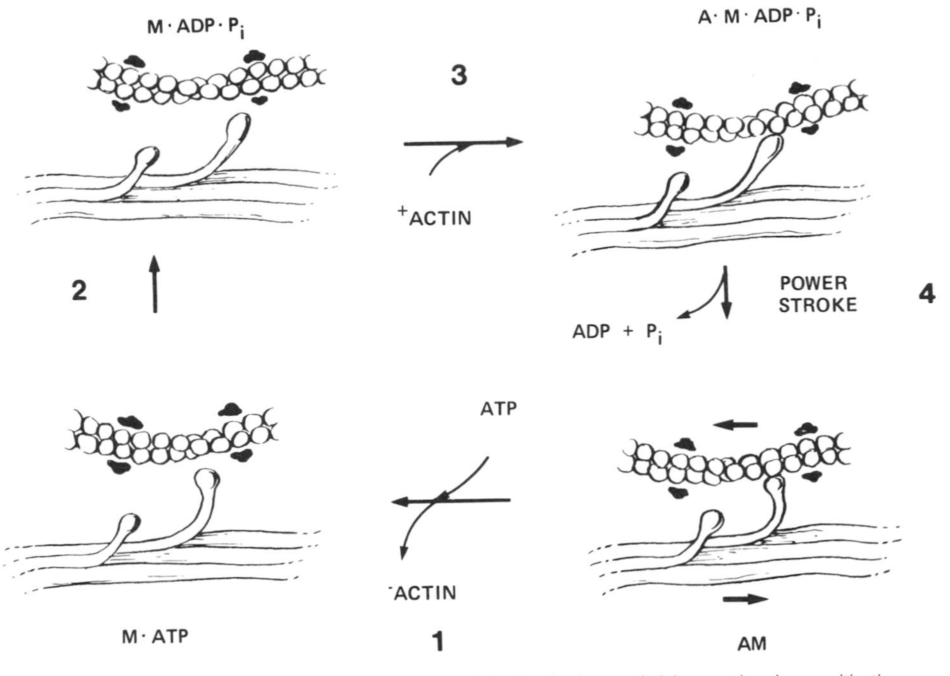

Figure 2.6. Schematic representation of the biochemical cross-bridge cycle along with the associated mechanical events. **1,** ATP binds to actin-myosin (AM) complex, displacing actin. **2,** ATP hydrolyzed on M·ATP complex. **3,** Actin reassociates with M·ADP·P_i complex. **4,** Attached cross-bridge generates force during power stroke. (After Lymn and Taylor, 1971.)

MECHANICAL PROPERTIES OF MUSCLES WITH DIFFERENT ARCHITECTURES

Congratulations! You have completed a complex discussion of two of the most important mechanical properties of skeletal muscle, as well as of the cross-bridge cycle. We are now in a position to further discuss the physiologic significance of muscle architecture, which was introduced in Chapter 1. We will see that even though we might understand the details of sarcomere structure and cross-bridge action, it is absolutely impossible to explain the force-generating property of a whole muscle without an understanding of muscle architecture. The important point to remember regarding muscle architecture is that *muscle force is proportional to physiologic cross-sectional area (PCSA), and muscle velocity is proportional to muscle fiber length.* We should note that by stating that velocity is proportional to fiber length, it is implicit that the total excursion (active range) of a muscle is also proportional to fiber length. Thus increasing fiber length results in both increased muscle velocity and excursion. It is probably apparent to you, based on the discussion in Chapter 1, that neither fiber length nor PCSA can easily be deduced based on gross muscle inspection. The detailed methods of Chapter 1 are required for architectural determination. However, after you have determined these architectural properties, you are in a position to understand how much force a muscle generates and how fast it contracts (or how far it contracts). Let's look at two specific architectural examples and their impact on the length-tension and force-velocity relationships.

Comparison of Two Muscles with Different Physiologic Cross-Sectional Areas

Suppose that two muscles had identical fiber lengths and pennation angles, but one muscle had twice the mass (equivalent to saying that one muscle had twice the number of fibers and thus twice the PCSA). What would be the difference in their mechanical properties? How would the length-tension and force-velocity curves be affected?

The schematic in Figure 2.7 demonstrates that the only effect is to increase maximum tetanic tension so that the length-tension curve has the same basic shape but is simply amplified upward in the case of the stronger muscle. Similarly, the force-velocity curve simply changes the location of P_O, but the curve retains the same basic shape. Note that if both curves are plotted on *relative* scales (*i.e.,* percent maximum tension instead of absolute tension), the two muscles of different architecture appear to have identical properties. This demonstrates that while architectural properties profoundly affect the extrinsic muscle properties (*i.e.,* the properties that vary with absolute muscle size, such as PCSA or mass), they have no affect on its intrinsic properties (*i.e.,* the properties that are independent of absolute muscle size, such as fiber length/muscle length ratio).

Comparison of Two Muscles with Different Fiber Lengths

Let us consider the effects of architecture using an example of two muscles with identical PCSAs and pennation angles but different fiber lengths. Before reading ahead, try to draw the appropriate length-tension and force-velocity curves.

As shown in Figure 2.7, the effect is to increase the muscle velocity (or, stated identically, to increase the muscle excursion). The peak absolute force of the length-tension curves is identical, but the absolute muscle active range is different. Did I say active range? That sounds a lot like active range of motion (ROM), a measurement that is extremely important in clinical evaluation. In fact, it is directly related to ROM. I will have more to say on this in Chapter 3. Suffice it to say that ROM is a direct result of muscle architecture and the joint properties on which the muscle acts.

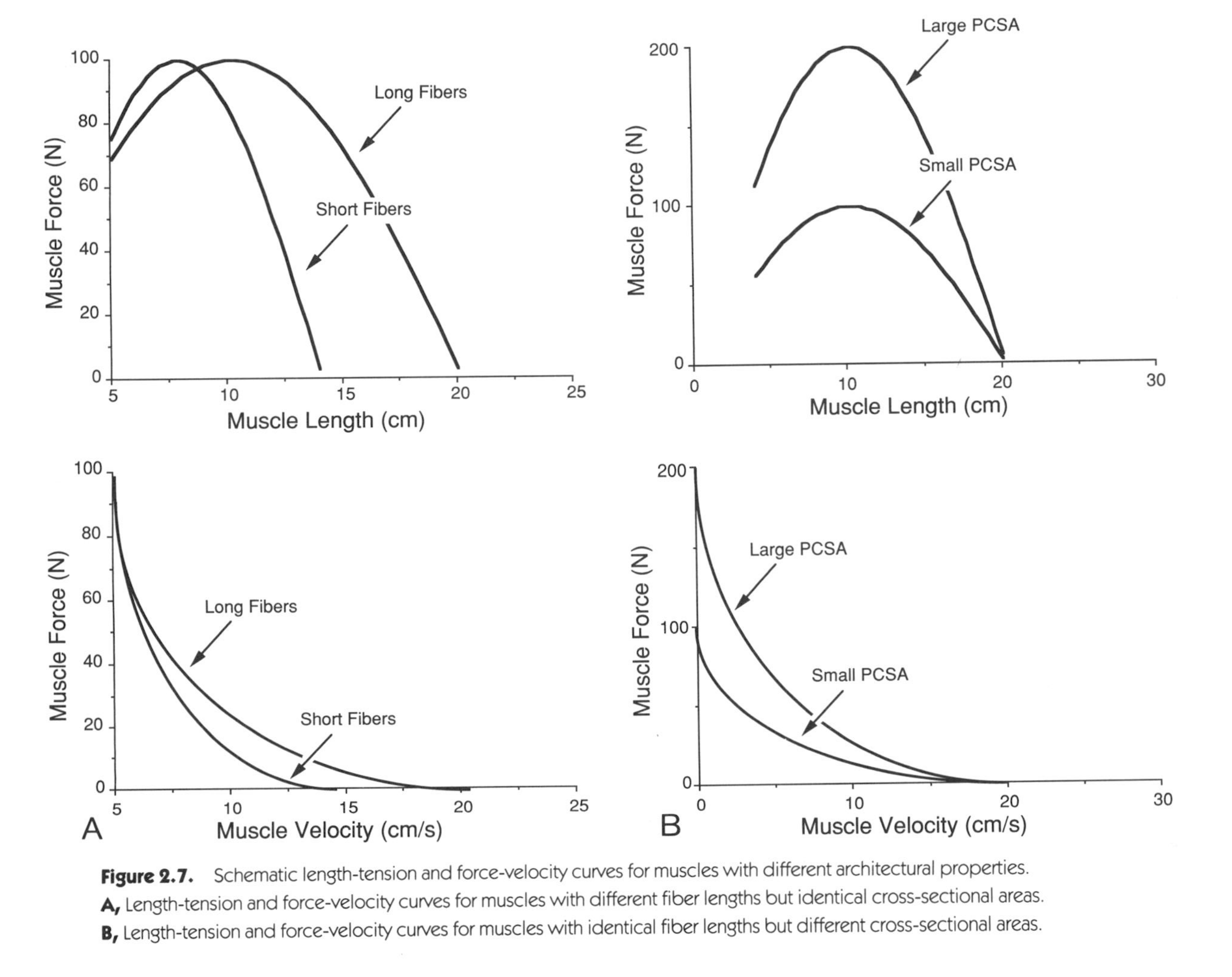

Figure 2.7. Schematic length-tension and force-velocity curves for muscles with different architectural properties. **A,** Length-tension and force-velocity curves for muscles with different fiber lengths but identical cross-sectional areas. **B,** Length-tension and force-velocity curves for muscles with identical fiber lengths but different cross-sectional areas.

For the same reason that fiber length increases the active muscle range of the length-tension relationship, it causes an increase in the muscle's absolute maximum contraction velocity (V_{max}). Again, while the fiber length increase causes an increase in these absolute properties, it has no effect on the intrinsic properties of the muscle. A similar exercise can be performed comparing muscles with different PCSAs and fiber lengths. Try predicting force-velocity and length-tension curves for the case where both architectural parameters are changed.

We have just seen vivid examples of the profound influence of muscle architecture on functional properties. You may want to return to the original discussion of architecture given in Chapter 1 and peruse Tables 1.3 and 1.4 in order to become more familiar with the architecture of several common muscles. Note that muscles with relatively long fibers (hamstrings, extensor carpi radialis longus, tibialis anterior) are muscles with high contraction velocities and large excursions. Conversely, muscles with large PCSAs (quadriceps, flexor carpi ulnaris) generate very large tensions. We now see that the architectural specialization observed in the numerous muscles has profound functional consequences. Muscles are able to perform a large range of tasks largely as a result of their intrinsic design rather than a specific set of command signals from the CNS. This design allows the CNS to act more as a coordinator of tasks rather than a definer of the particulars of the task. Interestingly, this is the same trend that is occurring in the microcomputer world regarding the tasks of computer central processing units (CPUs) and peripheral devices such as printers and video display terminals.

PART 3: MUSCLE FIBER TYPES AND MOTOR UNITS

MUSCLE FIBER TYPES

Historical View of Skeletal Muscle Fiber Types

To this point, I may have implied by omission that all fiber types are created equal. While this is true in terms of value, it is clearly not true in terms of properties. I have hinted at some of the differences between fiber types already in my discussion of developmental anatomy. In fact, by understanding the anatomic differences between fiber types, we gain insights into the rationale for the anatomic features themselves. This is much the same approach that the comparative anatomist takes in understanding skeletal function by observing the same skeletal feature across a variety of species.

To retain the view that the muscle fibers are the same is an oversimplification in light of the overwhelming evidence that skeletal muscle fibers are heterogeneous. In the early 1800s it was observed that the gross appearance of different skeletal muscles ranged in color from pale white to deep red. In fact, one of the earliest classification schemes for muscle was based on color, and thus muscles were classified as "red" or "white" (Table 2.2). However, as experimental methods became more sophisticated, it became clear that numerous other differences existed between muscles. For example, certain muscles contracted very rapidly, while others contracted more slowly. Certain muscles could maintain force for a long period of time, while others fatigued very rapidly. Certain muscles generated large forces, while others generated very small forces. Thus muscles were also classified as "fast" or "slow," and "fatigable" or "nonfatigable." In addition, with the advent of light microscopy and histochemistry, it was possible to classify individual fibers based on their appearance following a particular staining protocol. Many of these schemes did not simply correlate with the muscle color. In fact, many of them did not correlate at all with one another. Table 2.2 presents a few of the classifications schemes used historically and the bases for them. The main problem with fiber type classification schemes was that the classification only worked for one property but had no relationship to others. Thus while anatomists identified type 1 and type 2 fibers, physiologists spoke of "fast twitch" and "slow twitch" muscles. In addition, it was not clear whether muscle fibers were mutable over their life span.

Our current view of muscle fiber types is that skeletal muscle fibers possess a wide spectrum of morphologic, contractile and metabolic properties. The appropriate view of any classification scheme, therefore, is that it is an artificial system superimposed on a continuum for our convenience. In my view, the most useful scheme is one that can be related to other types of measurements (*e.g.,* physiologic and biochemical) in order to more fully understand muscle's normal and adaptive properties.

Skeletal Muscle Metabolism

Before discussing the physiologic properties of the various fiber types and the experimental methods for fiber type identification, a brief review of metabolism

Table 2.2.
Fiber Type Classification Schemes[a]

Basis for Scheme	Fiber Type Spectrum			Authors
Metabolic	SO	FOG	FG	Peter *et al.,* 1972
Morphology and physiology	Slow red	Fast white	Fast white	Ranvier, 1873
Z-line width	Red	Intermediate	White	Gauthier, 1969
Histochemistry	III	II	I	Romanul, 1964

[a]SO, slow oxidative; FOG, fast oxidative glycolytic; FG, fast glycolytic.

is presented. In this section, we will only present the relevant features of carbohydrate and lipid metabolism, which form the basis of our later discussions.

Muscle cells, like any other body cell, require energy to perform their normal functions. In contrast to other cells that only have to generate enough energy to maintain normal cellular processes, muscle cells have the additional burden of providing energy for force generation. Interestingly, cellular metabolism was discovered in skeletal muscle. We will see (page 101) that these metabolic pathways are important in determining a cell's ability to perform work under a variety of conditions.

The six-carbon carbohydrate molecule known as glucose serves as the major energy source of the cell. There are two main processes by which glucose can be oxidized to yield energy that is useable by the cell. Of these, one of them does not require oxygen (glycolysis) and one of them does (oxidative phosphorylation). Glycolysis occurs within the soluble cytoplasm of the cell, while oxidative phosphorylation occurs within the mitochondria (Figure 2.8).

GLYCOLYSIS: ENERGY WITHOUT OXYGEN

Glycolysis is the cellular process by which the glucose molecule is enzymatically broken down into two three carbon molecules known as pyruvate (Figure 2.9**A**). The chemical reaction can be represented as

$$\text{Glucose} + 2\,\text{ADP} + 2\,\text{P}_i \rightarrow 2\,\text{Pyruvate} + 2\,\text{ATP} \qquad (2.11)$$

This breakdown of glucose occurs as a series of chemical reactions that permits cellular control of the rate and amount of glucose metabolized. Note that for every molecule of glucose metabolized, two ATP molecules are created. If the glycolytic process is to occur continuously without oxygen, **several intermediate molecules must be regenerated by further reducing the pyruvate molecule to lactate:**

$$\text{Pyruvate} + \text{NADH} \rightarrow \text{Lactate} + \text{NAD}^+ \qquad (2.12)$$

where NADH and NAD^+ refer to the reduced and oxidized versions of nicotinamide adenine dinucleotide, a common reaction coenzyme. As we have seen, ATP is the primary energy molecule used in the cross-bridge cycle. While glycolysis can supply the energy needs of the cell, it is not extremely efficient.

OXIDATIVE PHOSPHORYLATION: ENERGY WITH OXYGEN

Lactate buildup after anaerobic glycolysis within the cell has two main drawbacks. First, lactate is an acidic molecule, and accumulation can alter the

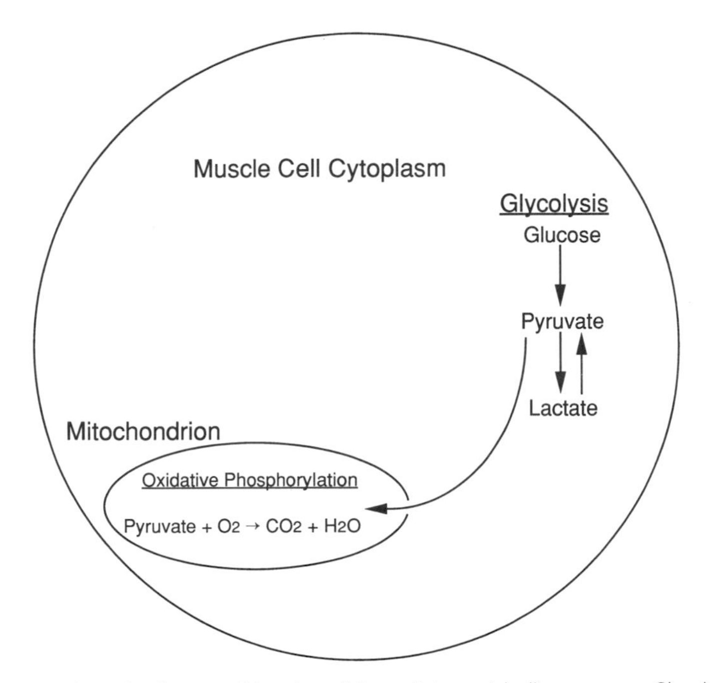

Figure 2.8. Schematic diagram of location of the cellular metabolic processes. Glycolysis and anaerobic glucose metabolism occur in the cytoplasm, while oxidative phosphorylation, which requires oxygen, occurs in the mitochondrion.

intracellular pH, thus altering subsequent cellular contractile and metabolic activity. In addition, lactate clearance from the cell requires further bodily energy to transport it via the blood and further metabolizing elsewhere—say, in the liver. When oxygen is present, the preferable metabolic pathway for glucose metabolism is to completely oxidize glucose to pyruvate and then, within the mitochondria, to further oxidize pyruvate into CO_2 and H_2O, which are easily cleared from the cell via diffusion into the blood and exhaling of CO_2 via the lungs (Figure 2.9**B**). Oxidative phosphorylation can thus be further represented as

$$\text{Pyruvate} + 15\,\text{ADP} + 15\,\text{P}_i + 4O_2 \rightarrow 3\,CO_2 + H_2O + 15\,\text{ATP} \quad (2.13)$$

Thus whereas anaerobic metabolism (glycolysis) of glucose yields 2 ATP per glucose molecule, oxidative metabolism yields 32 ATP per glucose molecule (15 ATP for each pyruvate and 2 ATP for glucose to pyruvate oxidation; Figure 2.9**C**). Thus when oxygen is available, it is a much more energy efficient mechanism to generate cellular energy.

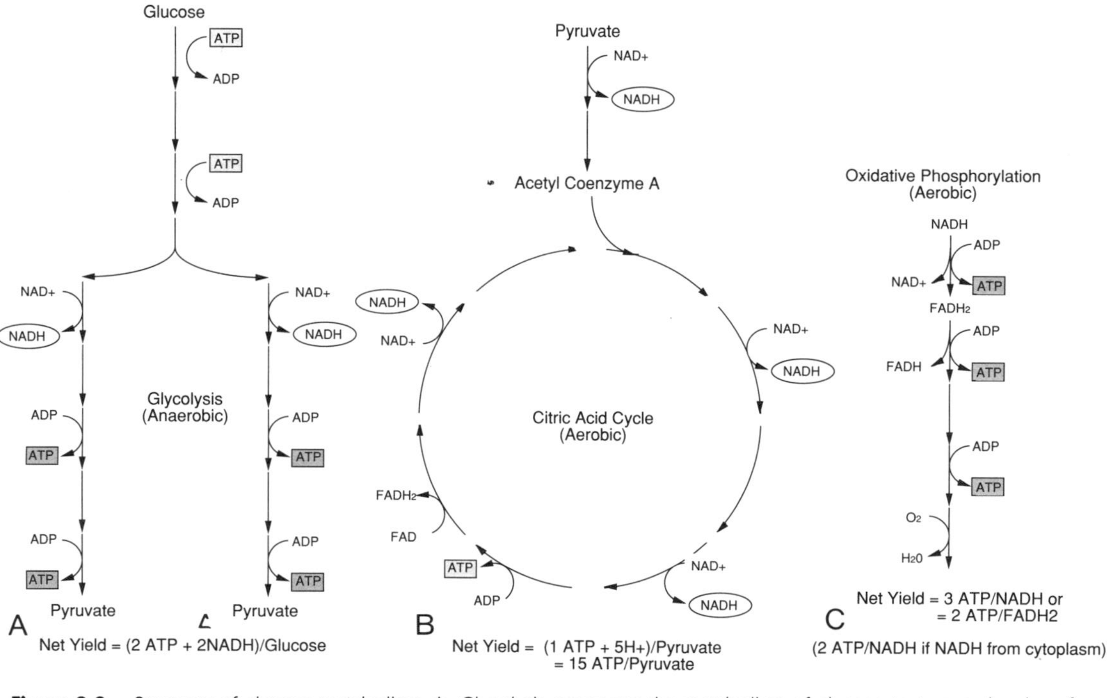

Figure 2.9. Summary of glucose metabolism. **A,** Glycolysis represents the metabolism of glucose to two molecules of pyruvate. This process yields two ATP molecules and two NADH molecules (shown in shaded squares and ellipses) and occurs without oxygen. **B,** The citric acid cycle represents the conversion of pyruvate into acetyl coenzyme A, which then passes through various intermediates (not shown) and yields one molecule of ATP, four molecules of NADH, and one molecule of $FADH_2$. The NADH and $FADH_2$ are available for further oxidation by the electron transport system, which yields much ATP. This process requires oxygen. **C,** Oxidative phosphorylation used to generate ATP from either NADH (three molecules ATP per NADH) or $FADH_2$ (two molecules ATP per $FADH_2$). If the NADH is derived from the cytoplasm via glycolysis, only two ATP are produced net, since one ATP is required to transport NADH into the mitochondrion.

FATTY ACID OXIDATION

Glucose is available from at least three sources (Figure 2.10): blood glucose, liver glycogen, and muscle glycogen. Glycogen is a large polymer of chemically linked glucose molecules, which are stored in muscle cells to provide energy under anaerobic conditions. However, another source of energy—fats—is also available to meet cellular needs. Fats, or more properly, fatty acids, are metabolized by a process called β-oxidation. β-Oxidation occurs in the mitochondria as with oxidative phosphorylation. From a simple 16-carbon chain of fatty acids (palmitate), two unit carbons are sequentially cleaved off to yield ATP:

$$\text{Palmitate} + \text{Acetyl CoA} + 7O_2 \rightarrow 7\,CO_2 + 4H_2O + 129\,\text{ATP} \qquad (2.14)$$

This is a whopping yield! We now see that oxidative metabolism can yield great quantities of ATP from glucose and fatty acids, while metabolism under anaerobic conditions has a much lower energy yield. We will return to these concepts in our discussion of motor unit identification by glycogen depletion (page 89) and muscle fatigue (page 101).

METHODS FOR TYPING MUSCLE FIBERS

The Metabolic Classification Scheme

As mentioned above, many schemes were proposed to classify skeletal muscle histochemical, physiologic, and morphologic properties. To date, many agree that the so-called metabolic classification scheme is the most useful. That is, the metabolic scheme allows one to switch back and forth between anatomic

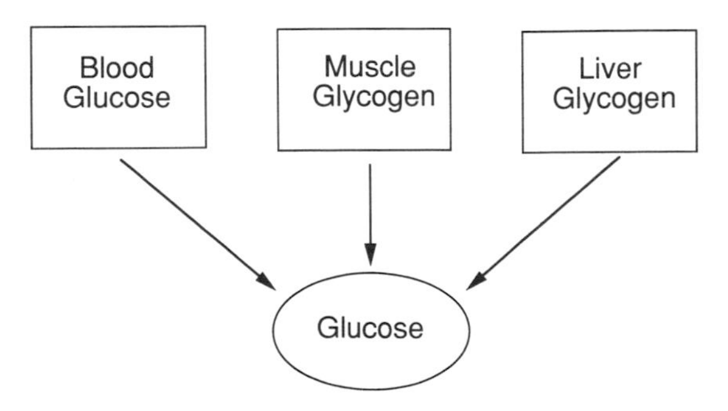

Figure 2.10. Three major sources of glucose are available for metabolism.

and physiologic studies and to understand one type of study based on the other. (Unfortunately, this scheme is not the one most widely used for study of human muscle.) This classification scheme was developed by Reggie Edgerton in the late 1960s and early 1970s in an iterative fashion. For example, Edgerton and colleagues published early work describing the glycogen content of red and white muscle fibers (Gillespie *et al.,* 1970). Later, based on his and others' physiologic observations that fast-contracting muscles were not always white, the scheme was expanded to include intermediate (in color) muscle fibers. Finally, again largely based on his own work, along with that of James Peter, the metabolic classification scheme was born (Peter *et al.,* 1972)! The strength of this scheme is that physiologic, biochemical, and histochemical experiments are combined to develop a scheme consistent across methodologies.

Histochemical Methods—Visualizing Chemicals and Enzymes in Tissue

In the metabolic method, three fundamental muscle fiber properties are identified using histochemical methods. ("Histochemical" ["histo" = tissue] implies that a chemical reaction is occurring in the tissue itself.) These histochemical methods rely on the fact that enzyme located in thin (6–8 μm) frozen sections of muscle fibers can be chemically reacted with certain products in order to visualize the activity of the enzyme. The most modern methods are actually able to quantify the activity of the enzymes in standard units (similar to *in vitro* biochemical studies) based only on measurements of optical density in frozen tissue sections (Figure 2.11**A**). Remember that a 6-μm section is still two to four sarcomeres thick!

The basic requirement for the histochemical assay is similar, at least in principle, to the requirement for any biochemical assay. First, a *substrate* (fuel) is provided for the enzyme to be studied. Second, an *energy source* is provided that allows the enzyme to utilize the substrate. Finally, a *reaction product* is linked to another product that can be visualized microscopically. Of course, the entire reaction takes place in the tissue itself, not in a test tube. Let us now proceed to the three histochemical assays typically used to determine muscle fiber types. These three assays are the myosin ATPase (MATPase) assay, the succinate dehydrogenase (SDH) assay, and the α-glycerophosphate dehydrogenase (αGP) assay.

MYOFIBRILLAR ATPASE—IDENTIFYING FAST AND SLOW FIBERS

In a classic example of combined physiologic and biochemical experimentation, Michael Barany measured V_{max} in several skeletal muscles (Barany, 1967). He then biochemically isolated the myosin from these muscles and

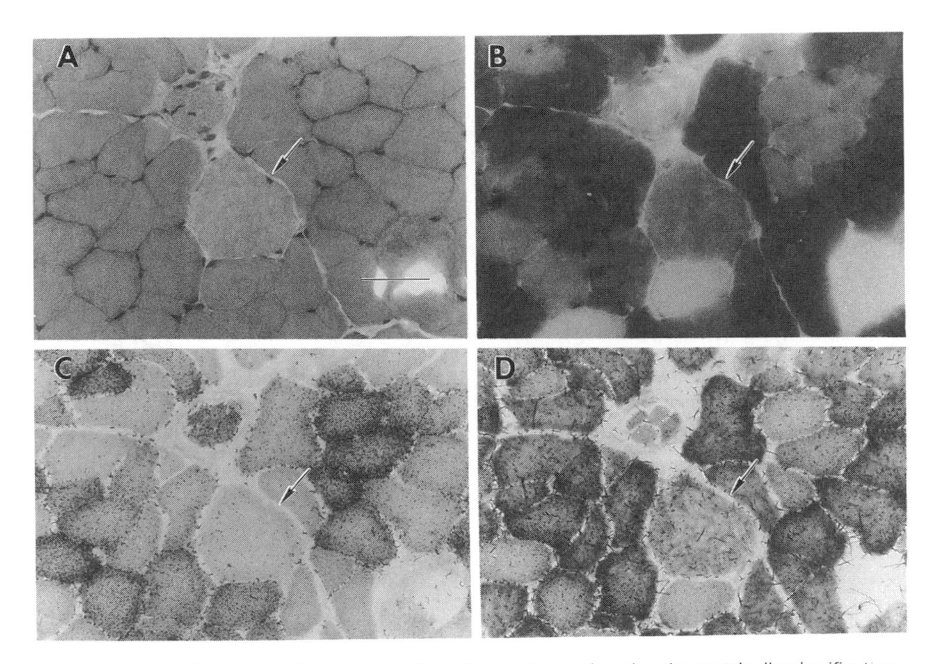

Figure 2.11. Histochemical characterization of skeletal muscle using the metabolic classification scheme. Each panel represents serial sections of rabbit tibialis anterior muscle. **A,** Hematoxylin and eosin stain for general fiber morphology. **B,** Myofibrillar ATPase stain to distinguish between fast and slow fibers. **C,** Succinate dehydrogenase stain to distinguish between oxidative and nonoxidative fibers. **D,** α-Glycerophosphate dehydrogenase stain to distinguish between glycolytic and nonglycolytic fibers. Arrow points to a fast glycolytic fiber.

measured their myosin ATPase activities (an "activity" is a rate expressed for an enzyme in units of moles of product per unit mass per time interval—for example, millimoles per milligram protein per minute in the case of the myosin ATPase assay). Barany found that V_{max} and myosin ATPase activity were directly proportional. Barany demonstrated that the activity of the myosin molecule limited the rate of muscle contraction and thus provided a window by which myosin activity could be used to infer contractile speed. In other words, he provided the expected link that could be used to combine biochemical and physiologic studies.

The histochemical assay for myofibrillar ATPase activity is used to distinguish between fast- and slow-contracting muscles fibers (although, strictly speaking, one should not histochemically type a fiber and call it fast or slow since the classification method does not actually *measure* speed but a histochemical appearance). Recall that the myosin molecule itself binds and hydrolyzes ATP during force generation. Because myosin ATPase activity is positively correlated with muscle contraction velocity, measures of ATPase activity can be interpreted in terms of contraction speed. The simplified cross-bridge cycle was presented as Equation 2.2. ATP is the reaction *energy source* and *substrate,* P_i is the reaction *product,* and myosin is the *enzyme.* Histochemically, P_i is invisible. Thus the assay requires that P_i be chemically reacted with calcium (Ca) in order to produce $CaPO_4$ (limestone), a white precipitate. Subsequent steps in the process convert $CaPO_4$ into CoS_2 which is brownish-black, more easily viewable, and less soluble. Thus, effectively, as P_i is released, a brownish-black product is deposited on the section. Fast-contracting muscle fibers hydrolyze ATP faster than slow-contracting fibers. Thus when given equivalent times, fast-contracting fibers appear dark histochemically, and slow-contracting fibers appear light (Figure 2.11**B**). In this manner, the myosin ATPase assay can be used to distinguish between fast- and slow-contracting muscle fibers.

SUCCINATE DEHYDROGENASE: IDENTIFYING OXIDATIVE POTENTIAL

The histochemical assay for SDH is used to distinguish between oxidative and nonoxidative (actually, "less" oxidative) fibers. Recall that fibers with a high oxidative capacity generate ATP via oxidative phosphorylation. This sequence of reactions occurs in the mitochondria, which is one reason that highly oxidative fibers have a high volume fraction of mitochondria (Eisenberg, 1983). The SDH enzyme is located in the inner membrane of the mitochondrion, bound to the cristae. SDH is responsible for oxidizing succinate to fumarate in the citric acid cycle:

$$\text{Succinate} + \text{NAD}^+ \rightarrow \text{Fumarate} + \text{NADH} \qquad (2.15)$$

Thus as this reaction proceeds, succinate is oxidized, and the reduced form of NADH is produced. Succinate is therefore the substrate, NADH is the reaction product (actually, a different electron acceptor is used for practical reasons), and SDH is the enzyme. The electron acceptor is chemically reacted in a second step with a purple tetrazolium salt in order to clearly visualize the location of the SDH enzyme. Similar to the ATPase assay, fibers rich in SDH (and thus rich in mitochondria) stain with a speckled pattern of the mitochondria, proportional to the number of mitochondria and the SDH activity within them (Figure 2.11**C**). Oxidative fibers have a relatively dense, purple speckled appearance, while nonoxidative fibers have only scattered purple speckles. Therefore, this histochemical assay reflects the relative oxidative potential of muscle fibers.

α-GLYCEROPHOSPHATE DEHYDROGENASE—IDENTIFYING GLYCOLYTIC POTENTIAL

The enzyme αGP is used to distinguish fibers based on their relative glycolytic potential. Recall that glycolysis is used to generate ATP in the absence of oxygen (anaerobically). The chemical reactions involved in glycolysis take place in the muscle cell cytoplasm (myoplasm). As such, the αGP stain is not confined to a specific cellular organelle as is the SDH stain, and the appearance is much more continuous across the cell. αGP is not actually involved with direct steps in glycolysis. Rather, the αGP enzyme is responsible for shuttling the NADH produced by glycolysis into the mitochondria where ATP can be produced. It is thus *related* to glycolytic activity in the sense that the more NADH that can be shuttled into the mitochondrion, the more energy that can be produced. It would be preferable to directly and selectively stain a rate-limiting glycolytic enzyme (such as phosphofructokinase [PFK], an important glycolysis regulatory enzyme), but technically, this is rather difficult to achieve. αGP is involved in the reaction

$$\text{reduced}\,\alpha\text{GP} + \text{NADH} + \text{H}^+ \rightarrow \text{into mitochondrion} \rightarrow \text{NAD}^+ + \text{oxidized}\,\alpha\text{GP} \qquad (2.16)$$

As such, αGP activity is directly related to the energy production of the glycolytic pathway. As with SDH, αGP is an enzyme that "dehydrogenates" or oxidizes its substrate, glycerol-1 phosphate. The histochemical method is completely analogous to the SDH assay where the reaction product is linked to the purple tetrazolium salt. However, visually, the reaction appears more homogeneous across the entire cell since the αGP enzyme is not organelle bound (Figure 2.11**D**). This assay can thus distinguish between glycolytic and nonglycolytic fibers.

Fiber Type Classifications Using Histochemical Methods

In their most basic form, the three histochemical methods described above can classify muscle fibers into fast or slow, oxidative or nonoxidative, and glycolytic or nonglycolytic. Potentially, if we stain a given muscle fiber for all three properties, we *could* obtain any of the *eight* (2^3) fiber types shown in Table 2.3. In reality (and fortunately), however, over 95% of normal muscle fibers can be classified into one of only *three* categories shown in Table 2.4.

Thus we have arrived (the long way!) at a classification scheme that fulfills the criteria set forth above: It can classify most muscle fibers, and it can be related to physiologic, biochemical, and morphologic measurements. While classification schemes are, by definition, artificial, this one is less so in that it interleaves well with many different experimental methodologies.

Table 2.3.
Potential Fiber Types Based on Histochemical Assay

Potential Fiber Type Designation	ATPase Activity (F or S)	SDH Activity (O)	α-GP Activity (G)
F	High	Low	Low
FO	High	High	Low
FG	High	Low	High
FOG	High	High	High
S	Low	Low	Low
SO	Low	High	Low
SG	Low	Low	High
SOG	Low	High	High

Table 2.4.
Actual Three Fiber Types Obtained By Histochemical Assay

Fiber Type Designation	ATPase Activity (F or S)	SDH Activity (O)	α-GP Activity (G)
FG	High	Low	High
FOG	High	High	High
SO	Low	High	Low

Classification by Immunohistochemistry

While histochemical fiber type identification can yield valuable insights into muscle function, it is limited in its ability to identify specific cellular proteins. For example, suppose two separate muscles were frozen, stained, and incubated for determination of myofibrillar ATPase activity. In both cases, let us suppose that light and dark fibers appeared, representing slow and fast fibers, respectively. Can we conclude that the fast fibers from one muscle are identical to fast fibers in the other muscle? Of course not, for the same reason that two individuals who run the same speed are not identical. Sometimes it is very important to know exactly which type of protein is contained in the cell. For this purpose, we must use very specific identification methods that identify proteins with the same specificity that fingerprints have for humans.

Antibodies to specific cellular proteins are produced by the immune system, which recognize specific portions of these proteins. While the details of antibody formation are beyond the scope of this discussion, the concept of antibody specificity is quite simple. Antibodies provide specific recognition of cellular proteins. Even small differences in protein structure (say, between two different forms of a "fast" myosin) can be seen. If the antibody is linked to a visible molecule, the location of specific cellular proteins can be determined. This is the principle of immunohistochemistry, which is in widespread use in current muscle biology research.

OTHER MUSCLE FIBER CLASSIFICATION SCHEMES

Before leaving the topic of muscle fiber classification schemes, and having laid the foundation for the most widely used scheme, let us take a closer look at some of the other schemes that were developed but proved to be less useful.

I mentioned the "red" and "white" classification scheme, which was one of the earliest developed. The earliest observations wee that muscles had a red appearance contracted slowly while muscles with a more pale, white appearance contracted rapidly. It was thus concluded that "red" was the same as "slow contracting" while "white" was the same as "fast contracting." However, it was soon determined that many red muscles were indeed fast contracting when measured physiologically! Thus while "red" and "white" were distinctions that could be applied to most muscles, they did not uniquely correlate with other measured muscle properties and thus were not useful in describing muscle (aside from color, of course!). We now know that the red appearance many muscles have is due to a rich muscle blood supply and myoglobin. Since high capillary density is correlated with muscle oxidative capacity, muscle composed of either slow oxidative (SO) fibers or fast oxidative glycolytic (FOG) fibers can possess a reddish appearance. Obviously, SO are slow contracting while FOG are fast contracting, thus explaining the observed discrepancy.

A second scheme used often in human muscle pathology and even in animal experiments is the so-called ATPase-based classification scheme (Brooke and Kaiser, 1970). In this scheme, several repetitions of the ATPase assay mentioned above are carried out on serial sections of muscle (serial sections are consecutive 8-μm sections in which the same fiber can be identified on one section and then the next, *etc.*). However, while the routine ATPase assay is carried out under alkaline conditions (pH 9.4), in the ATPase-based classification scheme, several other assays are performed under increasingly acidic conditions (around pH of 4), and optical density is measured. Thus the assay determines the sensitivity of the ATPase enzyme to the pH of the medium. It turns out (for reasons that are not clear) that fast muscle myosin has a different pH sensitivity than slow muscle myosin. Thus at acid pH, slow fibers stain more darkly than fast fibers, while at alkaline pH the opposite is true. In fact, the scheme takes this differential pH sensitivity a step further. It turns out that fast fibers themselves can be subdivided based on differential pH sensitivity over the range of pH 4.3–4.6. This classification scheme can thus differentiate between fast fibers (termed type 2) and slow fibers (termed type 1), and between (at least) two fast fiber *subtypes* (termed type 2A and type 2B). Table 2.5 presents the definitions of type 1, 2A, and 2B fibers based on the ATPase scheme. Note that this technique must be fine tuned to accurately and repeatedly obtain valid results on various tissues. The values presented are those obtained by Brooke and Kaiser for human muscle (1970).

It has been indirectly concluded that type 2A fibers have a greater oxidative capacity than type 2B fibers, and therefore, many equate them (incorrectly) with type FOG fibers. A number of studies have directly demonstrated that the metabolic scheme does not correlate well with the ATPase-based scheme. This should not be surprising based on an understanding of what the two schemes measure. Why should the pH sensitivity of the ATPase molecule be related to the oxidative or glycolytic capacity of the cell? A relationship *may* exist in general, as most cellular metabolic processes are complimentary. However, the relationships may not hold following a perturbation of the cellular environment and, therefore, must be used with great caution. It seems most prudent to measure directly the property of interest (*e.g.,* oxidative capacity) rather than relying on an indirect measure associated with the property of interest. (Enough said. Have I made my bias abundantly clear?)

PHYSIOLOGIC PROPERTIES OF MUSCLE FIBER TYPES

What are the differences between muscle fibers in terms of their physiologic properties? This is a difficult question to answer because it is currently not technically possible to perform all physiologic measurements on isolated, intact mammalian skeletal muscle fibers in the way in which they were performed by Gordon, Huxley, and Julian on frog muscle fibers since the large amount of interfibrillary connective tissue precludes single fiber isolation. Our best information comes from physiologic experiments on muscles composed *mainly* of one fiber type. Many animal muscles fulfill this criterion, while very few human muscles do. The problem with this approach is that it assumes that a muscle's properties are simply the sum of all the available fibers in the muscle and that each fiber exerts the same relative

Table 2.5.
Fiber Types Histochemical Appearance Using the ATPase Assay

Preincubation pH	Type 1	Type 2A	Type 2B
9.4	Light	Dark	Dark
4.6	Dark	Light	Medium
4.3	Dark	Light	Medium

influence. This approach also assumes that, for example, an SO fiber from a muscle with several fiber types (a "mixed" muscle) is the same as an SO fiber from a muscle composed entirely of SO fibers (a "homogeneous" muscle). Thus we proceed in our discussion of fiber type-specific properties obtained from whole muscle studies with the understanding that we may have to qualify the results to some extent as more data become available.

Maximum Contraction Velocity of Different Muscle Fiber Types

The force-velocity relationship described previously provides a convenient tool for muscle fiber type-specific characterization of "speed." The parameter V_{max} can be compared between muscles that have large differences in fiber type distribution in order to measure fiber type-specific values for V_{max}. As we have seen, muscle architecture has a profound influence on absolute contraction velocity, and therefore, all absolute velocities measured experimentally must be expressed in terms of a normalized velocity such as fiber lengths per second or sarcomere lengths per second in order to determine the intrinsic value of V_{max} for a fiber. I should again state at this point that comparison between muscles of different fiber types without correcting for architectural differences can lead to (and has led to) grossly errant conclusions.

Assuming that we perform the correct type of experiment, we will find that fast-contracting muscle fibers shorten approximately two to three times faster than slow-contracting fibers at V_{max} (Close, 1972). This is actually not a large difference, and, as we shall discuss in the next chapter, probably has very little influence on performance in sports or in rehabilitation. In spite of the popular discussion of fiber type distribution, it probably has very little to do with performance.

Maximum Tension Generated by Different Muscle Fiber Types

In a manner similar to that used for measurement of V_{max}, maximum tetanic tension (P_O) can be measured in muscles of different fiber type distributions. Again, one must account for differences in architecture in order to attribute differences in force to fiber type differences and not to architectural differences (*i.e.*, PCSA). This value is then normalized to the PCSA of the muscle studied, to yield the value known as "specific tension," or force of contraction per unit area of muscle.

This is a controversial area of current muscle physiology, and so it is likely that this discussion will require modification as more data are obtained. However, in measuring the specific tension of whole skeletal muscle, most investigators find that muscles composed mainly of fast fibers have a greater specific tension than muscles composed mainly of slow fibers. The typical value for specific tension of fast muscle is approximately 22 N/cm^2 while that for slow muscle is 10–15 N/cm^2. The common interpretation of these whole-muscle experiments has been that fast muscle fibers have a greater specific tension than slow muscle fibers. Of course the problem with this interpretation is that it assumes that a muscle fiber from a mixed muscle has the same properties as a muscle fiber of the same type that is in a homogeneous muscle. This assumption is probably not true.

The best estimates of specific tension come from isometric contractile experiments of single motor units (discussed below). I will discuss these experiments in more detail later in the chapter, but I will hint at the results since they are relevant to the present topic. Muscle fibers within a motor unit can be classified histochemically, and it turns out that they are generally the same type. Thus if the force generated by a motor unit is measured, and the *motor unit* PCSA is determined, specific tension of different motor unit types (and therefore muscle fiber types) can be calculated. The advantage of this method is that the contractile properties are all measured from the same fiber type. The problem with this scenario is that measurement of motor unit PCSA is *extremely* technically difficult. Generally, methods used for motor unit PCSA determination are extremely indirect, relying on a series of questionable assumptions (see motor unit section). In only one experiment have all of the fibers belonging to a motor unit been identified and summed to yield PCSA. These experiments, performed by Sue Bodine in Reggie Edgerton's laboratory, showed that fast muscle fibers develop just *slightly* more tension than slow muscle fibers (Bodine *et al.,* 1987). This is the best information available to date.

Specific tension of isolated, skinned mammalian muscle fibers has been determined. However, it is so difficult to estimate cross-sectional area of a single fiber and the experimental methods themselves have such potential problems that these measurements probably add little to the discussion already presented.

Endurance of Different Muscle Fiber Types

The endurance (or its opposite, fatigue) of muscle fibers is more difficult to precisely define than speed or strength. This is because endurance depends on the type of work the muscle is required to perform. For example, if the workload is extremely light, there is almost no difference between fiber types. If the workload is extremely heavy, the muscle fibers themselves do not fatigue; rather, the neuromuscular junction fatigues, and again, there is no difference between types. Because excitation-contraction coupling involves a chain of events, it is possible to produce fatigue by interrupting any point in the chain.

Thus a danger exists in simply ascribing a drop in force to muscle fiber fatigue without understanding the basis for the drop. The currently used method for fatigue measurement was developed for classification of single motor units and will thus be deferred until the motor unit presentation. Suffice it to say that the endurance of the various motor units (and muscle fiber types) differs considerably. However, it is difficult to give a quantitative difference unless the work conditions are known. Generally, SO fibers have the greatest endurance, followed by FOG fibers, and, lastly fast glycolytic (FG) fibers. This is not surprising in that we have seen that FG fibers have a very low oxidative capacity.

MORPHOLOGIC PROPERTIES OF DIFFERENT MUSCLE FIBER TYPES

If we understand the histochemical and physiologic differences between muscle fiber types, presented above, structural differences between the various fiber types generally follow logically (Table 2.6).

Contractile Protein Differences between Fiber Types

We have seen repeatedly that myosin differs considerably between fast and slow muscle fibers. Although this difference is profound functionally, there is really not a large structural difference between the different myosins as determined by electron microscopy and X-ray diffraction (Haselgrove, 1983). In fact, in terms of sarcomere force-generating components, while the proteins have very different functional properties, structurally they are quite similar. Fast and slow myosin have approximately the same mass and shape. Muscles composed of either fast or slow sarcomeres have approximately the same filament spacing and cross-bridge density.

Metabolic Differences between Fiber Types

Clearly, the large difference in oxidative and glycolytic capacity is represented in the cell as large differences in the concentration of the metabolic enzymes.

Table 2.6.
Differences between Fiber Types

Parameter	SO	FOG	FG
T/SR system quantity	Little	Much	Much
Z-disk width	Wide	Intermediate	Narrow
Contractile speed	Slow	Fast	Fast
Mitochondrial density	High	Moderate	Low
Lipid droplets	Many	Few	None
Glycogen granules	Few	Many	Many

For example, in the fast fibers, the cytoplasm has a much higher concentration of all of the glycolytic intermediates. Similarly, all of the oxidative fibers (FOG and SO) have a much higher concentration of oxidative enzymes. Since oxidative phosphorylation occurs in the mitochondria, oxidative fibers have a higher mitochondrial density than nonoxidative fibers. In a detailed quantitative study of the ultrastructure of the various fiber types, Brenda Eisenberg confirmed that highly oxidative fibers have a high concentration of mitochondria (up to 25%!) and may contain twice the volume fraction of lipid (Eisenberg, 1983). This alone is one reason why it is difficult to compare the specific tension of the various fiber types even if we were able to isolate intact single fibers. Not all of the space within the fiber is contractile material, and the difference between the fibers in the amount that is contractile material is type specific!

Membrane Differences between Fiber Types

Recall that the T-system and SR are involved in the excitation portion of excitation-contraction coupling. It makes sense that muscles are required to respond rapidly (fast fibers) would have a well-developed membranous activation system as shown biochemically. This is exactly what Eisenberg showed in her quantitative studies. The SR and T-system of fast fibers may occupy two to three times more volume in fast fibers than slow fibers. Thus differences in speed between fast and slow fibers result from differences in cross-bridge cycling rates and differences in activation speed (Gonzalez-Serratos, 1983).

Other Structural Differences between Fiber Types

A final interesting structural difference between fiber types that is very useful but poorly understood is the difference between muscle fiber type Z-disk thickness. Again, in her pioneering work, Brenda Eisenberg showed that FG fibers have the most narrow Z-disks (60 nm) while SO fibers have the widest Z-disks (150 nm). The thickness of the Z-disk in FOG fibers is intermediate (80 nm; Eisenberg, 1983). As stated, the reason for this difference is not clear, but it is interesting to note that in eccentric contraction–induced muscle injury, the Z-disk appears to be the weak link that is most susceptible to injury.

In their excellent anatomic ultrastructural studies, Lars-Eric Thornell and Michael Sjöström have demonstrated differences between fiber types in their M-band structure (Sjöström *et al.,* 1982; Thornell *et al.,* 1976). The M-band of type 1 fibers has five distinct bridges, while that from the 2B fiber has only three bridges (Figure 2.12). The 2A fiber M-band has three prominent central bridges and two faint outer bridges.

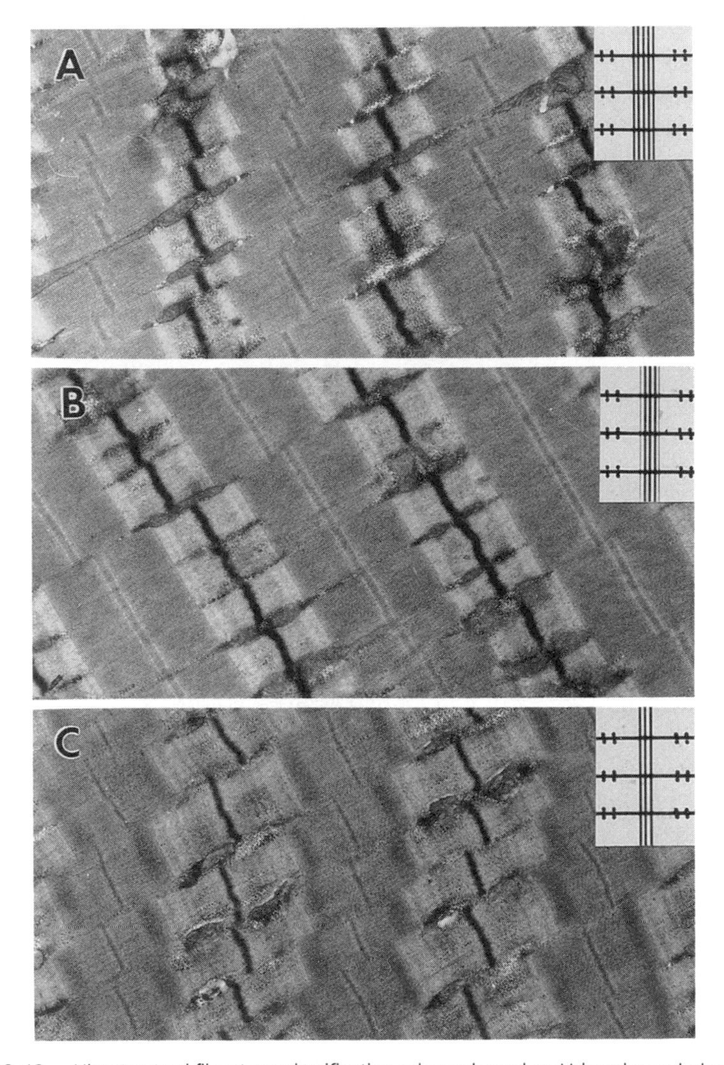

Figure 2.12. Ultrastructural fiber type classification scheme based on M-band morphology. Fibers with M-bands (magnification × 20,000), showing all five M-bridges with equal density, are classified as type 1 fibers (**A**). All other fibers are termed type 2. Of these fibers, those with M-band with the three middle M-bridges clearly visible but the two outer ones relatively less distinct are termed type 2A fibers (**B**). Fibers with only the three middle M-bridges clearly visible are allocated to type 2B (**C**). (Micrographs courtesy of Dr. Jan Fridén.)

While both of these ultrastructural differences hold true generally, again, there is a continuum of Z-disk widths and M-band bridging pattern, so that it is not a simple task to measure a Z-disk width or observe an M-band and then to unambiguously type the fiber. It appears that in about 80% of the fibers, M-band width provides a good indication of fiber type.

THE MOTOR UNIT

Introduction

To this point, we have discussed the properties of whole skeletal muscles and their composite muscle fibers. However, during normal activities, muscle fibers are activated by their composite nerves. What determines the distribution of muscle fibers that are normally activated during a particular task? How does the nervous system determine the force generated by a particular muscle? How are muscle fiber properties tailored to the task at hand? The answer to all of these questions requires an understanding of the anatomy and physiology of the motor unit (Burke, 1981).

Motor Unit Anatomy

While the functional unit of force generation is the sarcomere (actually, the half-sarcomere due to sarcomere symmetry), the functional unit of *movement* is the motor unit. A motor unit is defined as an α-motoneuron plus the muscle fibers it innervates. Motoneurons have their cell bodies in the ventral root of the spinal cord. The cell body is responsible for synthesis of the various nutrients responsible for maintenance of neuronal integrity. The long projection that extends from the cell body is known as the axon. Each cell body projects one axon through the ventral root, and this axon extends along with many other axons projecting from other cell bodies (together these axons are known as a peripheral nerve), to innervate a particular muscle. As the axon (or neuron) approaches the muscle, it branches many times (from just a few to hundreds of branches), and normally each small terminal branch innervates a single muscle fiber (see Chapter 1 for a description for the way in which this anatomic arrangement arises).

Thus a whole muscle contains many motor units, each of which contains a single motoneuron and its composite muscle fibers. The number of muscle fibers belonging to a motor unit (*i.e.,* the innervation ratio) as well as the number of motor units within a whole muscle vary widely. We will discuss the significance of innervation ratio and motor unit number in the next section.

One might guess that all of the muscle fibers within a motor unit might be located in a cluster within the muscle. This in not the case and actually would represent a pathologic condition. Although muscle fibers belonging to a particular motor unit are scattered over subregions of the muscle, fibers from one motor unit are interspersed among fibers of other motor units. The functional consequence of this dispersion is that the forces generated by a unit will be spread over a larger tissue area. This probably minimizes mechanical stress in focal regions within the muscle.

Identification of Muscle Fibers Belonging to a Motor Unit

A prerequisite to any discussion of motor unit properties is a general understanding of the methods used for identification of muscle fibers within a motor unit. Currently, it is not possible to stain for various motor units in the same way that is done for muscle fibers. Motor unit identification methods must identify muscle fibers that are all innervated by the same α-motoneuron. Thus one logical place to begin identifying motor units is the ventral root of the spinal cord, where the motoneurons originate.

Experimentally, the spinal cord can be surgically exposed, and the many motoneurons that exit the ventral root delicately teased apart. These ventral root filaments (actually motor neuron axons) can then be activated individually to stimulate only the fibers belonging to that unit. If the entire peripheral nerve is stimulated, many units and fibers are activated. Even when isolating ventral root filaments, it is possible to isolate, for example, two very small axons that appear to be the same axon. Thus certain tests are performed to ensure that, indeed, a single axon has been isolated.

The main test is to stimulate the filament and record the tension generated by the fibers in that unit. If a single axon were isolated, all muscle fibers belonging to the unit would contract at a single stimulation intensity. If the intensity is increased, and muscle force increases, then more than one axon has been isolated. This is because different axons have different thresholds for activation (see below). This criterion is known as the all-or-none response. If a unit demonstrates an all-or-none response, it is assumed that all of the fibers belong to a single motor unit. In an alternate method, the motoneuron cell body can be impaled by a microelectrode and stimulated to activate all terminal axon branches along with the motor unit muscle fibers.

Now that we are certain that a single axon and its composite fibers have been isolated, how do we identify those fibers? The ideal method would be to somehow "see" fibers that were actively contracting and ignore those that were not. Presently, this is not possible. However, we can trick the active fibers into making themselves visible by stimulating them in a certain way. We can't see active fibers, but we can sometimes see the results of what active fibers do. For example, as a muscle fiber is repetitively stimulated in a way that forces it to generate force anaerobically, it uses intracellular glycogen preferentially as a fuel source. It is a straightforward procedure to stain a muscle cross-section for glycogen. Thus if we force muscle fibers to perform anaerobic metabolism, glycogen will be depleted from the activated fibers and will remain in the nonactivated fibers. This "glycogen depletion" method for isolating motor units was pioneered by Edström and Kugelberg (1968) and has been of critical importance in current motor unit studies.

It is clear that the glycogen depletion method is not without ambiguity. The most obvious problem is developing stimulation protocols that force those muscle fibers that have a choice of aerobic or anaerobic metabolism to choose anaerobic metabolism. Thus the method itself tends to select for the easily identified fibers (the FG fibers) and select against identification of the highly oxidative fibers which have little anaerobic capacity (the SO fibers). This has implications in determining, by glycogen depletion, how many fibers belong to a motor unit. Values for innervation ratio obtained by glycogen depletion would tend to overestimate the number of FG fibers and underestimate the number of SO fibers belonging to a unit.

As a side note, an alternative method is to "feed" radioactive glucose to the muscle and then stimulate it repetitively. The active cells transport the glucose into the cell, thereby labeling them. This method has the advantage that it is not necessary to selectively activate the glycolytic pathway.

Motor Unit Physiologic Properties

Many of the classic motor unit physiology experiments were performed in the late 1960s and early 1970s. The work often cited is that of Bob Burke and his colleagues (Burke, 1967 and 1981). These investigators isolated single cat hindlimb motor units (using intracellular motoneuron stimulation) and measured numerous electrophysiologic properties of the motoneuron and mechanical properties of the motor units within the whole muscle. Interestingly (and fortunately), they found that motor units could generally be classified into three categories based on several physiologic properties of the contracting fibers (Figure 2.8). In other words, motor units were most easily classified based on the physiologic properties of their muscle fibers. These physiologic properties were (*a*) the motor unit twitch tension, (*b*) the fatigability of the unit in response to a specific stimulation protocol, and (*c*) the behavior of the tetanic tension record at an intermediate stimulation frequency. These will be discussed sequentially.

MOTOR UNIT TWITCH TENSION

Early motor unit studies revealed that in response to a single electrical impulse, some units developed very high twitch tensions while others developed relatively low twitch tensions and still others generated intermediate tensions. The exact basis for this difference was not clear. However, the units with low twitch tensions also tended to have slow contraction times while those with higher tensions tended to have fast contractions. This provided some of the first evidence that the different properties of motor units might have profound physiologic significance (Figure 2.13).

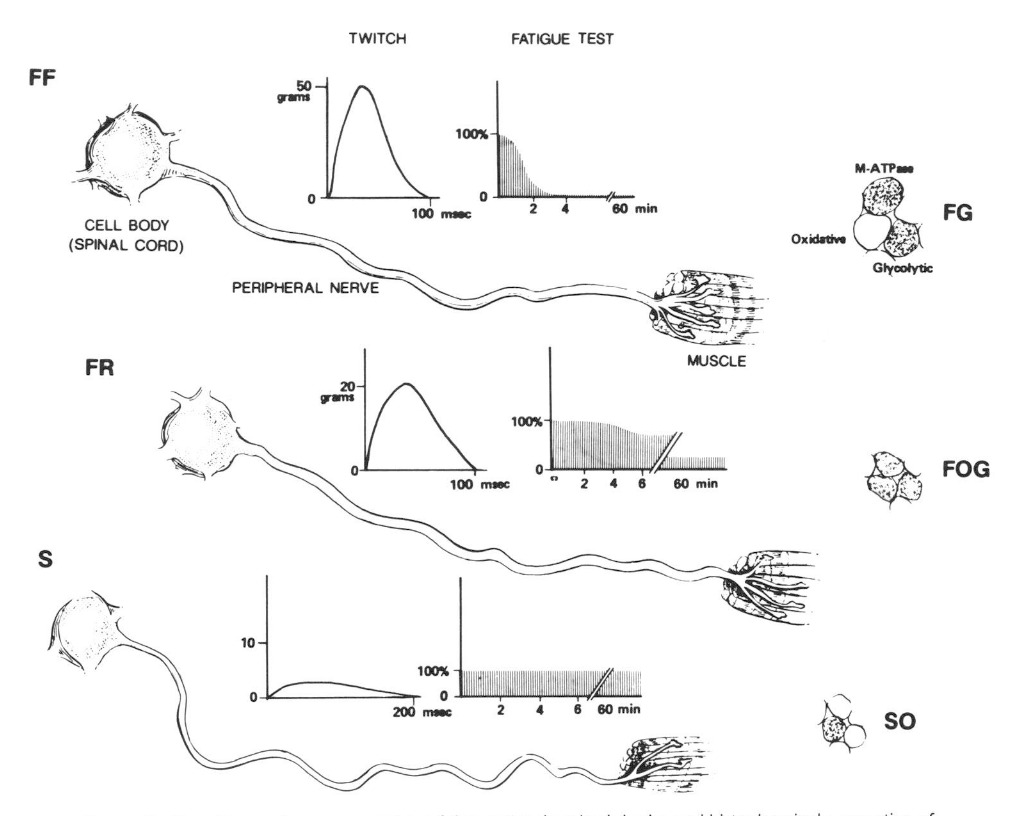

Figure 2.13. Schematic representation of the anatomic, physiologic, and histochemical properties of the three motor units types. FF units (top) have large axons that innervate many large muscle fibers. The units generate large tensions but fatigue rapidly (tension record insets). FR units (middle) have moderately sized axons that innervate many muscle fibers. The units generate moderate tensions and do not fatigue a great deal. S units (bottom) are composed of small axons that innervate a few small fibers. The units generate low forces but maintain force for a long time. Schematic diagram of histochemical staining pattern is shown on right of figure (*c.f.,* Figure 2.11). (Adapted from Edington and Edgerton, 1976.)

MOTOR UNIT FATIGUE INDEX

A second functional property used to distinguish between the various motor units was their "fatigue index," or how much the muscle tension declined upon repetitive stimulation. It was important in these early studies to choose a stimulation frequency that fatigued the muscle fibers themselves. As mentioned, the long chain of events in excitation-contraction coupling can be interrupted at any point, resulting in a force decrease. If our purpose is to identify muscle fibers of a motor unit, we must be sure that the fatigue measured is muscle fiber fatigue and not fatigue, say, of the nerve or neuromuscular junction. Experimentally, the electrical activity of the muscle fibers was measured during repetitive stimulation to guarantee that the activation pulses were reaching the muscle fibers. The fatigue index test required stimulation of the motor unit at approximately 40 Hz (generating about half-maximum tension) for one-third of a second, allowing the muscle to relax for two-thirds of a second, and then repeating the sequence. Thus every second, the motor unit received a burst of 40-Hz pulses. This sequence has been slightly modified by other investigators but is still basically used by many to identify the fatigue index of muscles and motor units. The stimulation protocol was continued for 2 minutes and the muscle tension measured. If the motor unit was highly fatigable, the tension dropped significantly compared to the initial tension. If the unit was not fatigable, the tension dropped only slightly or not at all. Using this approach, it was observed that units could be classified as highly fatigable (defined as units that generated less than 25% of the initial tension after 2 minutes), fatigue resistant (units that generated over 75% of the initial tension after 2 minutes), and fatigue intermediate (units that generated between 25% and 75% of the initial tension after 2 minutes; Figure 2.13).

MOTOR UNIT "SAG" PROPERTY

At this point, we have two classification criteria for motor units: twitch tension and fatigue index. A final and less well understood criterion for motor unit classification is based on the nature of the tension record in response to an unfused tetanic contraction. In some units, the tension was observed to increase smoothly, while in other units, the tension record first increased, and then decreased or "sagged" slightly. The presence or absence of "sag," while not clearly understood in origin, became the final classification criterion. In a manner completely analogous to muscle fiber type classification, these three properties when measured in motor units result in eight potential motor unit types. However, again in a manner analogous to muscle fiber types, only three types of motor units were commonly observed. A summary of these types is shown in Table 2.7.

As we demonstrated with muscle fiber types, we see that motor units also come only in three flavors: Those that have a fast contraction time, a low fatigue index, and sag are known as fast fatigable units (abbreviated FF). Those that have a fast contraction time, a high fatigue index, and demonstrate sag are known as fast fatigue resistant units (abbreviated FR). Finally, those that have slow contraction times, a high fatigue index, and no sag are known as slow units (abbreviated S).

Motor Unit Histochemistry

The alert student may already have an idea of what is to come. The issue to be addressed at this point is, What determines the physiologic properties of the motor units? (Hint: Compare Tables 2.4 and 2.7.) Using the glycogen depletion method, Burke and his colleagues identified the muscle fibers belonging to various motor units using the histochemical procedures previously described (Figure 2.13). It was determined that motor units of different types were composed of muscle fibers of different types. The correspondence between motor units and muscle fibers was as would be expected based on their physiologic properties. The FF motor units were composed of FG muscle fibers, the FR motor units were composed of FOG muscle fibers, and the S units were composed of SO muscle fibers (Table 2.8). I must add a caution following this discussion. It is often stated in the motor unit literature that muscle fibers within a motor unit are *exactly* the same. While it is true that they are the same fiber type, we know that all fibers of a given type are not exactly the same. Recent quantitative studies of the oxidative capacity of

Table 2.7.
Three Motor Unit Types Obtained Using Physiologic Measurements

Motor Unit Designation	Twitch Tension	Twitch Contraction Time	Fatigue Index	"Sag" Present?
FF	High	Fast	Low	Yes
FR	Moderate	Fast	High	Yes
S	Low	Slow	High	No

Table 2.8.
Correspondence between Motor Unit and Muscle Fiber Types

Motor Unit Designation	Muscle Fiber Type in the Motor Unit
Fast fatigable (FF)	Fast glycolytic (FG)
Fast fatigue-resistant (FR)	Fast oxidative glycolytic (FOG)
Slow (S)	Slow oxidative (SO)

different muscle fibers within the same unit reveal a surprising degree of variability between fibers (Martin *et al.,* 1988). These data suggest that while the α-motoneuron certainly influences motor unit properties, it does not absolutely determine them. This result has significant implications in studies of muscle plasticity (adaptation), which we will discuss in Chapters 4 and 5.

Determinant of Motor Unit Tension

As previously described, different motor unit types develop different tensions. Generally, fast motor units develop higher tensions than the slow motor units. Why is this? We might presume that because fast motor units are composed of fast muscle fibers that fast muscle fibers generate more tension than slow muscle fibers. On the other hand, perhaps fast and slow fibers generate the same tension, but fast units simply have a greater *number* of fibers than slow units. Perhaps still, fast and slow units have the same number of fibers of equal intrinsic strength, but the fast fibers are *larger* and therefore generate more tension. Which (if any) of these possibilities is the reason for the differences in muscle tension?

As you might imagine, determination of the number of fibers belonging to a motor unit (innervation ratio) is very difficult experimentally. We mentioned that experimental identification of muscle fibers belonging to a unit requires the glycogen depletion method, which tends to select for FG fibers (FF units) and against SO fibers (S units). This is the first problem. However, even after these fibers have been glycogen depleted, it is technically difficult to find them all within the muscle, especially if the muscle has a pennated architecture. Burke and others used a series of indirect calculations that attempt to account for the various anatomic features (innervation ratio, specific tension, and fiber size). They concluded that fast muscle fibers within a motor unit have a much larger specific tension than slow muscle fibers and have a somewhat higher innervation ratio (Burke, 1981). Unfortunately, it has not been possible to explain the difference in specific tension of fast and slow muscle fibers based on known structural features. However, using a different approach, Sue Bodine, working in Reggie Edgerton's laboratory, *directly measured* innervation ratio in a muscle with longitudinally oriented fibers (the cat tibialis anterior) and, using a stepwise regression model, demonstrated that the major reason that motor units generate different tensions is that high-tension motor units have a greater number of fibers (Bodine *et al.,* 1987). In addition, these high-tension units tend to have larger fibers within them (Figure 2.13). These two factors taken together suggest that motor unit tension is determined primarily by the number and size of the fibers within the unit and not as much by intrinsic differences (specific tension) between the fibers themselves.

The final chapter in this story has not yet been written. The take-home lesson is that our best evidence to date is that fast and slow muscle fibers within a motor unit have about the same specific tension, but that fiber size and innervation ratio differ significantly between motor unit types.

MOTOR UNIT RECRUITMENT

In our discussion of temporal summation, we mentioned that the nervous system can vary muscle force output by varying the stimulation frequency to the muscle fibers. This phenomenon is termed temporal summation. However, muscle force can also be varied by changing the number of motor units that are active at a given time. For relatively low-force contractions, few motor units are activated, while for higher force contractions, more units are activated. The process by which motor units are added as muscle force increases is known as recruitment. What factors determine the point during a contraction that a motor unit is recruited?

A classic study was performed in the 1960s by Elwood Henneman and colleagues whereby motoneuron electrical activity was measured as a muscle was slowly stretched, and therefore tension slowly increased. The increase in passive tension applied to a muscle caused more motor units to be recruited (Binder and Mendell, 1990); Henneman *et al.,* 1965; Figure 2.14). Henneman found that, at very low forces, electrical spikes were observed on the nerve, which were very low amplitude. (It was already known at the time that the amplitude of the spike is related to the size of the axon.) As muscle force increased, the size of the spikes also increased in a very orderly fashion. In other words, as force continued to increase, the units recruited always had larger and larger spikes. The entire process was reversed as force decreased. Henneman and colleagues interpreted this result to mean that at low muscle force levels motor units with small axons were first recruited, and, as force increased, larger and larger axons were recruited. This became known as the "size principle" and provided an anatomic basis for the orderly recruitment of motor units to produce a smooth contraction. Based on other studies, it was determined that, generally, small motor axons innervated slow motor units and larger motor axons innervated fast motor units. In fact, the FF units had the largest axons of all.

VOLUNTARY MOTOR UNIT RECRUITMENT

Essentially all of the data presented above were obtained from animal studies of isolated motor units. What evidence is there that human motor unit properties and recruitment patterns are similar? Obviously, it is not possible

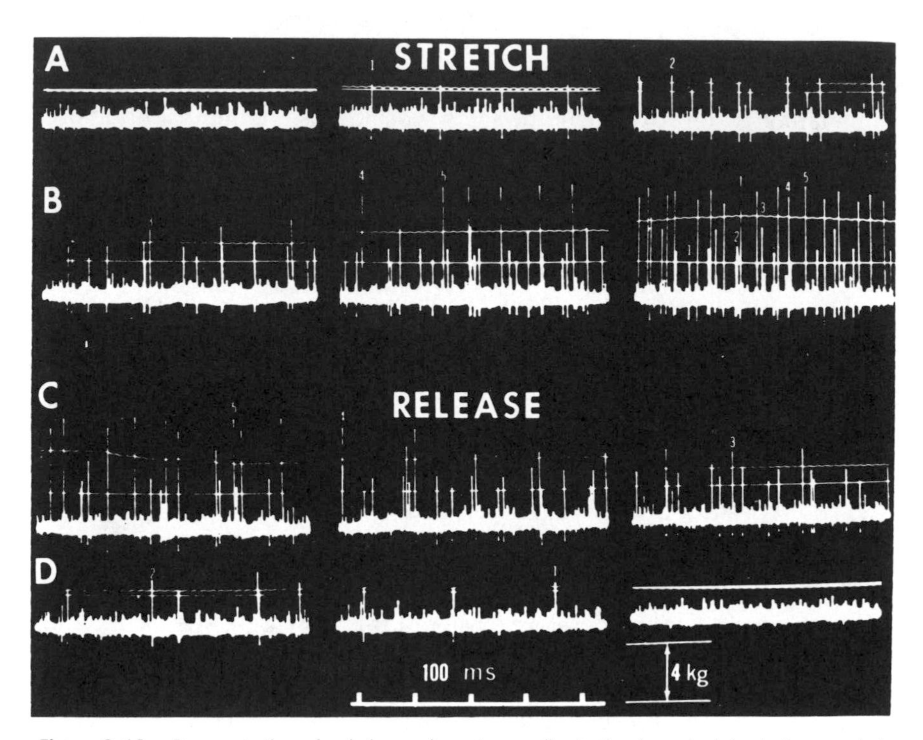

Figure 2.14. Demonstration of orderly recruitment according to the size principle. As the muscle is passively stretched, axons of various sizes (labeled **1** to **5**) are recruited. A continuous line across each trace represents muscle passive tension. As tension decreases (**bottom panel**), units drop out in the reverse order of recruitment. (From Henneman E, Somjen G, Carpenter DO. Functional significance of cell size in spinal motorneurons. J Neurophysiol 1965;28:560–580.)

to perform the identical experiments in humans. However, one method has been developed to study human motor unit properties that has validated many of the results from animal studies.

In the early 1970s, Milner-Brown and his colleagues, developed an ingenious method for measuring the contractile properties of human motor units (Milner-Brown *et al.,* 1973). The experimental apparatus consisted of small needle electrodes placed in the muscle of interest, a force transducer placed on the joint of interest, and surface electrodes to measure muscle electrical activity (Figure 2.15).

After placing the small electrodes in the muscle of interest, Milner-Brown *et al.,* asked the subject to attempt to activate voluntarily a single motor unit! With a little practice and feedback, this task can be performed. During these low-level voluntary activations, motor unit spike trains were recorded from

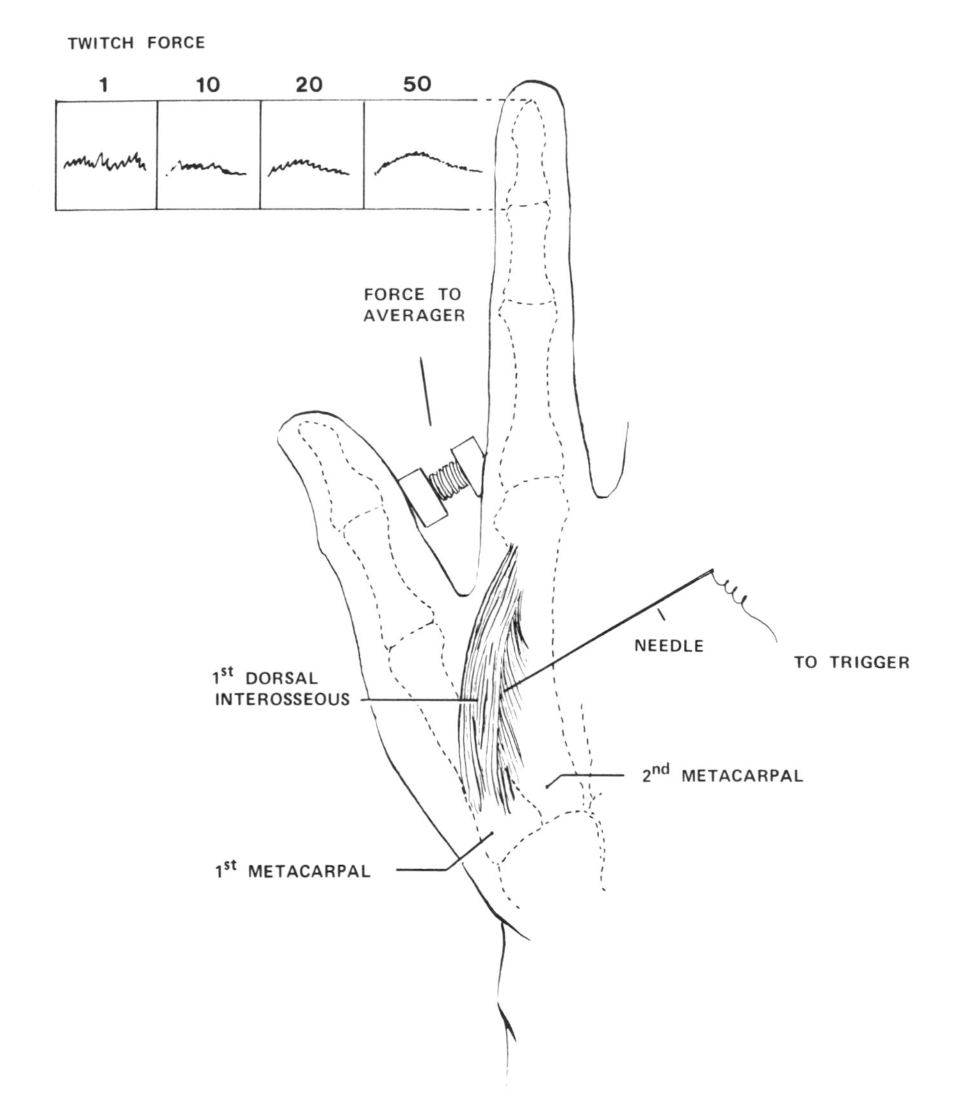

Figure 2.15. Experimental method for demonstration of human motor unit properties according to the spike-triggered averaging method. A needle is placed into the first dorsal interosseous muscle to record motor unit spikes and connected to a trigger. After each spike, finger abduction force is recorded. Many records are recorded (**upper panels**), which decreases background noise and reveals motor unit twitch tension.

the intramuscular electrodes. As voluntary activation level increased, the size of the motor unit spikes also increased. However, the really slick part of the experiment was the manner in which the investigators measured motor unit tension. At very low contraction levels, the force recorded was a very noisy force record—nothing like the smooth twitches recorded from animal motor units. Milner-Brown *et al.,* thus synchronized the force recording with the intramuscular electrical spikes recorded. Thus each time a spike of a particular size was recorded, they triggered their force recording equipment to measure tension. As more and more spikes triggered the recording equipment, the force records were averaged to yield records that looked like muscle twitches (Figure 2.16)! This technique was named spike-triggered averaging for obvious reasons. Using this technique, it has also been shown that at low levels of voluntary effort, slow contracting motor units with low tensions are recruited. As effort increases, faster motor units with higher tensions are recruited (Figure 2.17). Numerous subsequent experiments on a variety of muscles have essentially verified these initial studies. Thus it appears that the size principle is applicable to human as well as animal subjects.

Using all of this information, the following scheme was proposed for the manner in which motor units are recruited voluntarily (Figure 2.18): At very low exertion levels, the smallest axons (which have the lowest threshold to activation) are first activated. Most of these small axons innervate SO muscle fibers within S units. As voluntary effort increases, most of the next-larger axons are recruited, which activates the FOG fibers belonging to FR units. Finally, during maximal efforts, the largest axons are activated, most of which

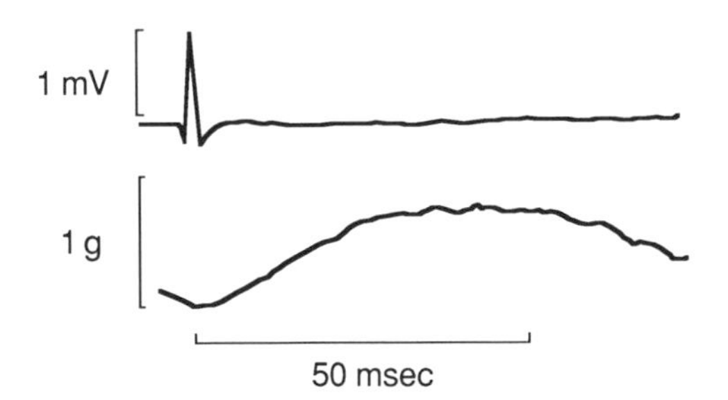

Figure 2.16. Single twitch recorded from human first dorsal interosseous muscle using spike-triggered averaging. Top record is the action potential, and bottom record is averaged twitch record. Not that this twitch looks very much like that obtained from isolated muscle (*c.f.,* Figure 2.2). (Adapted from Milner-Brown HS, Stein RB, Yemm R. The contractile properties of human motor units during voluntary isometric contraction. J Physiol [Lond]. 1973;228:285–306.)

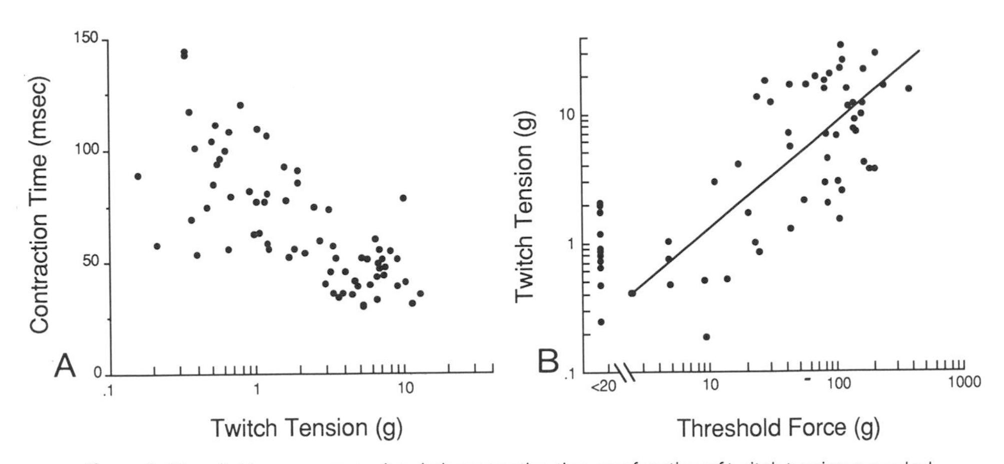

Figure 2.17. **A,** Human motor unit twitch contraction time as a function of twitch tension recorded by spike-triggered averaging. Note that as contractile tension increases, contraction time decreases. This suggests that the motor units with larger tension have faster contractile speed, as predicted by the size principle. **B,** Human motor unit twitch tension as a function of threshold voltage. As threshold increases, larger units are recruited, as predicted by the size principle. (Data from Milner-Brown *et al.,* 1973.)

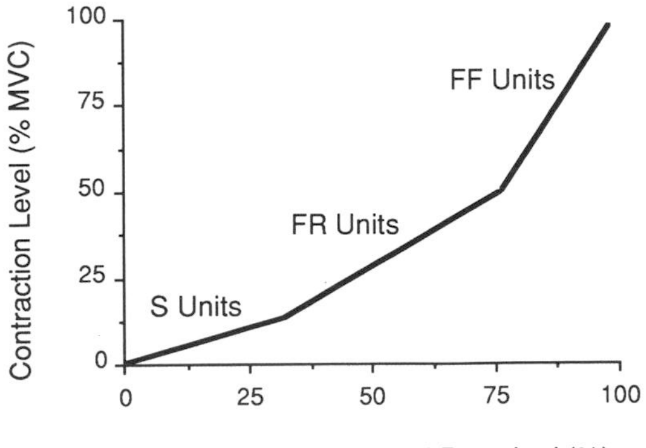

Figure 2.18. Schematic demonstration of predicted orderly recruitment of motor units during voluntary activity as a function of contractile force. At lower forces S units are recruited, while as force increases FR and FF units are recruited. (Adapted from Edgerton VR, Roy RR, Bodine SC, Sacks RD. The matching of neuronal and muscular physiology. In: Borer KT, Edington DW, White TP, eds. Frontiers of exercise biology. Illinois: Human Kinetics Publishers, 1983.)

innervate FG fibers and make up FF units. An appealing aspect of this hypothesis is that the units most often activated (S units) are those with the greatest endurance. The FF units, which are rarely activated, have the lowest endurance. In addition, the S units develop the slowest tension, and thus as contractions begin, low tensions are generated. This provides a mechanism for smoothly increasing tension as first S, then FR, and then FF units are recruited. This exquisite interrelationship of anatomic specialization and physiologic function is just one more structure-function relationship, which is the hallmark of the neuromuscular system.

PHYSIOLOGIC BASIS OF FATIGUE

Now that we have discussed fiber types, metabolism, and motor units, we are in a position to discuss the issue of fatigue. Nearly everyone is familiar with the feeling of muscle fatigue following prolonged exercise. However, a strict definition of fatigue has been more difficult to establish. This is due, in part, to the complex nature of voluntary contractions themselves. At least three major components are involved in the production of voluntary contractions (Figure 2.19): the CNS, the peripheral nerve and neuromuscular junction, and the skeletal muscles. *A priori* any one of these systems might be involved in the

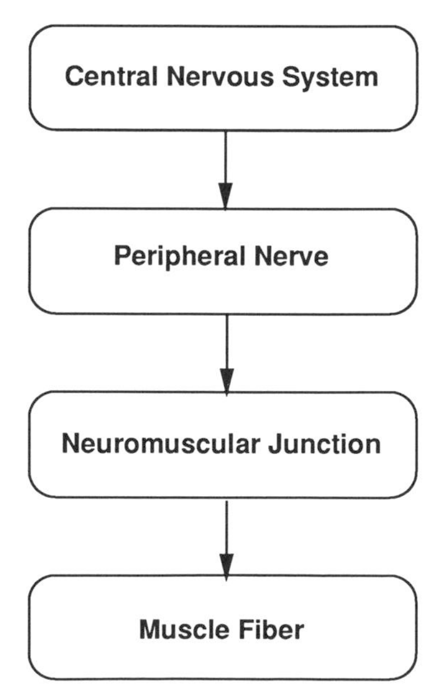

Figure 2.19. Chain of events involved in normal contraction and potential sites for fatigue. Contractile force can decrease if any portion of this chain is interrupted.

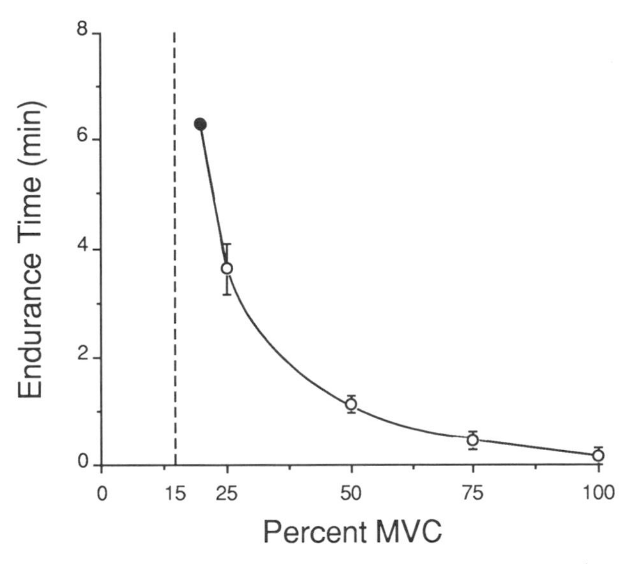

Figure 2.20. Relationship between endurance time and contraction intensity (% MVC). Note that forces lower than approximately 15% MVC can be maintained indefinitely (here defined as >45 minutes). As force increases, endurance time rapidly decreases. (Data from Rohmert, 1960.)

fatigue process. We will examine several of the classic fatigue studies that add to our current understanding of muscle fatigue.

Intuitively, it is obvious that low forces can be maintained longer than high forces. In fact, this relationship was quantified for a number of human muscles (Rohmert, 1960). Subjects were asked to maintain a target force ranging from 5% to 100% of their maximum voluntary contraction (MVC) level. For contraction levels less than 15% MVC, subjects could maintain the target level indefinitely (>45 minutes). However, as the target force increased, endurance time rapidly decreased (Figure 2.20). How can these changes in endurance time be explained? Why does muscle force decrease? Which of the different systems (Figure 2.19) changes in response to prolonged contraction?

Substrate Depletion in Fatigue

We have seen that ATP is the immediate energy source for force generation in muscle. However, under normal conditions, skeletal muscle only contains enough ATP to fuel two or three maximal contractions! What happens as ATP levels suddenly drop following contraction? An ATP regenerating system is present in muscle that is composed of the high-energy molecule creatine phosphate and the enzyme creatine phosphokinase (CK). Immediately after ATP depletion begins, it is replenished according to the reaction

$$\text{Creatine Phosphate} + \text{ADP} \rightarrow \text{Creatine} + \text{ATP} \qquad (2.17)$$

which is catalyzed by CK. As contraction proceeds, ATP levels remain relatively constant while creatine phosphate levels steadily drop (Bergström, 1967). The greater the workload, the greater the decrease in creatine phosphate. As cellular ATP levels continue to drop, cellular glycogen and fat are mobilized to replenish energy stores. The relative degree of glycogen and fat mobilized depends largely on the intensity of the exercise and the capability of the muscle fiber.

In an experimental investigation of glycogen metabolism, the Scandinavian physician Eric Hultman measured glycogen content in muscle biopsies obtained from the vastus lateralis muscles of cross-country skiers (Hultman, 1967). Hultman found, as expected, that as exercise proceeded, muscle glycogen levels dropped dramatically. In fact, the amount of time these skiers could pedal a bicycle ergometer at an intense level was directly related to the amount of glycogen in their muscles (Figure 2.21**A**). He also found that, following glycogen depletion, if the subjects were fed a high-carbohydrate diet, the amount of glycogen restored in the muscle actually exceeded the original amount. This overshoot was not seen if subjects were fed a high fat and protein diet following glycogen depletion. (Figure 2.21**B**). The overshoot was also not seen in the contralateral leg, which indicates that the "control" of glycogen storage was at the level of the muscle and not something that would affect all muscles in the body. In intense anaerobic exercise, substrate availability can thus limit performance.

Neural Fatigue Mechanisms

However, when muscle force declines, how can we be sure that the central drive from the CNS has not decreased? In a now classic study, Merton measured muscle force decline during fatiguing contractions (Merton, 1954). He was interested in addressing this question of whether muscle force declined because of a decrease in drive or intensity from the CNS (*i.e.,* CNS fatigue). To answer this question, periodically during the person's voluntary effort, Merton superimposed a massive electrical stimulation onto the voluntary contraction. He hypothesized that if muscle force decreased due to CNS fatigue, that electrical stimulation superimposed on the voluntary contraction would increase muscle force. In fact, no force increase was observed, and Merton concluded that CNS fatigue was not the cause for the muscle fatigue. This type of experiment has been confirmed by others, and it

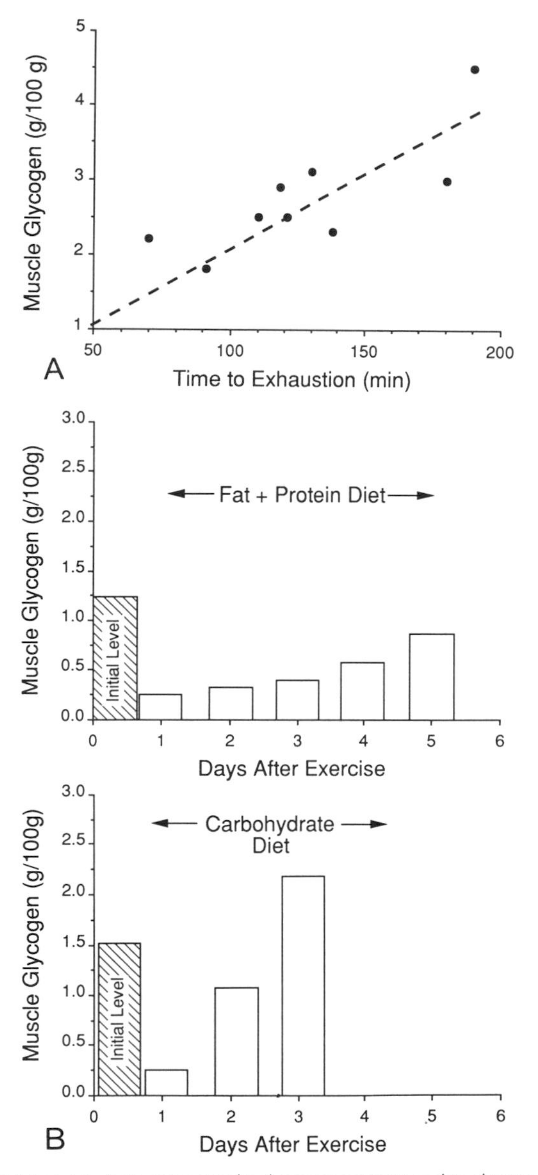

Figure 2.21. **A,** Relationship between muscle glycogen content and endurance time for intense bicycle ergometry. Increased muscle glycogen enables greater exercise time before exhaustion occurs. **B,** Muscle glycogen content before exercise (hatched bar) and after exercise (open bar). Data are shown after a fat and protein diet (**upper panel**), and after a carbohydrate diet (**lower panel**). Note that after exercise and a high carbohydrate diet, muscles store more glycogen than prior to exercise while a fat plus protein diet results in glycogen levels below the original level even for six days. (From Hultman E. Physiological role of muscle glycogen in man, with special reference to exercise. Circ Res 1967;21[suppl 1]:99–112.)

is generally agreed that muscle fatigue in highly motivated, trained subjects is not due to CNS fatigue.

Similar experiments have been performed to determine whether fatigue of the neuromuscular junction or sarcolemma occurs during muscle fatigue. In these experiments, during fatiguing MVCs, electrical stimuli were again superimposed and the muscle mass action potential measured (*i.e.*, the M-wave). Brenda Bigland-Ritchie and her colleagues demonstrated that almost no change in the M-wave occurred in spite of the force decrease (Bigland-Ritchie *et al.,* 1982). These data suggested that the weak link in fatigue was also not the neuromuscular junction or muscle sarcolemma.

Fatigue in Isolated Nerve-Muscle Preparations

In contrast to human studies, many investigators have studied the behavior of the muscle action potential and muscle force during repetitive electrical stimulation to the point of fatigue in isolated nerve-muscle preparations (*e.g.,* Pagala *et al.,* 1984). Experimentally, muscles are indirectly stimulated via the motor nerve and muscle force and action potential are measured. Several investigators have shown that during muscular fatigue, force and M-wave changes do not follow the same time course. For example, at moderate stimulation frequencies (*e.g.,* 30 Hz), muscle force declines faster than the action potential, which suggests that both excitation and contraction can be altered during fatigue (refer to the above discussion of excitation-contraction coupling). However, it is not clear that these same mechanisms are functionally significant during voluntary activity in humans. These studies do emphasize that the fatigue process may involve several different links in the neuromotor chain.

Neuromotor Control and Muscle Fatigue

Recently, Bigland-Ritchie and her colleagues succeeded in measuring single muscle fiber action potentials in human adductor pollicis longus muscles (Table 1.1) during MVCs (Bigland-Ritchie *et al.,* 1983a and 1983b). They found that as muscle force declined during the MVC, the average motor unit firing rate decreased from approximately 30 Hz to approximately 15 Hz (Figure 2.22). However, based on our previous discussion of temporal summation (see page 52), doesn't decreased firing frequency result in decreased muscle force? Interestingly, these investigators hypothesized a reasonable neuromotor control strategy for such a frequency decrease. By measuring the time course of the muscle twitch during fatigue, they found that the muscle actually contracted and relaxed more slowly, with the time-to-peak tension (TPT) and half-relaxation time (HRT) dramatically *increasing.* What would be the effect

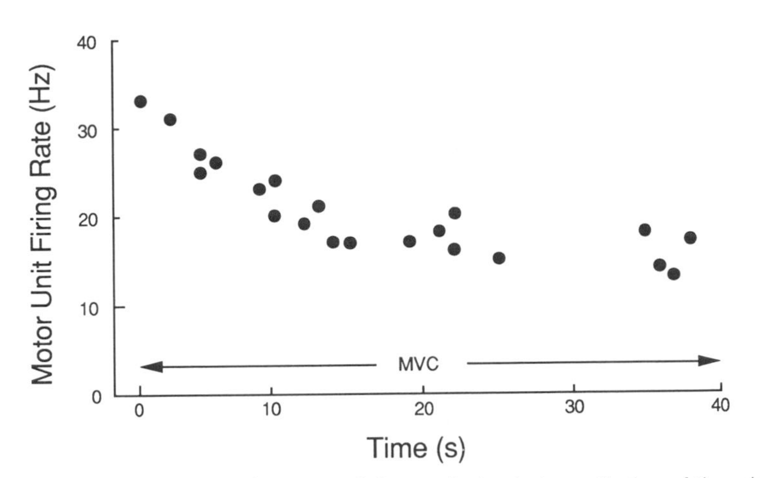

Figure 2.22. Motor unit firing frequency during maximal voluntary activation of the adductor pollicis longus muscle. As fatigue occurs, firing rate decreases. (From Bigland-Ritchie B, Johansson R, Lippold OCJ, Smith S, Woods JJ. Changes in motoneuron firing rates during sustained maximal voluntary contractions. J Physiol 1983;340:335–346.)

of muscle slowing? Slower muscles generate higher forces at lower frequencies since the muscle is not able to keep up with the incoming impulses, much as a slow-contracting muscle shows tetanic fusion at lower frequencies than a fast muscle. Therefore, if firing rate decreased and TPT and HRT increased, the net effect would be to maintain approximately the same degree of fusion in the contractile record. It is as if the nervous system has a feedback system that senses the muscle speed and drives it with the appropriate stimulation frequency. Support for this idea was obtained by comparing the average firing rate of fast muscles (*e.g.*, biceps brachii) to slow muscles (*e.g.*, soleus). They found that while the biceps firing rate was approximately 31 Hz, the soleus firing rate was only about 10 Hz. These firing rates varied in proportion to the relative speeds of the two muscles, providing support for the idea that firing rates are tailored to muscle contractile properties.

THE ELECTROMYOGRAM

The Electromyogram—an Electrical Activation Signal

We are now at the point in our tour of muscle physiology that we might begin to wonder how these various neuromuscular components function in daily activities, not simply in voluntary isometric contractions under experimental conditions. Before moving into that most exciting arena, let us pause and take a moment to understand a measurement often made from muscles in order to understand their normal function. This measurement is the electromyogram (EMG).

Recall from the discussion of excitation-contraction coupling that each time a muscle fiber is activated, an action potential is conducted from the nerve, along the muscle and into the fiber. Thus an electrical signal is emitted by the muscle fiber during each activation. If a suitable receiver is placed near the fiber (for example, an EMG electrode), this electrical activity can be measured (Figure 2.23). As such, the raw EMG signal is useful to determine when a particular muscle is active. In fact, one of the best uses for the EMG is to determine the relative timing of muscle activation during gait (discussed in Chapter 3). Some have taken the measurement of EMG a step further—EMG quantification.

Quantifying the EMG—How Much Is Enough?

If the raw EMG signal is observed, it is difficult to quantify the activity of the muscle. Theoretically at least, the EMG contains electrical signals from every muscle fiber in the vicinity of the EMG electrode. Can this be counted? Again, theoretically, it has been shown that it is *reasonable* to process the EMG signal into a "rectified, integrated" signal that is more easily managed than the raw signal. The rectification step converts the EMG record into one that contains only positive voltages. The integration step electrically "adds up" all of these separate spikes to yield a number proportional to all of the spikes generated. Recently, Carl Gans and Jerry Loeb have presented an excellent discussion of practical EMG measurement methods (Loeb and Gans, 1986). In addition, John Basmajian and Carlo DeLuca have provided a comprehensive review of the electrical activity of different muscles during various movements as well as many theoretical aspects of EMG measurement (Basmajian and DeLuca, 1985). Suffice it to say that the EMG, if acquired and processed correctly, should represent the total number of fibers active at a particular moment in time.

EMG-Force Relationship

Much has been written about the predictability and reliability of the EMG-force relationship. There is general agreement on the utility of the EMG as an indicator of the activation pattern for a motor pool. However, it should be recognized that the *EMG cannot always be uniquely related to muscle force.* For example, suppose the EMG is measured from a muscle that is shortening at 5% V_{max}. The force generated by this muscle will be approximately 75% P_O. However, the EMG from the same muscle might be measured under conditions where it is shortening at only 1% V_{max}. In this case, much more tension will be generated (90% P_O *vs.* 75% P_O) due to the slower contraction velocity, but the EMG signal will be the same. The take-home

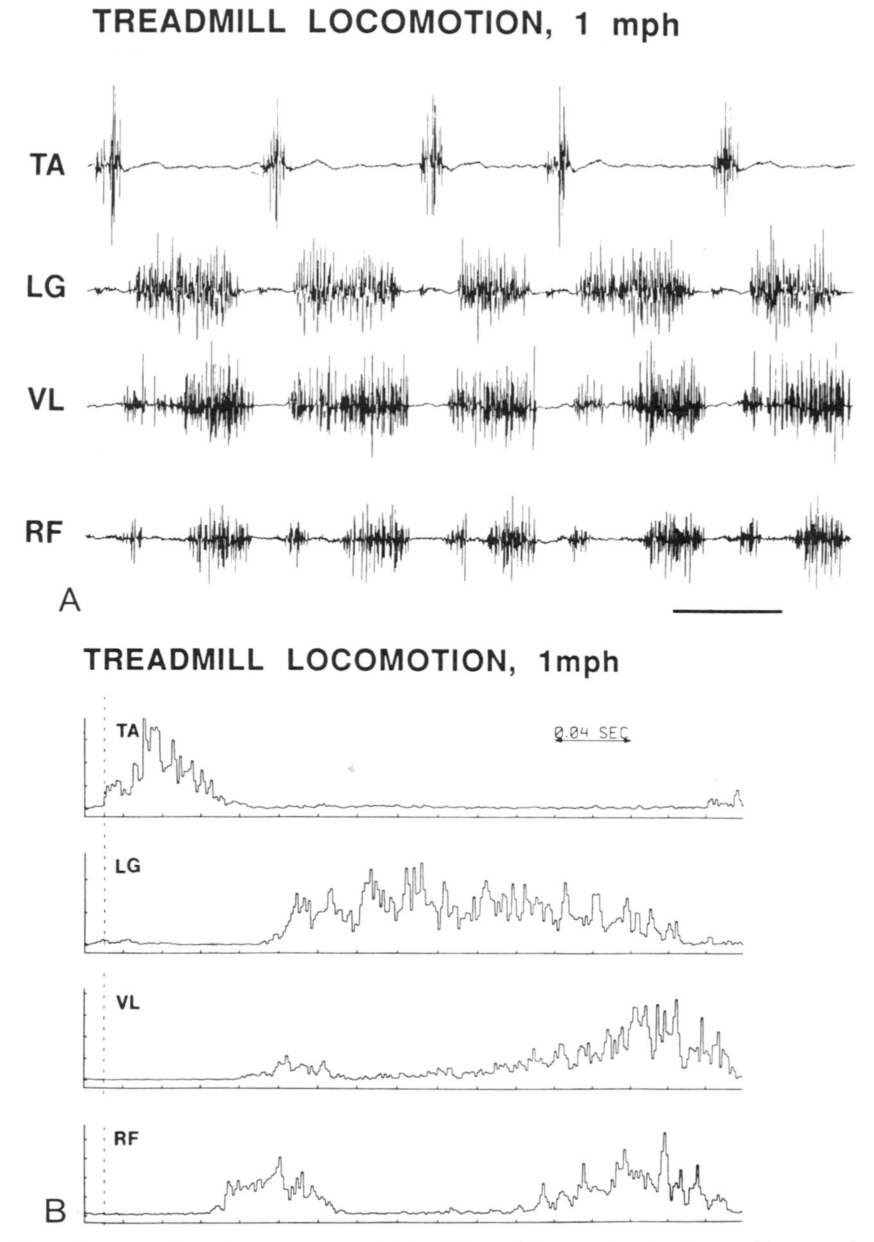

Figure 2.23. **A,** Raw electromyograms obtained from rat leg muscles at a slow walking speed. Note that muscles fire cyclically and relatively regularly during these cycles. Calibration bar = 0.25 sec. **B,** Data from **A** rectified and averaged over the cycles. This processed EMG permits description of the typical muscle activation signal. (Experimental records courtesy of Dr. Sue Bodine-Fowler)

lesson is, an EMG can give you information regarding muscle activation patterns but be careful when interpreting an EMG record in terms of muscle force. In isometric or perfectly eccentric or concentric conditions, forces might be estimated, but otherwise, be very careful!

SUMMARY

Skeletal muscle is activated sequentially by the process known as excitation-contraction coupling. Muscle force is generated by the cyclic interaction between actin and myosin, using ATP as the direct energy source. Isometric muscle force varies with length as explained by the length-tension curve. The force-velocity relationship describes isotonic behavior, muscle contraction velocity at constant load. Muscle fiber types and motor units come in three general types that can be characterized physiologically and histochemically. The properties of the fibers that make up the various motor unit types generally follow logically from a knowledge of their physiologic properties. Motor units are recruited generally in an orderly fashion according to Henneman's size principle. Although the size of a unit can be estimated in several ways, the order of recruitment seems to be closely matched with the unit's maximum tetanic tension. You're now about to enter unexplored territory. In the next chapter, we will synthesize much of this information in a form relevant to normal movement.

REFERENCES

Barany, M. (1967). ATPase activity of myosin correlated with speed of muscle shortening. J. Gen. Physiol. 50:197–216.

Basmajian, J.V. and DeLuca, C.J. (1985). Muscles Alive. Their Functions Revealed by Electromyography. 5th ed. Baltimore: Williams & Wilkins.

Bergström, J. (1967). Local changes of ATP and phosphorylcreatine in human muscle tissue in connection with exercise. Circ. Res. 21(suppl 1):191–198.

Bigland-Ritchie, B., Johansson, R., Lippold, O.C.J., Smith, S., and Woods, J. (1983a). Changes in motoneurone firing rates during sustained maximal voluntary contractions. J. Physiol. 340:335–346.

Bigland-Ritchie, B., Johansson, R., Lippold, O.C.J., and Woods, J.J. (1983b). Contractile speed and EMG changes during fatigue of sustained maximal voluntary contractions. J. Neurophysiol. 50:313–324.

Bigland-Ritchie, B., Kukulka, C.G., Lippold, O.C.J., and Woods, J.J. (1982) The absence of neuromuscular transmission failure in sustained maximal voluntary contractions. J. Physiol. 330:265–278.

Binder, M.D., and Mendell, L.M., eds. (1990). The Segmental Motor System. New York: Oxford University Press.

Bodine, S.C., Roy, R.R., Eldred, E., and Edgerton, V.R. (1987). Maximal force as a function of anatomical features of motor units in the cat tibialis anterior. J. Neurophysiol. 6:1730–1745.

Brooke, M.H., and Kaiser, K.K. (1970). Muscle fiber types: how many and what kind? Arch. Neurol. 23:369–379.

Burke, R.E. (1967). Motor unit types of cat triceps surae muscle. J. Physiol. 193:141–160.

Burke, R.E. (1981). Motor units: anatomy, physiology, and functional organization. In: Brookhart, J.M., Mountcastle, V.B., Brooks, V.B., Geiger, S.R., eds. Handbook of Physiology. Bethesda, MD: American Physiological Society, 345–422.

Close, R.I. (1972). Dynamic properties of mammalian skeletal muscles. Physiol. Rev. 52:129–197.

Ebashi, S., Maruyama, K., and Endo, M., eds. (1980). Muscle Contraction. Its Regulatory Mechanisms. Tokyo: Japan Scientific Societies Press.

Edington, D.W., and Edgerton, V.R. (1976). Biology of Physical Activity. Boston: Houghton Mifflin.

Edgerton, V.R., Roy R.R., Bodine, S.C., Sacks, R.D. (1983). The matching of neuronal and muscular physiology. In: Borer, K.T., Edington, D.W., White, T.P., eds. Frontiers of Exercise Biology. Illinois: Human Kinetics Publishers.

Edman, K.A.P. (1966). The relation between sarcomere length and active tension in isolated semitendinosus fibers of the frog. J. Physiol. [Long]. 183:407–417.

Edström, L., and Kugelberg, E. (1968). Histochemical composition, distribution of fibers and fatiguability of single motor units. J. Neurol. Neurosurg. Psychiatry 31:424–433.

Eisenberg, B.R. (1983). Quantitative ultrastructure of mammalian skeletal muscle. In: Peachy, L.D., Adrian, R.H., and Geiger, S.R., eds. Handbook of Physiology. Bethesda, MD: American Physiological Society, 73–112.

Entman, M.L., and Van Winkle, W.B., eds. (1986). Sarcoplasmic Reticulum in Muscle Physiology. Vol. 1. Boca Raton, FLA: CRC Press, 1986.

Gauthier, G.F. (1969). On the relationship of ultrastructural and cytochemical features to color in mammalian skeletal muscle. Z. Zellforsch. Mikrosk. Anat. 95:462–482.

Gillespie, C.A., Simpson, D.R., and Edgerton, V.R. (1970). High glycogen content of red as opposed to white skeletal muscle fibers of guinea pigs. J. Histochem. Cytochem. 18:552–558.

Goldman, Y.E. (1987). Kinetics of the actomyosin atpase in muscle fibers. Ann. Rev. Physiol. 49:637–654.

Gonzalez-Serratos, H. (1983). Inward spread of activation in twitch skeletal muscle fibers. In: Peachy, L.D., Adrian, R.H., and Geiger, S.R., eds. Handbook of Physiology. Bethesda, MD: American Physiological Society, 325–353.

Gordon, A.M., Huxley, A.F., and Julian, F.J. (1966). The variation is isometric tension with sarcomere length in vertebrate muscle fibers. J. Physiol. 184:170–192.

Haselgrove, J.C. (1983). Structure of vertebrate striated muscle as determined by x-ray-diffraction studies. In: Peachy, L.D., Adrian, R.H. and Geiger, S.R., eds. Handbook of Physiology. Bethesda, MD: American Physiological Society, 143–171.

Henneman, E., Somjen, G., and Carpenter, D.O. (1965). Functional significance of cell size in spinal motorneurons. J. Neurophysiol. 28:560–580.

Hill, A.V. (1938). The heat of shortening and the dynamic constants of muscle. Proc. R. Soc. Lond. [Biol]. 126:136–195.

Hill, A.V. (1964). The effect of load on the heat of shortening of muscle. Proc. R. Soc. Lond. [Biol]. 159:297–318.

Hill, A.V. (1970). First and Last Experiments in Muscle Mechanics. Cambridge: Cambridge University Press.

Hoffer, J.A., Loeb, G.E., Marks, W.B., O'Donovan, M.J., Pratt, C.A., and Sugano, N. (1987). Cat hindlimb motoneurons during locomotion. I. destination, axonal conduction velocity, and recruitment threshold. J. Neurophysiol. 57:510–573.

Hoffer, J.A., O'Donovan, M.J., Pratt, C.A., and Loeb, G.E. (1981). Discharge patterns of hindlimb motorneurons during normal cat locomotion. Science. 213:466–468.

Horowits, R., and Podolsky, R.J. (1987). The positional stability of thick filaments in activated skeletal muscle depends on sarcomere length: evidence for the role of titin filaments. J. Cell Biol. 105:2217–2223.

Hultman, E. (1967). Physiological role of muscle glycogen in man, with special reference to exercise. Circ. Res. 21(suppl 1):99–112.
Huxley, A.F. (1957). Muscle structure and theories of contraction. Prog. Biophys. 7:255–318.
Katz, B. (1939). The relation between force and speed in muscular contraction. J. Physiol. 96:45–64.
Loeb, G.E., and Gans, C. (1986). Electromyography for Experimentalists. Chicago: University of Chicago Press.
Lymn, R.W., and Taylor, E.W. (1971). Mechanism of adenosine triphosphate hydrolysis by actomyosin. Biochemistry. 10:4617–4624.
Magid, A., and Law, D.J. (1985). Myofibrils bear most of the resting tension in frog skeletal muscle. Science. 230:1280–1282.
Martin, T.P., Bodine-Fowler, S., Roy, R.R., Eldred, E., and Edgerton, V.R. (1988). Metabolic and fiber size properties of cat tibialis anterior motor units. Am. J. Physiol. 255:C43–C50.
Maruyama, K., and Gergely, J. (1962). Interaction of actomyosin with adenosine triphosphate at low ionic strength: dissociation of actomyosin during the clear phase. J. Biol. Chem. 237:1095–1099.
McCray, J.A., Herbette, L., Kihara, T., and Trentham, D.R. (1980). A new approach to time-resolved studies of ATP-requiring biological systems: laser flash photolysis of caged ATP. Proc. Natl. Acad. Sci. 77:7237–7241.
Merton, P.A. (1954). Voluntary strength and fatigue. J. Physiol. 123:553–564.
Milner-Brown, H.S., Stein, R.B., and Yemm, R. (1973). The contractile properties of human motor units during voluntary isometric contraction. J. Physiol. [Lond]. 228:285–306.
Moss, R.L. (1979). Sarcomere length-tension relations of frog skinned muscle fibers during calcium activation at short lengths. J. Physiol. 292:177–192.
Pagala, M., Namba, T., and Grob, D. (1984). Failure of neuromuscular transmission and contractility during muscle fatigue. Muscle Nerve. 7:454–464.
Peachey, L.D., and Franzini-Armstrong, C. (1983). Structure and function of membrane systems of skeletal muscle cells. In: Peachey, L.D., ed. Handbook of Physiology. Bethesda, MD: American Physiological Society, 23–73.
Peter, J.B., Barnard, R.J., Edgerton, V.R., Gillespie, C.A., and Stempel, K.E. (1972). Metabolic profiles on three fiber types of skeletal muscle in guinea pigs and rabbits. Biochemistry. 11:2627–2733.
Podolsky, R.J., and Shoenberg. (1983). Force generation and shortening in skeletal muscle. In: Peachy, L.D., Adrian, R.H., and Geiger, S.R., eds. Handbook of Physiology. Bethesda, MD: American Physiological Society, 173–187.
Rohmert, W. (1960). Ermittung von Erholung-spausen fur statische Arbeit des Menschen. Int. Angew. Physiol. 18:123.
Rüdel, R., and Taylor, S.R. (1971). Harderian gland: an extraretinal photoreceptor influencing the pineal gland in neonatal rats? Science. 167:884–885.
Sjöström, M., Kidman, S., Henriksson-Larsen, K., and Angquist, K.A. (1982). Z- and M-band appearance in different histochemically defined types of human skeletal muscle fibers. J. Histochem. Cytochem. 30:1–11.
Taylor, S.R., and Rüdel, R. (1970). Striated muscle fibers: inactivation of contraction induced by shortening. Science. 167:882–884.
Thornell, L.E., Sjöström, M., and Ringqvist, M. (1976). Attempts to correlate histochemical and ultrastructural features of individual skeletal muscle fibers. J. Ultrastruct. Res. 57:224.
Webb, M.R., and Trentham, D.R. (1983). Chemical mechanism of myosin-catalyzed ATP hydrolysis. In: Peachy, L.D., Adrian, R.H., and Geiger, S.R., eds., Handbook of Physiology. Bethesda, MD: American Physiological Society. 237–255.

Chapter

3

The Production of Movement

OVERVIEW

In this chapter, the "rubber meets the road!" The anatomy and physiology of nerves and muscles have already been presented. Now you will see how these systems, along with tendons and bones, produce movement. We will start with a discussion of the basic components of the movement generating system: the muscles, skeleton, and tendons. The pieces of the system are put together and then activated to produce movement. Muscles, tendons, and joints are complex in their interaction. Principles that apply to the determination of strength and range of motion along with implications for physical examination will also be presented. This is the home stretch of the basic science portion of the text.

INTRODUCTION

In the preceding two chapters, I have presented the parts that make up the motors driving the human movement machine. You know a great deal about skeletal muscle anatomy and physiology. You also have some idea of the relationship between the nervous system and the muscles (motors) it controls. We have discussed the way that muscle force is regulated by either altering activation frequency or altering the number of activated motor units. This is precisely the position many find themselves in following a good neuromuscular physiology course. These courses are taken by undergraduate and graduate students, medical students, and therapists who need to apply this information to the real world of weak or injured patients and elite or injured athletes. We must now ask: Does this basic information apply to the real world of patients and athletes? Can we use our newfound information to the benefit of our patients? The answer is, fortunately, yes. However, certain definitions must first be made, which are needed to describe the action of muscles located in the real-world environment. It is not possible to simply speak of isometric or isotonic muscle contractions in the real world since muscle activation is never truly performed at a constant length or under a constant load. In fact, before discussing movement *per se,* we must digress for a moment to further discuss the muscle-tendon motors and the joints on which they act to produce movement. Following this introduction, we will apply these ideas to the musculoskeletal system.

MUSCLE-TENDON INTERACTION

Most anatomy texts teach that tendons function to connect skeletal muscles to bones. However, tendons should not be considered simply as rigid linkage structures. It is well known that tendons are relatively compliant (stretchy) tissues.

Tendon Mechanical Properties

If a tendon is physically connected to force transducer and stretched under a load, it is possible to measure the lengthening of the tendon material that occurs during the stretching. The change in tendon length can be measured in absolute units (*e.g.,* mm) and can be expressed relative to the starting tendon length as strain (Equation 3.1):

$$\epsilon = \frac{l - l_o}{l_o} \qquad (3.1)$$

where l is the tendon length, l_o is the initial length, and ϵ is strain. Strain is, therefore, the change in tendon length relative to its starting length and is a convenient method for expressing the stretchiness or compliance of materials. When a tendon is loaded and strain measured, a curve such as that shown in Figure 3.1 is obtained. If, instead of tendon load, we normalized the load to the cross-sectional area of the tendon, we could calculate the stress on the tendon as

$$\sigma = \frac{F}{A} \qquad (3.2)$$

where F is the tendon load (in Newtons of load or force), A is the tendon cross-sectional area (in area units such as m^2), and σ is stress (in units of N/m^2, which are abbreviated as Pascals or Pa). By expressing load in terms of stress (which is normalized) and deformation in terms of strain (which is also normalized), it is possible to compare the properties of materials of different sizes and shapes. One interesting parameter to compare is the relative stiffnesses (Young's modulus) of materials, which is simply the change in stress for a given change in strain. In other words, Young's modulus (E) can be calculated as

$$E = \frac{\sigma}{\epsilon} \tag{3.3}$$

Since strain has no units, the units of modulus are the same as those for stress, namely, Pascals. Typical values for the modulus of muscle, tendon, and bone are 200 kPa, 1 GPa, and 20 GPa respectively. It is clear that bone is the stiffest and muscle the least stiff of the connective tissues.

Returning to our discussion of tendon, note that, at low loads, tendons strain quite a bit for a small change in load (Figure 3.1). In this low-load region, the tendon is thus very compliant, and this region is known as the "toe" region

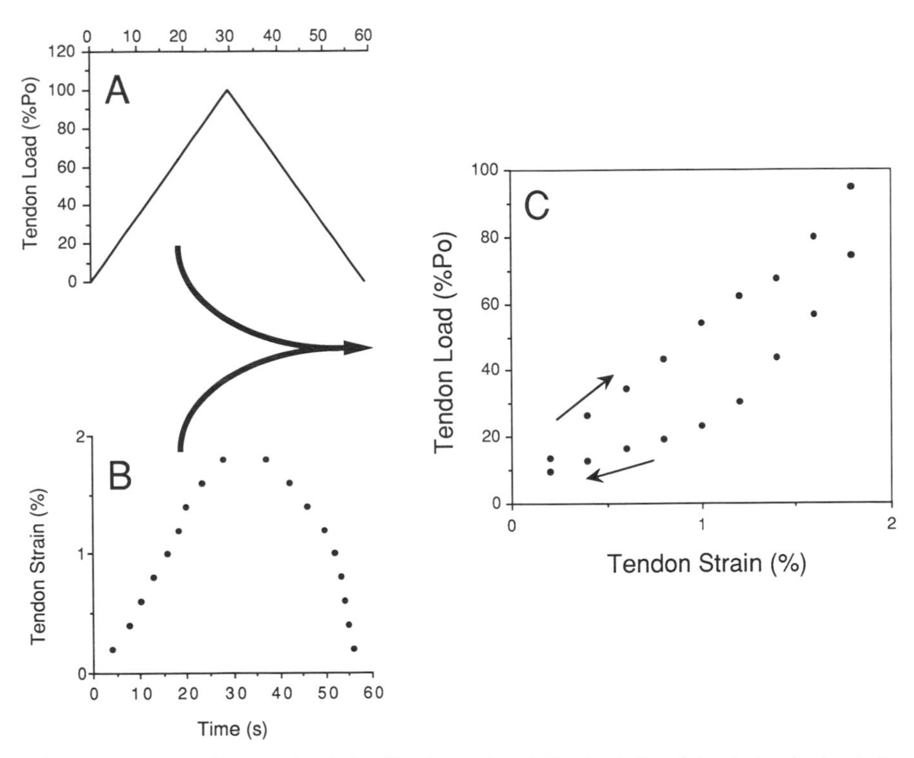

Figure 3.1. Typical load-strain relationship of a tendon. **A,** Tendon is linearly loaded and unloaded over time. **B,** Strain is measured in tendon during loading and unloading phases. **C,** Load and strain at corresponding time points are plotted to yield the load-strain relationship. Note that, at low loads, the tendon elongates (strains) a great deal for a given change in load. However, at higher loads, the tendon is stiffer and, therefore, elongates less for a given change in load. Shown in the figure are the loading and unloading portions of the curve (*arrows*). (From Lieber RL, Leonard ME, Brown CG, Trestik CL. Frog semitendinosus tendon load-strain and stress-strain properties during passive loading. Am J Physiol 1991;261:C86–C92.)

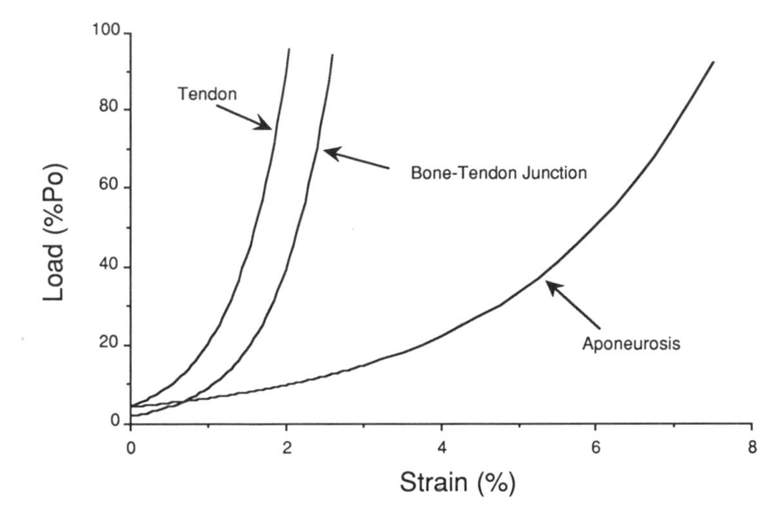

Figure 3.2. Relative load strain for different regions of the frog semitendinosus tendon. During loading, the aponeurosis strains more than the bone-tendon junction, which strains more than the tendon. (From Lieber RL, Leonard ME, Brown CG, Trestik CL. Frog semitendinosus tendon load-strain and stress-strain properties during passive loading. Am J Physiol 1991;261:C86–C92.)

(Butler *et al.,* 1978). While numerous measurements of connective tissue load-strain properties have been made, very few have been performed under physiologic conditions, *i.e.,* under loading conditions that would be expected based on the strength of the attached muscles. Estimates of tendon strain during muscle contraction have been made by Zajac (1989) based on literature values and recently directly measured during muscle contraction by Lieber *et al.,* (1991a and 1991b). The two approaches generally agree that tendons strain approximately 3% at muscle maximum tetanic tension. We thus should consider muscles as being connected to bones via relatively compliant connectors known as tendons, which strain during muscle contraction. In fact, the level of sophistication goes farther in that experimental measurements have demonstrated that different tendon regions have different properties—the tendon closer to the bone is about five times stiffer than the tendon connected to the muscle fibers (Figure 3.2).

Physiologic Significance of Tendon Compliance

What is the physiologic significance (if any) of tendon compliance? Intuitively, we can appreciate that as a muscle develops force, tendons will strain, allowing the muscle to shorten (Figure 3.3). This is one reason that muscle activation at a fixed joint angle will never produce a truly isometric contraction. However, less intuitive is that, because of tendon compliance, a muscle-tendon unit will have an increased operating range relative to the range attributed to muscle fibers alone. This is because some of the length change that would be required to be taken up by muscle fibers is actually taken up by tendons. Thus by attaching muscles to bones via compliant tendons, the operating range of the muscle-tendon unit is increased. Of course, the downside to this arrangement is that there will be some delay from the time when the muscle develops tension to when the bones begin to move. We thus have a trade-off between increased range and increased control. Recently, the British functional morphologist R. McNeil Alexander (1988) presented various biologic examples of compliant systems and their influence on normal function. We will return to the issue of tendon compliance in our discussion of the gait cycle (page 153). Suffice it to say that the relative amount of tendon compared to muscle fibers will determine the magnitude of the increase in operating range of the muscle-tendon unit. In fact, if tendon length is increased, the operating range and length corresponding to peak force changes even more (Figure 3.4**A,B**). In some ways, then, *tendon length* is a design parameter in much the same way as is muscle architecture.

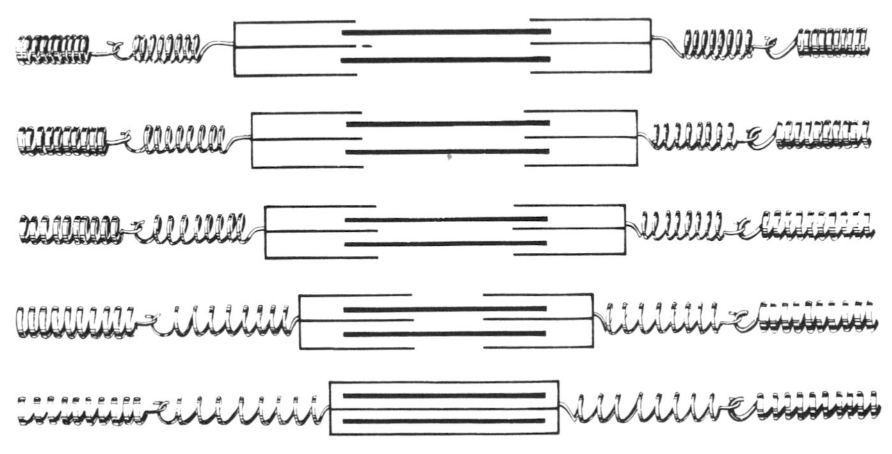

Figure 3.3. Schematic model of a muscle "isometrically" contracting while in series with a tendon. The muscle (schematically represented by the sarcomere) shortens at the expense of tendon lengthening (schematically represented by the springs). The springs strain to different extents due to differences in intrinsic compliance at the fiber-tendon and tendon-bone junctions.

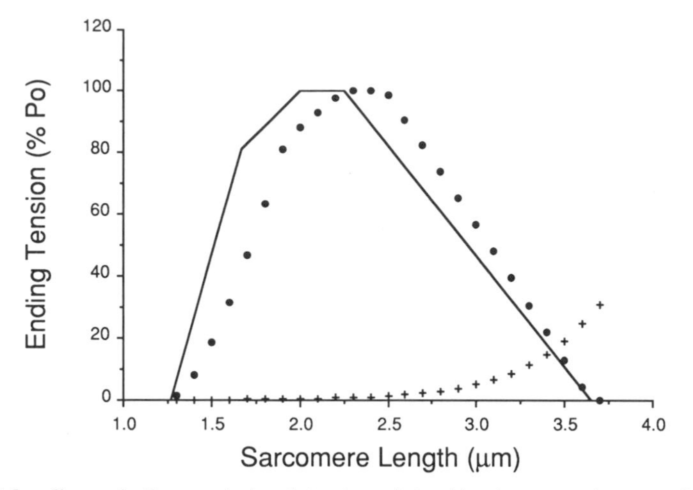

Figure 3.4. Change in the muscle length-tension relationship due to tendon compliance. The effect of tendon compliance is to skew the sarcomere length-tension curve (*dotted line*) relative to a muscle with extremely stiff tendons (*solid line*). (From Lieber RL, Brown CG, Trestik CL. Model of muscle-tendon interaction during frog semitendinosus fixed-end contractions. J Biomech, in press.)

JOINT MOMENTS

Definition of Torque

In the musculoskeletal system, muscles generate force and transmit the force, via tendons, to the bones. If muscles generate sufficient force, bones move. As you know, bones are not free to move in space but usually rotate about a joint axis. We thus have a mechanical system that can be quantitatively described using the concept of torque. The equation for torque is

$$\tau = \vec{r} \times \vec{F} \tag{3.4}$$

where τ is torque, r is the moment arm, and F is the applied force; $\times$ represents the vector cross-product of the two quantities (Figure 3.5**A**). Thus a force, F, is applied a distance r away from an axis (*open circle,* Figure 3.5**A**) that is free to rotate. The arrows above the moment arm and force variables signify that they are vector quantities. That is, both have magnitude as well as direction. This is another way of saying that the orientation between r and F is important. If we expand this equation, we can express torque as

$$\tau = |r| \cdot |F| \cdot \sin\theta \tag{3.5}$$

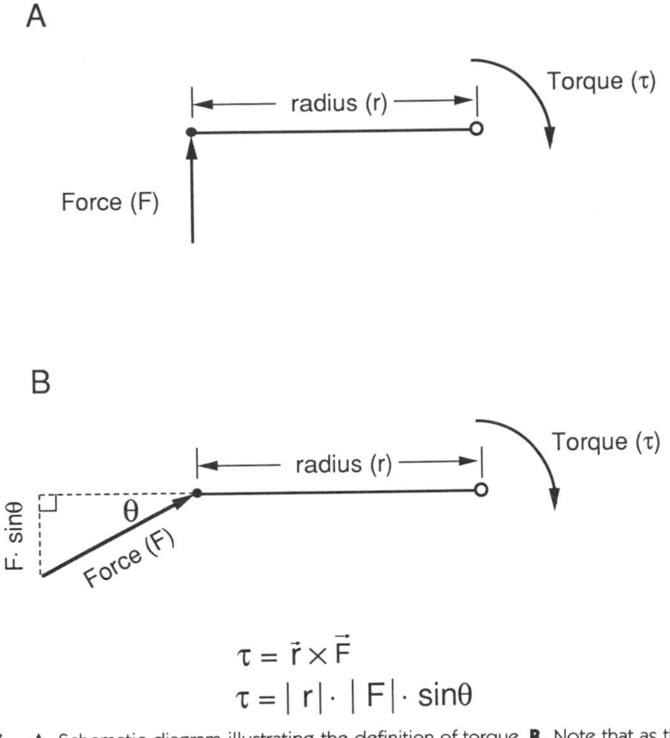

$$\tau = \vec{r} \times \vec{F}$$
$$\tau = |r| \cdot |F| \cdot \sin\theta$$

Figure 3.5. **A,** Schematic diagram illustrating the definition of torque. **B,** Note that as the force is applied to the axis at an angle θ, the component of force tending to rotate the axis if F·sinθ, which is equal to the moment arm.

where the vertical bars around *r* and *F* represent the vector magnitudes, and θ is the angle between the direction of force application and the axis of rotation. The basis for this expression can be seen in Figure 3.5**B** where the force, *F*, is applied at an angle θ, relative to the axis of force generation. The *component* of force that causes rotation is F·sinθ, the short side of the triangle shown in dotted lines. F·sinθ is the perpendicular distance between the point of force application and the axis of rotation, which is referred to as the moment arm.

Moment arm is measured in distance units such as meters. Because the sinθ term is dimensionless, torque has units of Newton-meters (force-distance units). Another common unit of torque is foot-pounds (such as are used in the automobile industry, and, interestingly, on the isokinetic dynamometer machines used in rehabilitation). Proper representation of torque always requires use of the correct units.

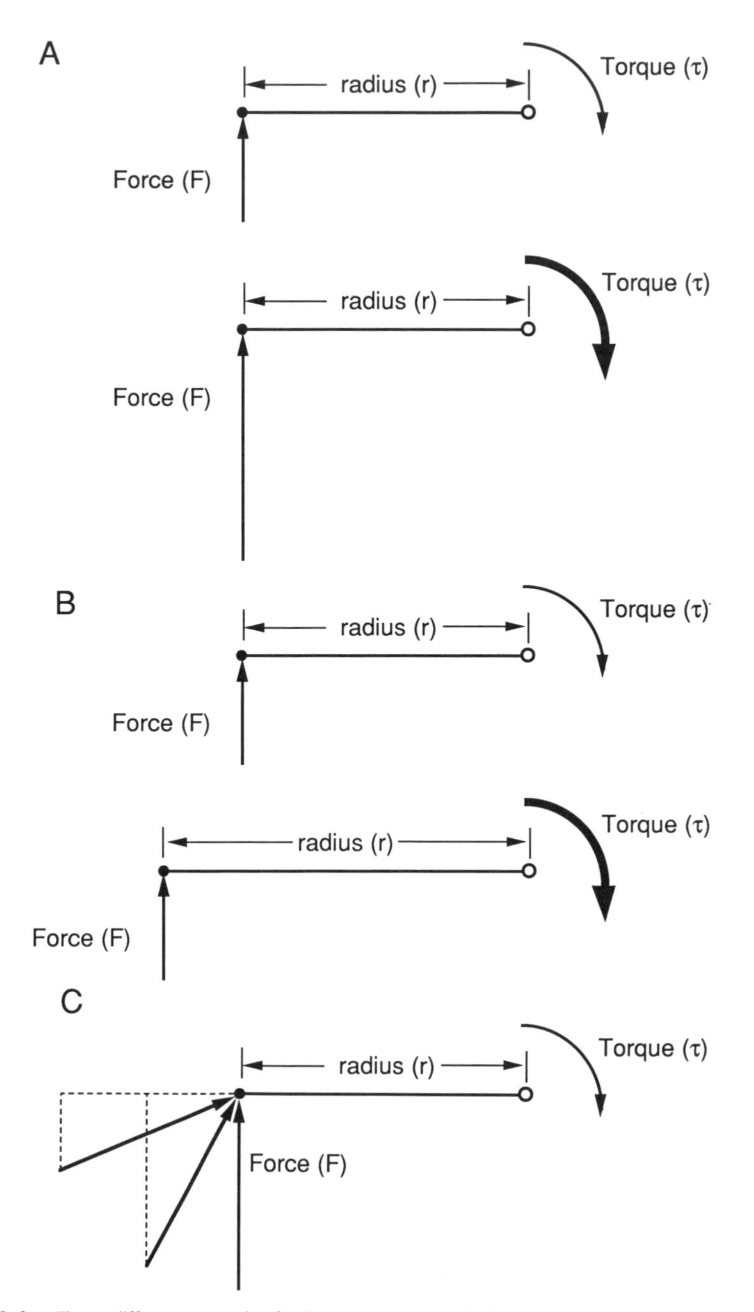

Figure 3.6. Three different strategies for increasing torque. **A,** Torque increases as force increases (*bold curved line* signifies increased torque). **B,** Torque increases as moment arm increases. **C,** Torque increases as the angle between force application and axis of rotation increases.

Applying this idea to the musculoskeletal system, we see that "strength" as we commonly use the term actually represents torque. Let's emphasize this point: All real-world performance (sprinting, lifting weights, getting out of bed, writing) represents manifestations of torque by the musculoskeletal system. If we grasp the implications of this statement, it is clearly incorrect to conclude that a person is strong simply because their muscles generate large forces. This may sound sacrilegious and it may even indeed be the case, but it is not *necessarily* required. It would be just as ridiculous to conclude that a person is strong because they have large moment arms! Based on the above discussion, it is clear that at least three strategies exist for changing torque: changing force, changing moment arm, or changing the angle between the two (all strategies illustrated in Figure 3.6).

Torque Examples

A common example of torque generation is seen when we need to generate a large force to remove the lid from a can of paint. We do so using a screwdriver, which, because of its long handle (moment arm) generates a large torque and thus a large upward force on the lid (Figure 3.7). Our downward imposed force is transferred via a fulcrum (the lip of the can) to produce an upward output force. Similarly, the human triceps muscle (Table 1.1) can generate a contractile force that is transmitted via the elbow joint to produce elbow extension. With this in mind, it is interesting to compare elbow extensor strength of various animals. In the animal kingdom, it is well known that baboons have elbow extension strength three to six times that of humans. However, analysis of baboon triceps muscles reveals that the muscles are approximately the same size of those of humans. Observation of the baboon's ulna reveals the basis for the great extension strength: The very long olecranon process (moment arm) extends 3–4 cm posteriorly from the elbow joint (axis of rotation)! Thus when we speak of real-world strength or performance, we must speak in terms of a torque or joint moment. When we discuss joint moments, we must consider both the muscle force producing the moment and moment arm about which that force is acting.

Muscle Force Producing the Moment

Determination of the muscle force that produces a joint moment is simple in principle but extremely difficult in practice. As discussed in Chapter 2, muscle force varies with muscle length (length-tension relationship), muscle velocity (force-velocity relationship), and activation level (fiber recruitment). If we know muscle length, velocity, and recruitment level, we can predict muscle force. In practice, this is easier said than done.

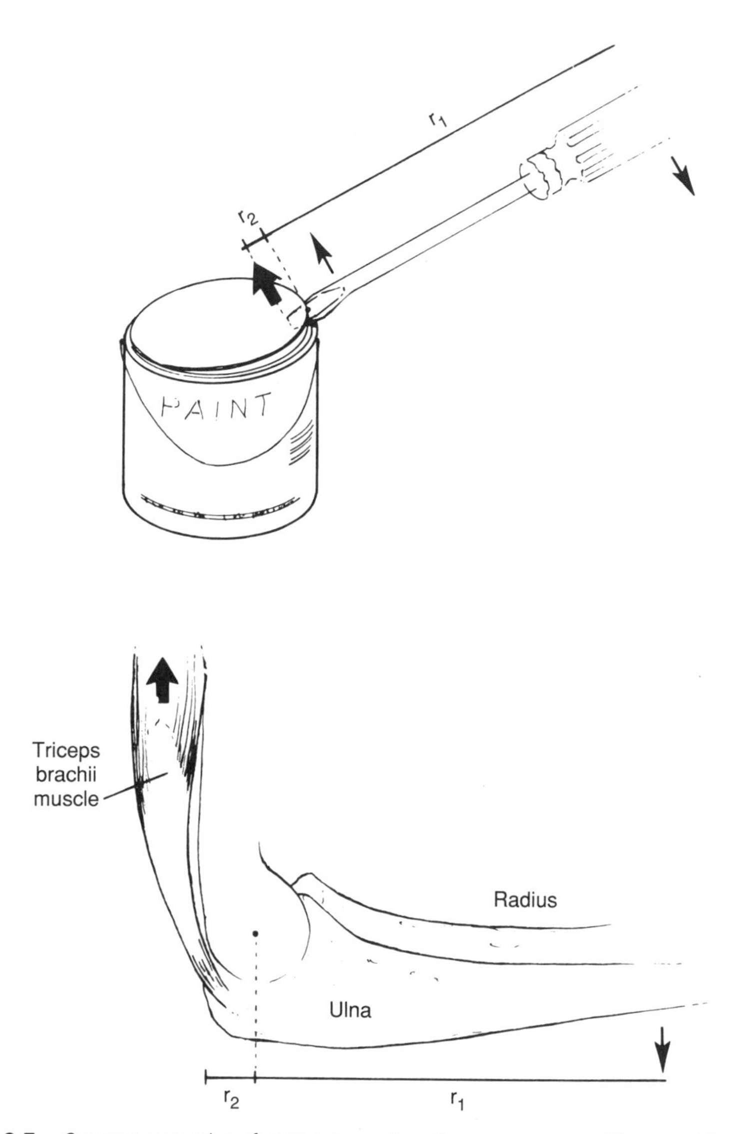

Figure 3.7. Common examples of torque generation. Torque on a screwdriver provides a large upward force to remove the lid of a paint can (**upper panel**). Input moment arm is r_1 while output moment arm is r_2 and the fulcrum is the edge of the paint can. Triceps muscle force is applied via moment arm r_2 to the olecranon process to generate an elbow extension torque via moment arm r_1.

Moment Arm Producing the Moment

Determination of joint moment arm requires an understanding of the anatomy and movement (kinematics) of the joint of interest. For example, some joints can be considered to rotate about a fixed point. A good example of such a joint is the elbow. At the elbow joint, where the humerus and ulna articulate, the resulting rotation occurs primarily about a fixed point, referred to as the center of rotation. In the case of the elbow joint, this center of rotation is relatively constant throughout the joint range of motion. However, in other joints (for example the knee), because the articulating surfaces of the tibia and femur are not perfect circles, the center of rotation moves in space as the knee joint rotates. In the case of the knee, it is not appropriate to discuss a single center of rotation—rather we must speak of a center of rotation corresponding to a particular joint angle, or, using the terminology of joint kinematics, we must speak of the instant center of rotation (ICR), that is, the center of rotation at any "instant" in time or space.

Identification of the joint ICR is difficult in practice. The most common method for ICR determination implements the technique of rigid body kinematics. Using rigid body kinematics, two points are identified on two rigid bodies before and after rotation, and the axis about which the bodies rotate is identified using simple algebraic methods. Often such measurements are taken from radiographs. Such methods have been used a great deal in orthopaedic surgery, for example, to define the normal kinematics of joints that are surgically replaced. The intention is to design artificial joints that mimic the kinematics of normal joints.

Having defined a joint ICR, the moment arm is defined as the perpendicular distance from point of force application to the axis of rotation. This is illustrated in Figure 3.8 for a simulated elbow joint. In Figure 3.8**A**, the elbow joint is almost fully extended. Let the angle, θ, between the brachialis muscle and the ulna be relatively small, *e.g.,* $\theta = 20°$. Let the distance between the brachialis insertion site and the elbow instant center be 5 cm. In this case, the *perpendicular* distance between the point of force application and the elbow ICR is shown by the dotted line in Figure 3.8**A** and is equivalent to 5 cm $\times$ $\sin(20°)$ = 1.7 cm. Thus because the joint is nearly fully extended, this presents an unfavorable mechanical advantage to the muscle—the moment arm is relatively small. Much of the force generated by the muscle will simply *compress* the joint, not rotate it. Contrast this situation with the conditions shown in Figure 3.8**B**, where the joint has now been flexed to $\theta = 50°$. Now, the moment arm equals 5 cm $\times$ $\sin(50°)$ = 4.3 cm. We see that for a simple hinge joint (a joint with a fixed ICR), the maximum moment arm is attained at $\theta = 90°$. If we plotted moment arm *vs.* joint angle for this simple hinge joint, we

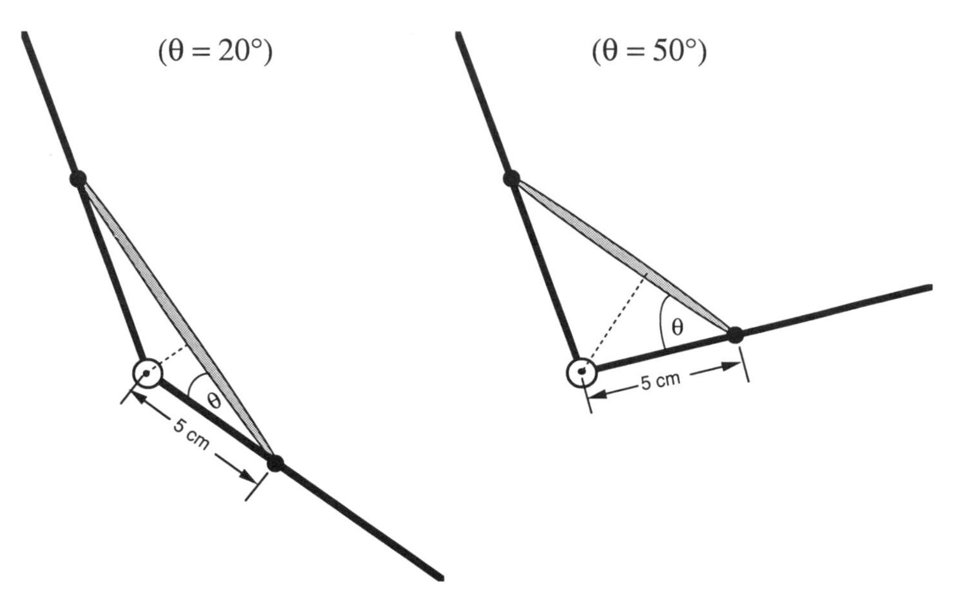

Figure 3.8. Schematic example of altered torque with change in joint angle. **A,** At an acute joint angle ($\theta = 20°$) moment arm is relatively small (1.7 cm). **B,** At larger joint angles ($\theta = 50°$), moment arm increases considerably to 4.3 cm. This is because the angle between the muscle force and point of force application changes.

would obtain a simple *sine* function that has a maximum of 5 cm occurring at $\theta = 90°$ of flexion (Figure 3.9**B**). Such a curve can be generated for any joint. In general, the curves are not quite as simple as the one previously discussed.

MUSCLE-JOINT INTERACTION DURING ISOMETRIC TORQUE PRODUCTION

We are now in a position to accurately discuss torque generation as it applies to the musculoskeletal system. For a given muscle-joint system, we can define muscle force-generating properties and joint kinematics. Unfortunately, the detailed relationship between human muscle and joint properties has not been thoroughly studied. To clearly understand the stated problem, refer to Figure 3.9. In the top portion of the figure, we have placed a hypothetical length-tension relationship, which is expressed as a tension-joint angle relationship. In Chapter 2, we detailed the length-tension curve for single frog muscle fibers (Figure 2.3). The muscle length-joint angle will simply be some portion of this curve. For our purposes, we will assume that a muscle increases force and then decreases as a trigonometric *sine* function as a joint rotates (Figure 3.9**A**). Note that the detailed form of both muscle and moment

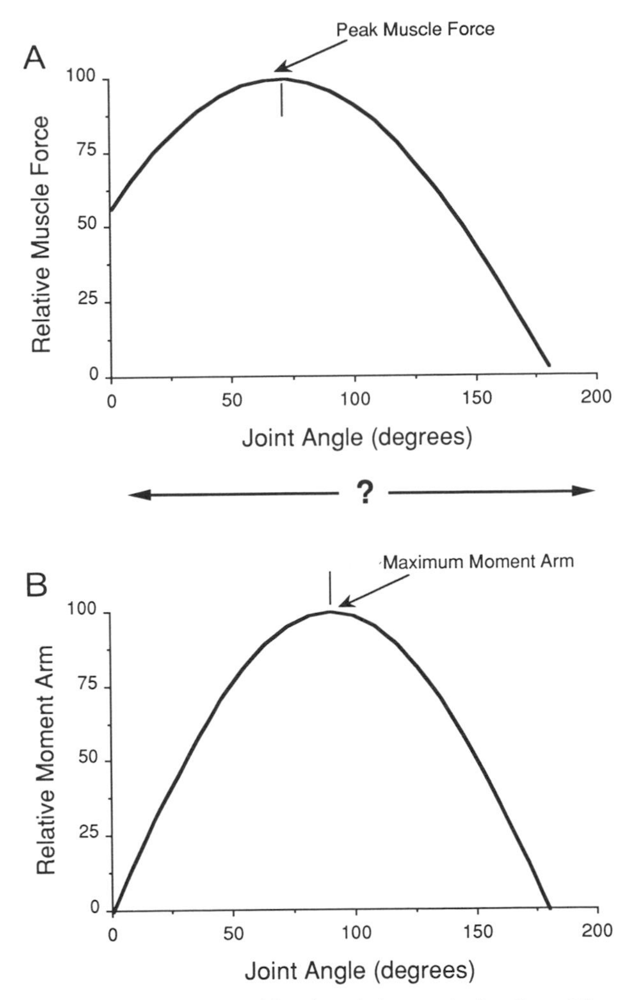

Figure 3.9. Conceptual question posed by the relative angular location of the angle at which maximum muscle force occurs and the angle corresponding to maximum moment arm. **A,** muscle force as a function of joint angle. **B,** moment arm as a function of joint angle. *Small vertical lines* represent peaks of each curve. *Horizontal arrows* emphasize that relative angular relationship between these curves is generally unknown.

arm curves is relatively arbitrary and only intended to illustrate muscle-joint interaction. In Figure 3.9**B**, we have placed a typical moment arm curve. Since we established that the moment is simply the product of muscle force and moment arm, the joint moment results from the product of these two curves. However, the key question that must be addressed is, What is the relative angular relationship between the muscle and joint properties (*arrows* in

Figure 3.9)? That is, is muscle force maximum at the same angle where moment arm is maximum? If not, at what joint angle is muscle force maximum? If the two curves are offset relative to one another (vertical dotted lines), their product will be dramatically altered. Understanding the relationship between muscle and joint properties is important from the point of view of understanding the normal design of the musculoskeletal system. It is also important in providing a scientific basis for surgical procedures that involve movement or transfer of muscles from one position to another. If we are to mimic the natural function of the musculoskeletal system, we must understand the relationship between muscles and joints. Specifically, we must define the relative angular relationship between the two curves in Figure 3.9.

Joint Angle Corresponding to Maximum Muscle Force

Examination of current physiology texts reveals a good deal of wishy-washiness regarding the definition of the joint angle that corresponds to maximum muscle force. Typically, it is stated that muscle force is maximum when the joint is in a neutral position, or when the muscle is at resting length. What is the basis for such statements? Unfortunately, there is very little scientific basis for such a statement. Recent studies of torque generation in animals and humans have produced conflicting results. Some studies have concluded that the joint angle at which muscle force was maximum was either outside the normal range of motion, or at one extreme end of the range of motion. What an interesting design!

Torque Generation in the Frog Hindlimb

We recently studied the relationship between sarcomere length and joint angle in the frog (Lieber and Boakes, 1988). Before you close the book, recall that frog muscle is the *only* skeletal muscle for which the sarcomere length-tension relationship is available (Figure 2.3). In other words, definitive determination of the relationship between sarcomere length and joint angle has only been reported for frog skeletal muscle.

In this study, the semitendinosus muscle was examined during knee rotation. In the frog, the semitendinosus is a biarticular hamstring that crosses the knee and hip. The frog pelvis, femur, tibia, and semitendinosus muscles were isolated and placed in a specially designed jig, surrounded by physiologic saline solution (Figure 3.10). With the hip at 90° of flexion, the knee joint was rotated throughout its range of motion, from 0° to 180° of flexion. During joint rotation, sarcomere length was measured by laser diffraction. (Laser diffraction is a method based on the fact that the A-I band periodicity acts as a

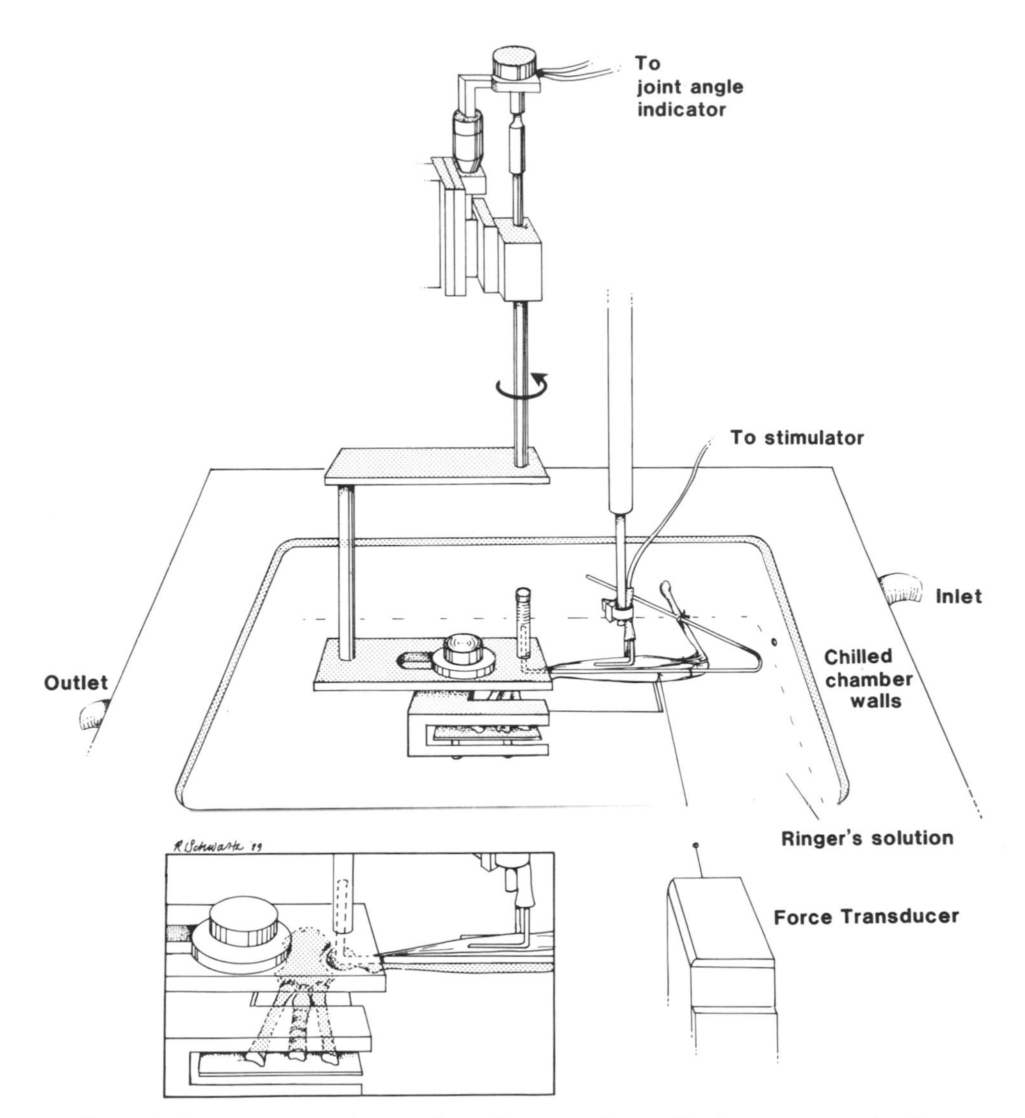

Figure 3.10. Apparatus used to experimentally measure frog semitendinosus sarcomere length, hip and knee joint angle, and joint torque. The frog bone-muscle complex is secured to the rotating arm of the jig. Stimulating electrodes flank the semitendinosus muscle. Joint angle is changed by rotating the arm, and joint angle is read directly from a goniometer. The femur and tibia are stabilized throughout the experiment. (From Mai MT, Lieber RL. Interaction between semitendinosus muscle and knee and hip joints during torque production in the frog hindlimb. J Biomech 1990;23:271–279.)

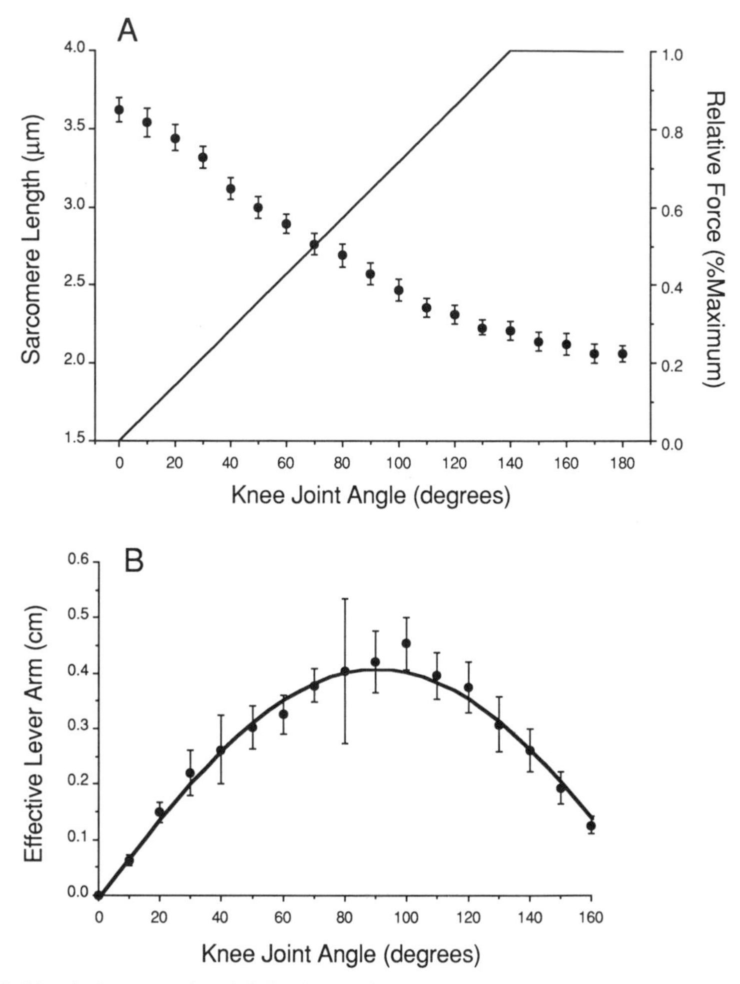

Figure 3.11. **A,** Sarcomere length (left axis in μm) *vs.* knee joint angle. All sarcomere lengths are on the descending limb of the length-tension curve. Solid line and right axis represent the approximate muscle force corresponding to the measured sarcomere length. **B,** Moment arm *vs.* knee joint angle. *Symbols* and *solid line* represent mean values ± standard error from 10 muscles. Note that muscle force and moment arm vary significantly throughout the joint range of motion. (Data from Lieber and Boakes, 1988).

diffraction grating to incident laser light. Because of destructive and constructive interference, a diffraction pattern is produced that is proportional to sarcomere length.) The relationship between sarcomere length and joint angle, shown in Figure 3.11**A**, presented a striking illustration of the manner in which muscle force can change during joint rotation. Note that with the joint fully extended (0°), the muscle was highly lengthened to a sarcomere length of 3.6 μm, the sarcomere length corresponding to zero active tension on the sarcomere length-tension curve (Figure 2.3). As expected, when the knee was flexed, the muscle shortened and sarcomere length decreased to approximately 2.0 μm, the sarcomere length corresponding to maximum tetanic tension. Isn't it interesting that in this muscle-joint system, the muscle would produce no force with the joint fully extended (sarcomere length 3.6 μm) and maximum force (sarcomere length 2.0 μm) with the joint fully flexed? What an intriguing design! In the same experimental system, the knee joint kinematics were also studied, and the knee was shown to behave as a simple hinge joint, as illustrated in Figure 3.11**B**. Finally, in the same system, the muscle was stimulated, the joint was rotated throughout the same range of motion, and the torque was measured. The torque-joint angle relationship shown in Figure 3.12 was obtained. The combined results of this study are illustrated schematically in Figure 3.13**A** and graphically in Figure 3.13**B**. Note that maximum torque (the optimal joint angle) was obtained at 120° of flexion, which was neither the angle at which muscle force was maximum (160°) nor the angle at which moment arm was maximum (90°). Thus torque production in this system (and probably other musculoskeletal systems) resulted from the

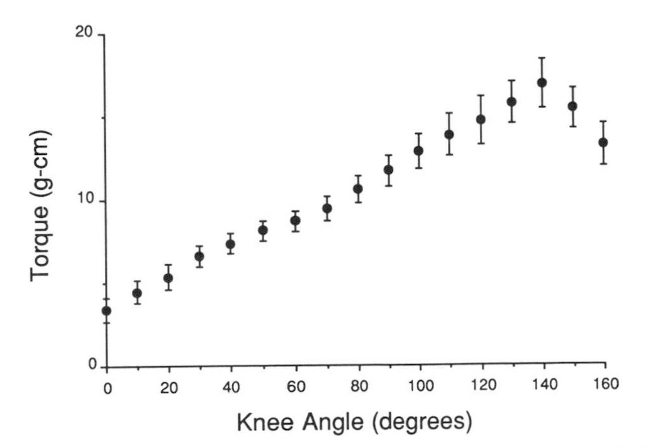

Figure 3.12. Knee joint torque *vs.* joint angle for the frog semitendinosus muscle during knee joint rotation. Data shown are ± mean standard error for all frogs. Note that torque increases relatively linearly from 0° to 140° and drops off sharply from 140° to 160°. (Data from Lieber and Boakes, 1988.)

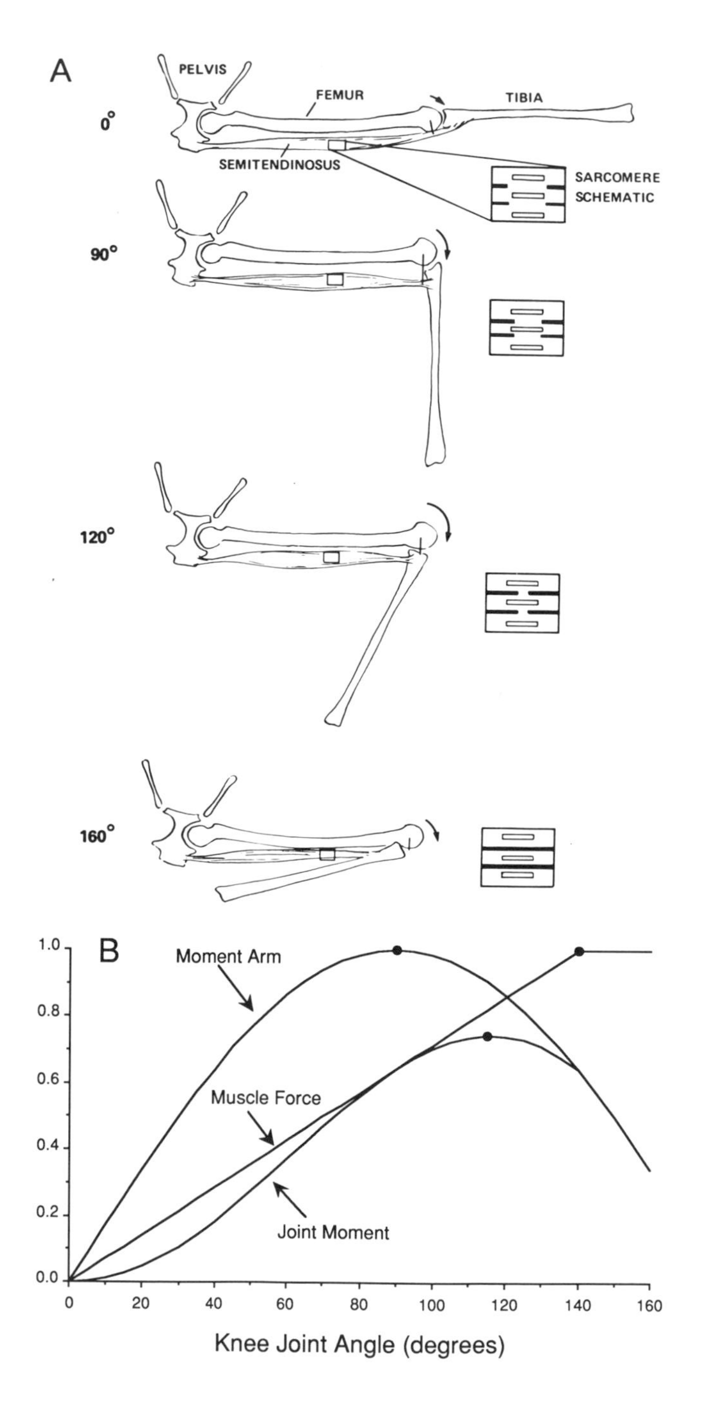
A
PELVIS
FEMUR
TIBIA
0°
SEMITENDINOSUS
SARCOMERE
SCHEMATIC
90°
120°
160°
B
Moment Arm
Muscle Force
Joint Moment
1.0
0.8
0.6
0.4
0.2
0.0
0
20
40
60
80
100
120
140
160
Knee Joint Angle (degrees)

interaction between muscle and joint properties and not either property alone.

Generalized Design of Musculoskeletal Torque Generators

Unfortunately, the details of this section cannot be fully written as yet. We are just now beginning to understand muscle-joint interaction as described above. However, several main points can be made, which provide a basis for understanding any musculoskeletal torque generating system. Recall that in the frog system above, as the knee flexed, sarcomere length decreased from 3.6 μm to 2.0 μm. Why? Did this have to be the case? The answer is, of course, no! The mechanical reason for the relatively large sarcomere length change (1.6 μm) during joint rotation was that the number of sarcomeres arranged in series along the fibers was such that when muscle-tendon length decreased from full extension to full flexion, in order to take up that length change, each sarcomere was required to shorten 1.6 μm. Clearly, if there had been twice as many sarcomeres arranged in series, each sarcomere would have only shortened half as much, or 0.8 μm. For example, suppose a muscle-joint system were configured such that, by extending from 40° to 80°, the muscle went from its minimum to its maximum length (Figure 3.14**A**). Now, suppose the muscle fiber length were significantly increased. What happens to joint range of motion? Clearly, since more sarcomeres are in series to take up the length change, joint range of motion increases. Now the muscle can extend from 70° to 145°—a total of 75° (compared to the previous 40°; Figure 3.14**B**). Therefore, by increasing fiber length, active range of motion has increased from 40° to 75° (Figure 3.15)! This demonstrates the intimate interaction between the muscle and the joint about which it rotates.

Based on this example, we can see that the ratio between muscle fiber length (number of sarcomeres in series) and moment arm will influence the amount of sarcomere shortening that will occur during joint rotation (Equation 3.6). This ratio can be calculated for any muscle joint system as

$$\text{Ratio} = \frac{\text{Fiber Length}}{\text{Moment Arm}} \tag{3.6}$$

Figure 3.13. **A,** Schematic relationship between relative muscle force, relative effective lever arm, and relative joint torque as a function of knee joint angle. Length of *curved arrow* at knee represents torque magnitude. Thin *vertical line* at knee represents moment arm. **B,** Graphical representation of relationship diagrammed in **A.** Note that throughout the range of motion, muscle force and effective lever arm interact to produce torque. Note that maximum muscle force and maximum effective lever arm (*filled circles*) do not coincide with the optimal joint angle. (Data from Lieber and Boakes, 1988.)

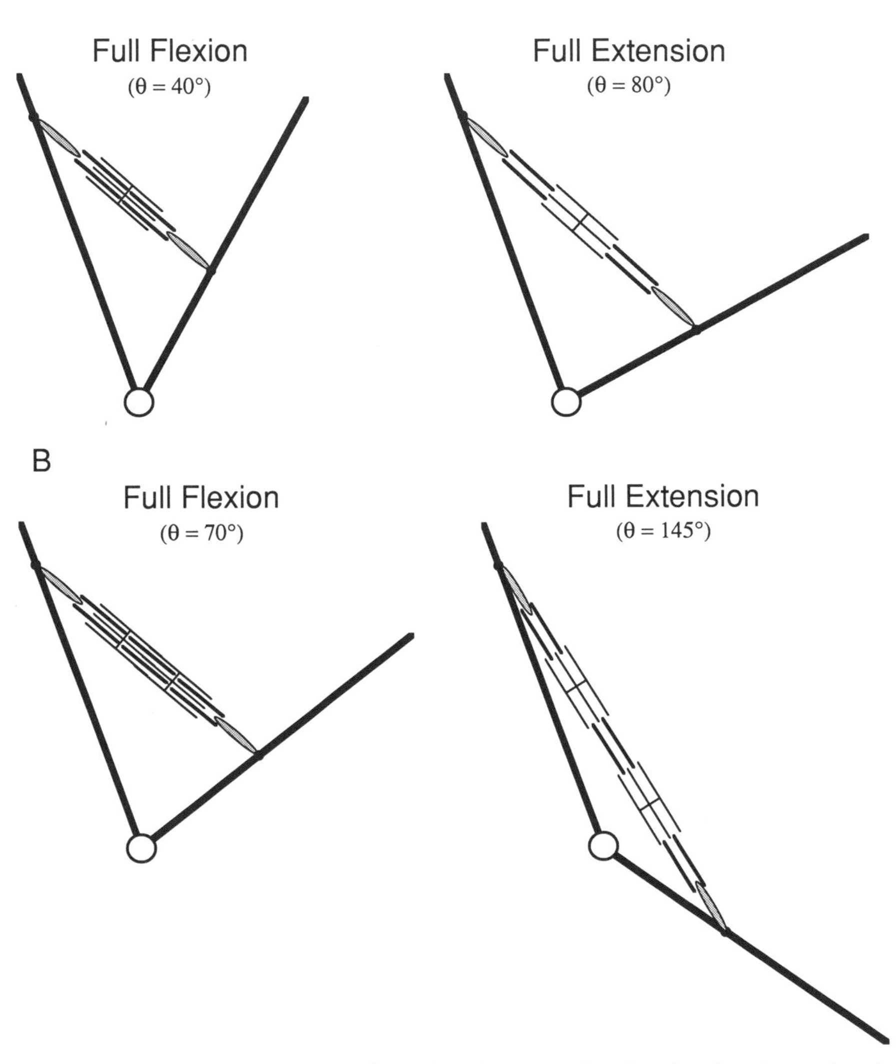

Figure 3.14. Schematic illustration of muscle active range of motion changing due to altered muscle fiber length. **A,** Short muscle fibers result in only 40° range of motion. **B,** Long muscle fibers result in a range of motion of about 75°.

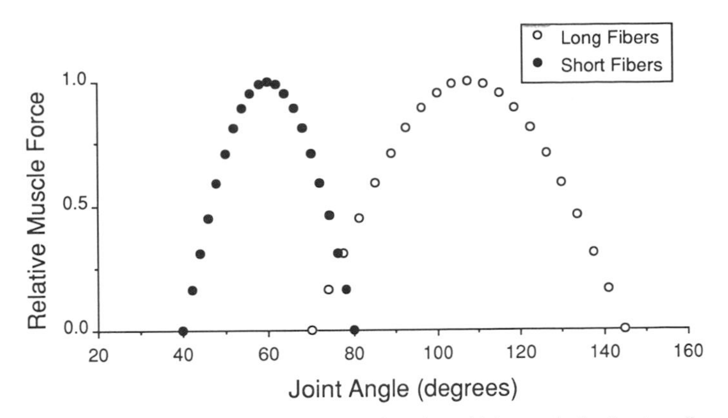

Figure 3.15. Comparison of muscle force as a function of joint angle for the two "muscles" in Figure 3.14. Note that the muscle with the long fibers (*open circles*) has a range of motion of 75° while the muscle with the short fibers (*filled circles*) has a range of only 40°.

and will determine the relative influence of the muscle on the muscle-joint torque generator. As we have seen, if fiber length is very long compared to moment arm, relatively little sarcomere length change will occur during joint rotation, and muscle force change will contribute little to the joint moment. If, however, fiber length is very short and moment arm is long, the sarcomeres will change length a great deal during joint rotation, and so will muscle force. This dramatically affects the muscle contribution to the joint moment.

Human Musculoskeletal Torque Generators

In the frog muscle-joint system presented above, we had a specific case where muscle, bone, and joint properties could be measured directly to determine their relative influence in torque generation. Obviously, such invasive experiments cannot be performed on humans. However, using ingenious assumptions and collecting the available experimental torque data from the literature for the hip, knee, and ankle joint, Melissa Hoy and Felix Zajac developed a model of torque production in the human lower limb (Zajac, 1989; Hoy *et al.,* 1990). They predicted muscle force based on architectural measurements from cadaver specimens (Figure 3.16**A**), moment arm from skeletal studies (Figure 3.16**B**), and predicted the torque produced by each muscle group (Figure 3.16**C**). By making small adjustments in the various model assumptions, they fit their data to actually measured torque data

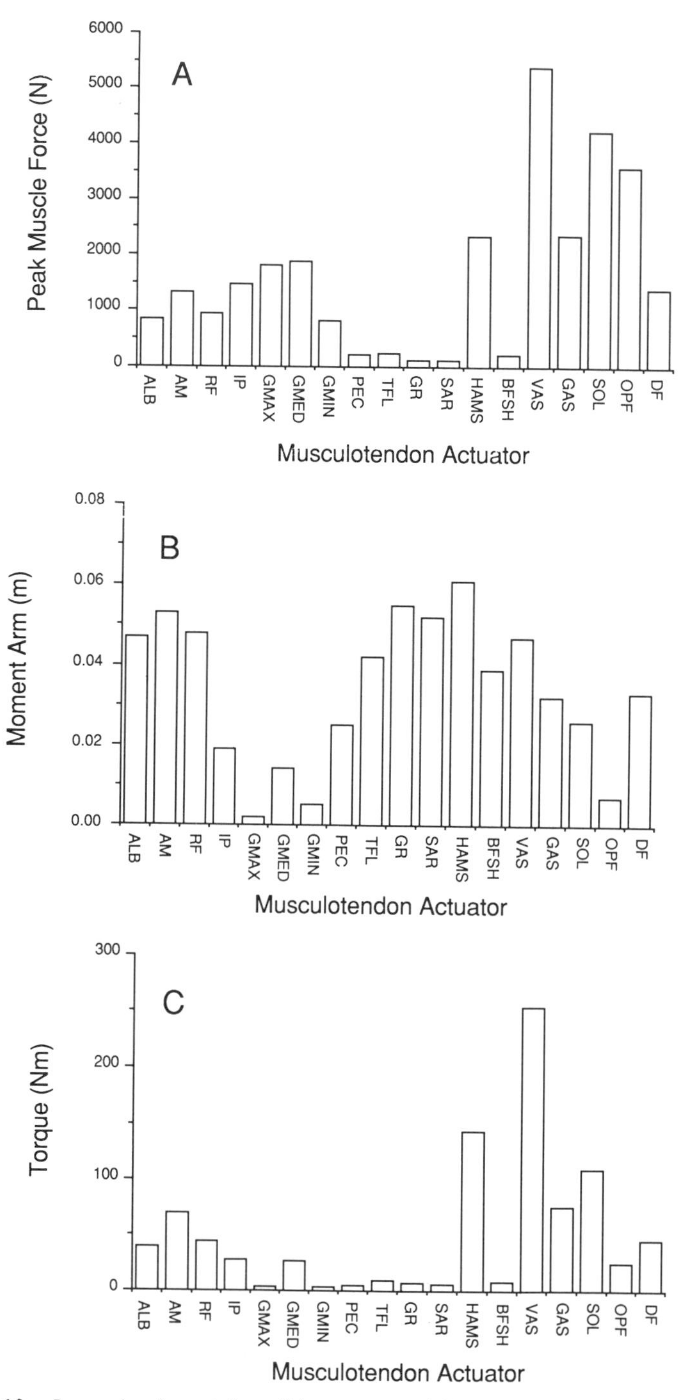

Figure 3.16. Bar graphs of muscle force (**A**), moment arm (**B**), and torque (**C**) for various muscles of the lower limb. (Data from Hoy *et al.*, 1990.)

obtained during human voluntary contraction. Comparison of the three components of Figure 3.16 immediately yields the conclusion that muscle force, moment arm, and torque are not simply scaled versions of one another. Inspect Figure 3.16**C** (torque), and determine which of its components most strongly influences it. Doesn't Figure 3.16**A** look more like 3.16**C** than does 3.16**B**? This leads us to conclude generally that muscle force is probably the major determinant of peak joint torque. However, note the many exceptions: Soleus torque and gastrocnemius torque are very similar in spite of the fact that the soleus produces almost twice the contractile force that the gastrocnemius does. Why? The gastrocnemius must have a larger moment arm (Figure 3.16**B**). As another example, the gluteus medius and maximus generate approximately the same force. However, the hip extension torque of the gluteus medius is over four times that of the gluteus maximus due to its large moment arm. Look for other relationships in the figure itself.

In their analysis of the lower limb, Hoy *et al.* also showed that the ratio between fiber length and moment arm varied considerably between muscle-joint systems. For example, in the lower extremity, fiber length/moment arm ratios varied from 80 (for the gluteus maximus) to about 1.0 (for the soleus). Other values are shown in Table 3.1. In the upper extremity, using the architectural data of Lieber *et al.* (1990) and the joint kinematic data of Horii *et al.* (1991), these ratios vary from 11 (for the extensor carpi radialis brevis) to about 2.0 (for the flexor carpi ulnaris). Again, other intermediate values are given in Table 3.1.

What we have demonstrated here is yet another design parameter of the musculoskeletal system. In addition to the previously described muscle architectural properties (fiber length, physiologic cross-sectional area, tendon length), we now have the fiber length/moment arm ratio (Equation 3.4) which determines the relative influence of muscle on the muscle-joint torque generator. In fact, if either fiber length or moment arm are altered (*e.g.,* due to surgery or trauma), the joint strength and range of motion can be dramatically altered.

Performance in the Musculoskeletal System

We should now review that, to this point, we have determined a number of intrinsic and extrinsic design parameters in the musculoskeletal system that are related to different aspects of performance. These are listed in Table 3.2. Note that it is possible to alter muscle, muscle-joint, muscle-tendon, and torque motor properties in numerous ways and in numerous combinations. It is not surprising to see such variation in design from muscle to muscle and joint to joint.

Table 3.1.
Fiber Lengths and Moment Arms of Limb Muscles[a]

Muscle	Fiber Length (m)	Fiber Length/ Moment Arm Ratio
Leg muscles		
Adductor longus	.132	2.80
Adductor magnus	.144	2.71
Rectus femoris	.082	1.70
Iliopsoas	.127	6.54
Gluteus maximus	.180	79.5
Gluteus medius	.081	5.79
Gluteus minimus	.064	13.9
Pectineus	.130	5.27
Tensor fascia latae	.118	2.80
Gracilis	.345	6.26
Sartorius	.566	10.8
Hamstrings	.107	1.76
Biceps femoris (short head)	.173	4.38
Vasti	.084	1.78
Gastrocnemius	.048	1.47
Soleus	.024	.926
Other plantar flexors	.038	5.77
Dorsiflexors	.101	3.08
Wrist muscles		
Extensor carpi radialis longus	.76	11
Extensor carpi radialis brevis	.48	4
Extensor carpi ulnaris	.50	8.3
Flexor carpi radialis	.51	3.4
Flexor carpi ulnaris	.41	2.6

[a]Data compiled from Hoy *et al.*, 1990; Lieber *et al.*, 1990; and Horii *et al.*, 1991.

PHYSIOLOGIC RANGE OF MOTION

The normal voluntary active range of motion (ROM) has been defined experimentally for a number of joints and is commonly used during physical examination to diagnose deficiency or disease. We have just discussed the mechanism of torque generation in the musculoskeletal system and are thus in position to discuss the physiologic basis for ROM. Before proceeding directly to the ROM discussion, we must first discuss the way in which muscle contractile properties vary as a function of architecture, because different architectural arrangements will have profound implications in determining voluntary ROM.

Table 3.2.
Design Parameters of the Musculoskeletal System

Parameter	Property Affected
Fiber length	Velocity or excursion
Fiber area	Fiber force
Physiologic cross-sectional area	Maximum muscle force and torque
Tendon length	Increased range, damping, and energy storage
Moment arm	Maximum torque
Tendon length/fiber length ratio	Relative stiffness of muscle-tendon unit
Fiber length/moment arm ratio	Relative muscle-joint influence
Fiber type distribution	Muscle speed and endurance
Motor unit distribution	Relative muscle control

Force-Generating Properties of Muscles with Different Architectures

We stated in Chapter 2 that muscle architecture had a profound influence on muscle force-generating properties. Recall that muscle force is proportional to physiologic cross-sectional area (PCSA) while muscle speed (or excursion) is proportional to fiber length. Let us illustrate these points by way of an example using two different muscles. Suppose we had two muscles with dramatically different designs but identical muscle fiber types, as illustrated in Figure 3.17. Both muscles have approximately the same amount of contractile material (mass), but the arrangement of this material is quite different. The muscle in Figure 3.17**B** has relatively long fibers that extend almost the entire length of the muscle and are parallel to the muscle's force-generating axis. In architectural lingo this is the classic parallel-fibered muscle. Contrast this with the muscle shown if Figure 3.17**A**, which has relatively short fibers that extend a very short length relative to the muscle and are tilted by about 30° to the muscle's force-generating axis. This is the classic pennated muscle. It must be emphasized that the intrinsic length-tension and force-velocity properties of these two muscles are identical; *i.e.,* the properties of the composite sarcomeres are identical. It is the *arrangement* of the sarcomeres that imparts the functional differences on the two muscles. In Figure 3.18**A**, we have plotted the length-tension curves of the two muscles. Note that muscle B, with its longer fibers, has a greater absolute working range than muscle A. This is because, for a given length change, the sarcomeres in muscle B lengthen less, the length change being distributed over a greater number of sarcomeres. However, note also that muscle B generates a lower tension than muscle A, because muscle A contains a much greater PCSA. We might say that muscle A is designed for force production while muscle B is designed for excursion. This

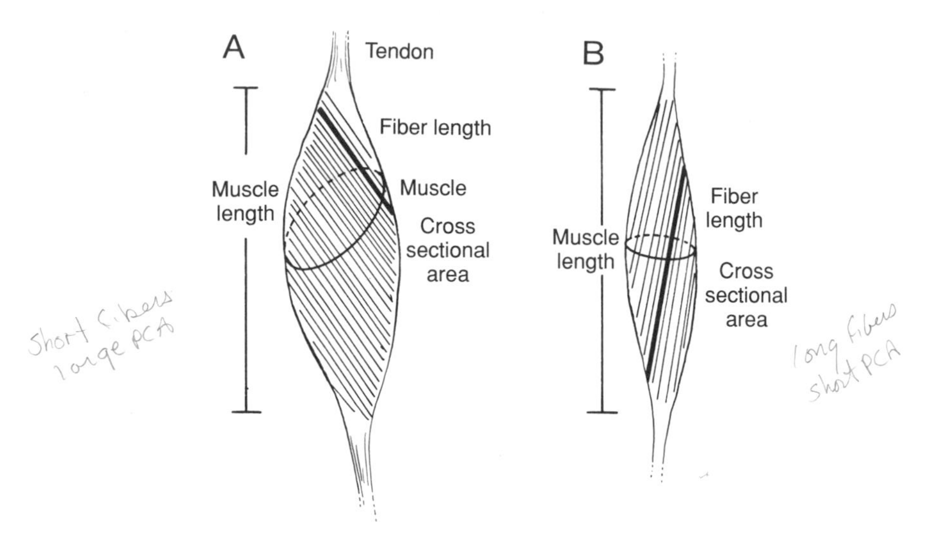

Figure 3.17. Schematic representation of muscles with different architectures. **A,** muscle with short fibers and a large PCSA, **B,** muscle with long fibers and a small PCSA.

concept is well illustrated by the force-velocity curves plotted in Figure 3.18**B**. Note that muscle B has a maximum contraction velocity (V_{max}) that is much greater than muscle A, again because of its very long fibers. Each sarcomere within the fiber contracts at the same velocity, whether in muscle A or muscle B. However, by placing more sarcomeres in series in muscle B, the overall muscle velocity is greater. Again, note that the maximum tetanic tension (P_o), for muscle A is much greater than that observed for muscle B, due to its greater PCSA.

Range of Motion as a Function of Architecture

Muscles with longer fibers thus have a longer functional ranges than muscles with shorter fibers. Returning to our previous discussion, does this imply that muscles with longer fibers are associated with joints that have larger ROMs? Think about it before reading on. The answer is no. It is true that a *muscle* with longer fibers does have a longer working range. However, the amount of muscle length change that occurs as a joint rotates is very strongly dependent on the muscle moment arm as described above. This idea is illustrated in Figure 3.19, where we have attached a simulated "muscle" using two different moment arms. In Figure 3.19**A**, the moment arm is much less than in Figure 3.19**B**. This means that in Figure 3.19**A**, the muscle will change length much less for a given change in joint angle compared to the same change in joint

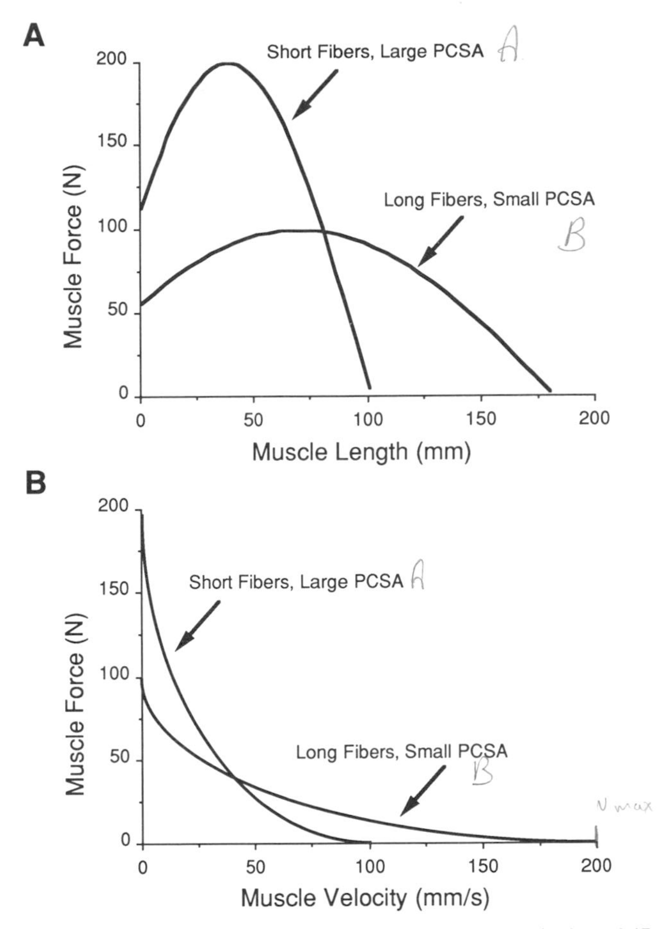

Figure 3.18. **A,** Schematic length-tension relationships of muscles shown in Figure 3.17. Note that the muscle in Figure 3.17**B** has a longer working range and lower maximum tetanic tension due to long fibers and small PCSA. **B,** Force-velocity relationships of muscles shown in Figure 3.17. Note that the muscle in 3.17**B** with longer fibers has a higher contractile velocity but a lower maximum tetanic tension.

angle in Figure 3.19**B**. As a result, the active ROM for the muscle-joint system shown in Figure 3.19**A** will be much greater than that which is shown in Figure 3.19**B** in spite of the fact that their muscular properties are identical. In fact, in the current example, increasing moment arm decreased range of motion from 40° (Figure 3.19**A**) to only 25° (Figure 3.19**B**)!

We should now qualify our statment about muscle design and architecture.

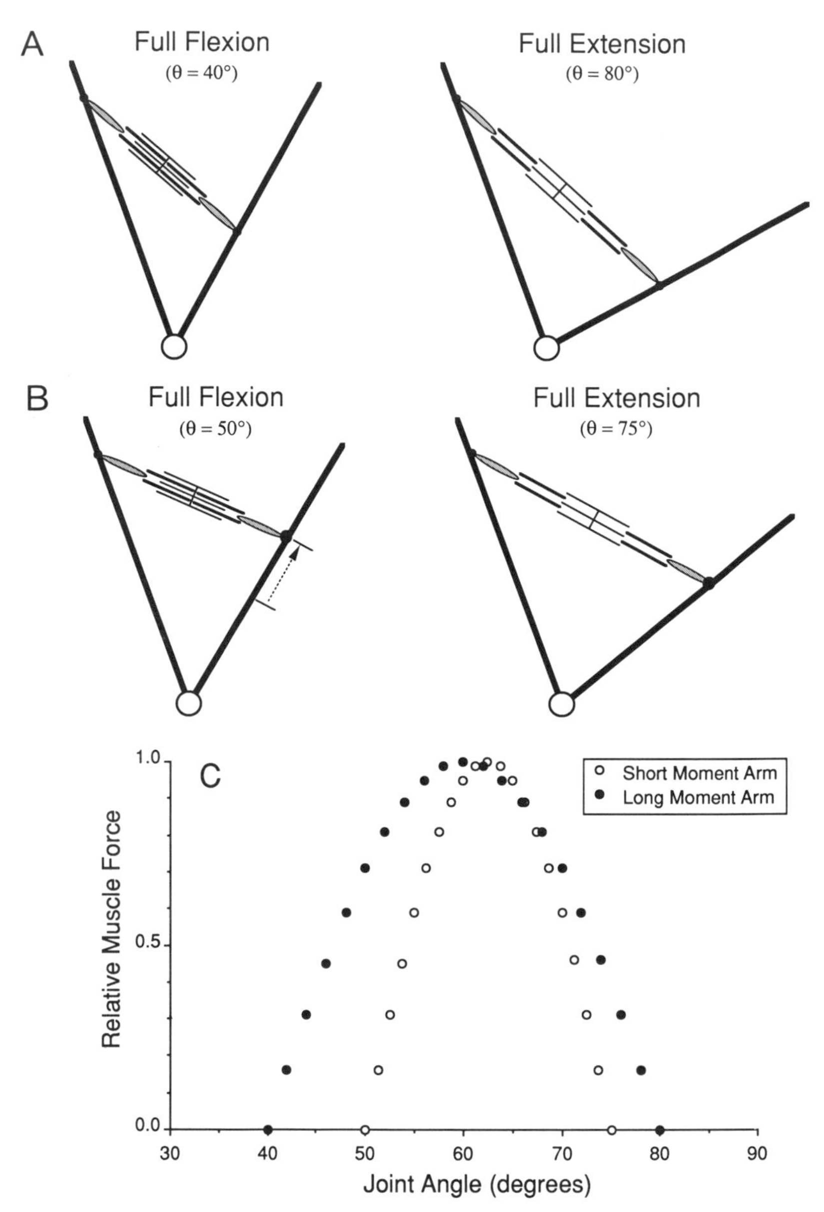

Figure 3.19. Effect of changing moment arm on active range of motion. In this example, the muscle in Figure 3.14**A** was attached with two different moment arms. **A,** 40° range of motion for "normal" muscle. **B,** Moment arm increase results in a decrease in range of motion to 25°. In **B,** the active ROM is smaller since the moment arm is greater, and therefore, more sarcomere length change occurs for a given angular rotation. **C,** Comparison of force *vs.* joint angle (range of motion) for muscles with short (*open circles*) or long (*filled circles*) moment arms.

Muscles that are, say, designed for speed because of their very long fibers may not actually produce large velocities if they are placed in position with a very large moment arm. The increased moment arm causes a greater joint moment, and the muscle is actually best suited for torque production. Similarly, a muscle that appears to be designed for force production due to the large PCSA, if placed in position with a very small moment arm, may actually produce high joint excursions or angular velocities.

To summarize, muscle design may or may not be a reflection of its actual use in the physiologic muscle-joint torque-generating system. It does seem, in general, that muscle fiber length and muscle moment arm are positively correlated. Thus muscles with long fibers tend to have long moment arms, but this is not necessarily the case. Muscle architectural features may represent muscle adaptation to kinematic criteria. However, definitive answers to these suggestions await further study.

ISOKINETIC DYNAMOMETERS USED IN PHYSICAL ASSESSMENT

It is of importance to be able to objectively characterize human performance in sports and rehabilitation, not only to evaluate patient progress but also to ascertain the efficacy of clinical treatment. Using objective criteria such as maximum joint moment and ROM, it is possible to evaluate the efficacy of many surgical and rehabilitative procedures. One of the most commonly used tools for musculoskeletal assessment is the isokinetic dynamometer. This is a device that measures joint moment while maintaining a constant joint velocity (isokinetic = "same motion"). If joint velocity is set to zero, an isometric joint moment *vs.* joint angle relationship is generated, which, as described above, represents the interaction between muscle and joint properties. It is clearly not appropriate to ascribe any portion of a moment *vs.* joint angle curve to either muscle or joint properties alone.

Physiologic Cross-Sectional Area and Fiber Length Influence on Isokinetic Torque

Let us consider the underlying physiologic processes that make up the isokinetic joint moment. As joint angular velocity increases, dynamic muscle contractile properties become important. We can assume that joint kinematics are velocity independent and, thus, that variations in the moment-angle curves as a function of velocity represent variations in muscle force. How should we interpret such isokinetic data? What should we expect to see? Let's answer these questions based on what we have learned about skeletal muscle. First, the moment achieved during isokinetic contraction (concentric muscle contraction) must necessarily be less than that achieved during isometric contraction due to the force-velocity relationship (Figure 2.4). Second, the muscle force generated during isokinetic contraction will be a function of the muscle's PCSA and its fiber length. The reason that isokinetic muscle force varies with PCSA is obvious since force is directly proportional to PCSA. The reason that muscle force varies with fiber length is less obvious although straightforward. Remember that isokinetic moment measurements are obtained while the joint is limited to a specific isokinetic movement. This means that the muscle is also forced to maintain a certain (not constant) shortening velocity. Since shortening velocity is fixed, the longer the muscle fibers, the higher the muscle force can be sustained during shortening. This is because longer muscle fibers have more sarcomeres in series and, with longer fibers, each sarcomere will have a slower absolute contraction velocity, allowing it to stay higher on its force-velocity curve (*i.e.,* closer to P_0). This idea is illustrated in Figure 3.20, where force is measured from two hypothetical muscles that are shortening at the same velocity. Note that, for a given shortening velocity (vertical dotted line), the muscle with longer fibers maintains a higher force compared to the muscle with shorter fibers. This means that the muscle with longer fibers generates the greater force.

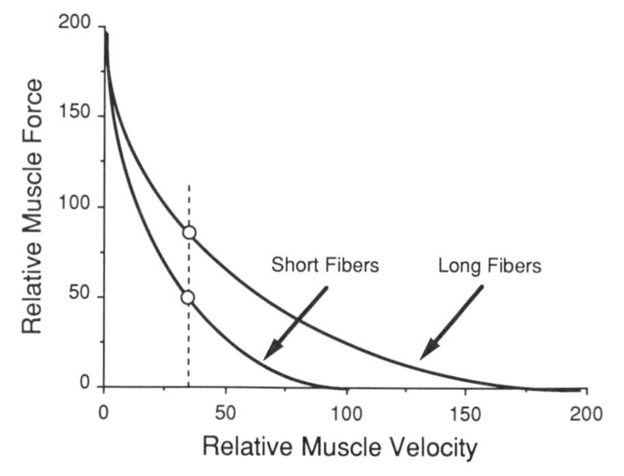

Figure 3.20. Force-velocity relationships of two muscles with different fiber lengths but identical strengths. Note that the muscle with longer fibers maintains a higher force at a given velocity (*dashed vertical line*) due to a smaller sarcomere shortening velocity. Difference between muscles becomes smaller as shortening velocity decreases.

Fiber Length and Physiologic Cross-Sectional Area in Force Production

We are now in a position to amplify a point made earlier. Remember that we stated that a muscle with long fibers was designed for speed or excursion, while a muscle with a high PCSA was designed for force production. We now see that, in muscle-joint systems, this is not necessarily the case. For example, we saw in Figure 3.20 that the muscle with longer fibers was actually designed for force production in the sense that it generated a greater force during movement. This equivalence of fiber length or PCSA as design strategies to increase force was recently explicitly stated by Gans and De Vree (1987). The take-home lesson in these types of discussions is that it is better to understand the underlying physiology than to try to memorize a few facts about muscle design. In this way, you can simply reason out a particular situation in terms of force or speed production.

The joint moment-isokinetic velocity relationship has been studied in a number of circumstances for variously aged populations of both sexes. These data served as normative values in evaluating limb status. Early on in the development of the isokinetic dynamometer, Perrine and Edgerton (1978) measured this relationship and attempted to interpret the data in terms of the muscle force-velocity properties (Figure 3.21). The torque-isokinetic velocity relationship looked (dangerously) similar to the isolated muscle force-velocity relationship. The main difference between the two was that the torque observed at very low velocities was much lower than was expected based on extrapolation of the torque-velocity relationship to zero velocity (dotted line). Perrine and Edgerton interpreted these data as representing a safety in the neural system that prevents overactivation of the muscle to dangerously high levels.

Limitations of Isokinetic Dynamometers

Before leaving the subject of isokinetic testing, we should mention several real-world issues that severely limit our ability to obtain meaningful muscular data using isokinetic dynamometers. The first issue is the time required to recruit muscle fibers. Since this time typically ranges from 50–200 msec, it is not wise to utilize any torque data obtained during this startup portion of the dynamic torque curve. At high angular velocities, this may represent a majority of the record itself! During recruitment and acceleration, the limb is accelerating from stationary to the selected speed and thus is not exerting maximum torque. Unfortunately, at the end of the limb acceleration phase, the limb strikes the testing bar (even if it was already in contact) and generates a force spike simply due to the impact. As a result, torques measured in this

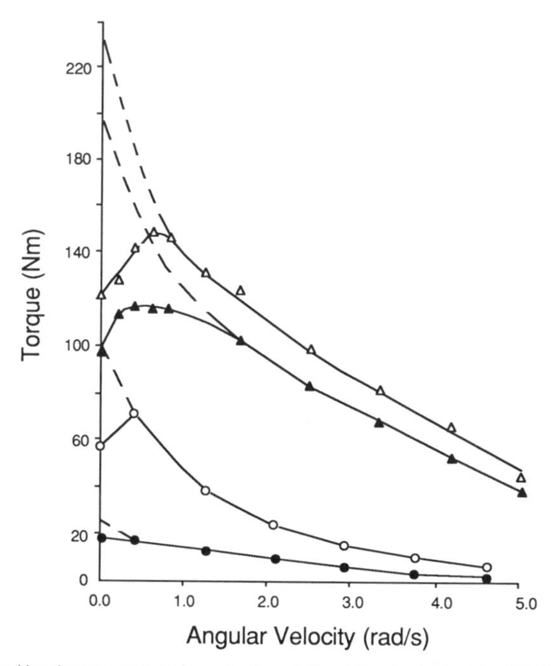

Figure 3.21. Isokinetic torque-angular velocity relationship elucidated by Wickiewicz *et al.* (1983) on quadriceps (*open triangles*), hamstrings (*filled triangles*), plantarflexors (*open circles*), and dorsiflexors (*filled circles*). *Dotted line* represents the theoretical force-velocity curve of Hill (1939). Note that forces at low angular velocities are much lower than predicted. This may represent an "override" of the nervous system to prevent very high muscle forces. (Data from Wickiewicz *et al.*, 1983.)

deceleration region can be artificially high. There are some who think this impact force is also dangerous to the subject.

As a result of limitations such as these, several manufacturers have installed electronic damping circuits to minimize the observation of these mechanical artifacts. One must be sure that the damping itself does not mask the intended measurement. Similarly, some investigators limit the range over which data are actually recorded or the time. Unfortunately, even with these precautions, it is not clear exactly which properties of the musculoskeletal system are expressed during isokinetic testing.

THE GAIT CYCLE

We have discussed muscle-joint interaction under very specific conditions: isometric, isokinetic, and isotonic conditions. However, what muscle forces and lengths are observed during normal movement? This activation of muscles and joints during normal movement has been studied both in animals and in humans (Basmajian and DeLuca, 1985). A complete description of muscle length, joint angles and gait speeds was presented by Goslow *et al.* (1973). In cats, they measured isolated bone dimensions and bone movement (using real-time radiography at the Harvard Museum of Comparative Zoology) during walking at various speeds. They then related the various muscle lengths and dimensions to the cats' movement and to the known pattern of muscle electrical activation (Engberg and Lundberg, 1969).

We will now apply much of our newfound knowledge of muscle and joint properties to a discussion of the typical action of muscles and joints during walking and running. As we shall see, the gait cycle is a wonderfully orchestrated sequence of electrical and mechanical events that culminate in the coordinated propulsion of a body through space (Figure 3.22). First, we will define the different phases of the gait cycle, and then we will discuss skeletal, electrical, and muscular changes that occur during the various phases. Finally, we will interpret this information in terms of what we have learned about muscle's static and dynamic properties.

Phases of the Gait Cycle

In its most general form, the gait cycle is divided into two phases: stance and swing (Table 3.3, Figure 3.22). These phases refer to the action of a limb that is either in contact with the ground and, therefore, providing some body support (stance) or a limb that is airborne, preparing for the next step (swing). Both stance and swing phases are subdivided into portions that refer to the action of the various muscular groups causing the movement. During swing phase, the limb *flexors* are sequentially activated at the knee, hip, and ankle to lift the limb off of the ground. The limb extensors are then activated to extend the knee and place the foot onto the ground. These two phases are referred to as F and E_1, respectively, referring to "flexion" and the first portion of "extension." Together, they comprise the swing phase (Table 3.3). Next, the extensor muscles are activated to higher levels, and as the foot strikes the ground, muscles and skeleton absorb the shock of ground impact, yielding somewhat under the weight of the body. These events comprise the E_2 or yield phase of gait. Finally, in a synchronized movement, all extensors propel the body until the foot finally leaves the ground to begin another swing phase. This final propulsive phase is known as E_3. It is obvious that these phases, referred to as the Philipson step cycle (Philipson, 1905), are named for the kinematics of functional muscle groups that are active during the various

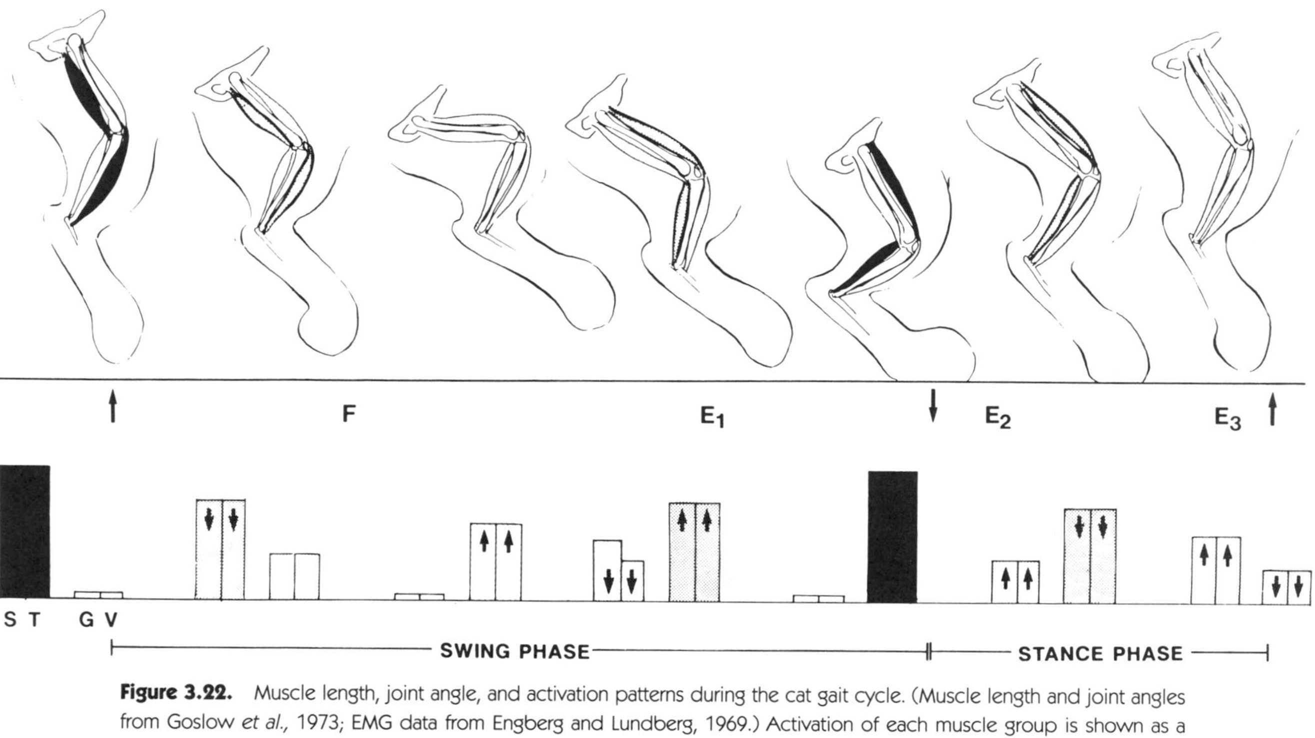

Figure 3.22. Muscle length, joint angle, and activation patterns during the cat gait cycle. (Muscle length and joint angles from Goslow *et al.*, 1973; EMG data from Engberg and Lundberg, 1969.) Activation of each muscle group is shown as a different shading pattern. Darker shading represents higher activation. Lower bar graphs represent activation patterns for muscles during that gait phase. Step cycle phase and toe off and toe strike are shown immediately beneath the line. **S,** semitendinosus; **T,** tibialis anterior; **G** gastrocnemius; **V,** vasti.

Table 3.3.
Components of the Philipson Gait Cycle

Swing Phase		Stance Phase	
F	E_1	E_2	E_3
Flexion phase as foot is lifted from ground	Extension of limb until foot contacts ground	"Yield" phase as foot contacts ground and supports body	Propulsive phase as limb extends and propels body into space

phases (Table 3.3). As we will see, there is much more to be considered during gait than simply which muscle is "on" and which is "off."

Muscle Group Activities during the Gait Cycle

The accomplishment of a step should be viewed as a coordinated effort between the neural control system (brain and spinal cord), the muscular system, and the skeletal system. A more detailed view of the muscle groups involved in gait is summarized in Figure 3.22 for one cycle. In this example, certain specific muscles will be discussed as typical for the function of that muscle group. The iliopsoas (IP) will be considered a typical hip flexor, the semitendinosus (ST) a knee flexor, and the tibialis anterior (TA) an ankle flexor. The semimembranosus (SM) will be considered a typical hip extensor, the vastus lateralis (VL) a knee extensor, and the medial gastrocnemius (MG) and soleus (S) ankle extensors. We are about to see that muscle actions are much more complex than might be anticipated. In fact, the notion that muscles simply contract and cause joint rotation is incorrect and oversimplified.

During the flexion phase (F), the limb flexors (IP, ST, and TA) are activated, resulting in active flexor muscle shortening, to elevate the foot off of the ground (Figure 3.22; Table 3.4). As a result, the hip, knee, and ankle flex. During this same period, because of active shortening of the flexors, the extensors (SM, VL, MG, S) are passively stretched. As the body mass moves forward, the limb extensors are activated, and the flexors shut off to initiate limb extension (E_1). During E_1, the extensors are activated, causing limb extension, and now the extensors actively shorten. Due to (low) electrical activity in the flexors, flexors actively lengthen. As the foot strikes the ground (E_2, the yield phase of stance), the extensors are highly active. However, the mass and momentum of the body are so great that in spite of this high extensor activity, the knee and ankle joints are forced to flex or yield! Obviously, forced flexion of the hip, knee, and ankle during active extensor

Table 3.4.
Muscle Actions during Different Phases of the Cat Gait Cycle[a,b]

Muscle Group	Swing		Stance	
Extensors	F	E_1	E_2	E_3
Hip	PL	AL	ISO	AS
Knee	PL	AS	AL	AS
Ankle	PL	AS	AL	AS
Flexors				
Hip	AS	ISO	AL	AL
Knee	AS/AL	AL	PS	AS/AL
Ankle	AS	AL	PL	AL

[a]Adapted from Goslow G, Reiwking R, Stuart D. The cat step cycle: hindlimb joint angles and muscle lengths during unrestrained locomotion. J Morphol 1973;141:1–42.
[b]PL, passive lengthening; AL, active lengthening; AS, active shortening; PS, passive shortening; ISO, isometric.

activity will result in active *lengthening* of the VL, MG, and S, respectively! We recall from the force-velocity relationship that active lengthening (eccentric contraction) results in high muscle forces. It is thus not surprising that muscle injury and soreness are sometimes associated with this type of muscle activity seen during this phase of the gait cycle (Chapter 6). It is also not surprising that during E_2 the flexors are completely silent, since gravity and momentum are doing what the flexors would do anyway. In the last portion of stance, the lengthening extensors have generated enough force to reverse the transient joint flexion back into extension, and the flexors dramatically change from active lengthening to active shortening. This transition from lengthening to shortening (the so-called stretch-shorten cycle) has significant energetic consequences. Active shortening of the extensors during late E_3 results in pronounced joint extension and propels the body forward into space. The flexors, which were electrically silent during E_2, begin to be activated during the late stages of E_3 (in preparation for F) and are, therefore, actively lengthened.

Muscle Force Modulation during the Gait Cycle

During gait most muscle groups thus experience periods of active shortening, active lengthening, and passive lengthening (Table 3.4). It is exciting to understand the underlying events of this complex sequence of events, which we routinely refer to as simply "walking" or "running." Let us not leave this realm without considering some of the fine points of gait, which relate these gross muscle and joint changes to the muscle's physiologic properties and to the joint's kinematic properties. Several questions come to mind upon gross inspection of the gait cycle. First, how much does muscle force modulation *really matter* in accomplishing the gait cycle? We showed that muscle force changes in response to length (length-tension relationship) and velocity (force-velocity relationship), but how do these relationships figure into physiologic action of muscles during gait? Several elegant studies of gait and mathematical modeling of gait suggest that *muscle length changes are relatively small during the time when muscles are active.* In other words, when muscle force is increasing in flexors during the E_3 phase of gait, it is not simply because sarcomere length is changing, resulting in greater overlap of thick and thin filaments and an increase in potential force. In this case, it is due primarily to increased muscle fiber recruitment. Most investigators agree that the total length muscle change that occurs during the gait cycle is only about 20%–30% of the total physiologic range of the muscle (the physiologic muscle range is that observed when the joint is forced to go from full flexion to full extension). For example, in the study of four ankle extensors previously mentioned, Goslow *et al.* (1973) measured skeletal joint angles and muscle physiologic properties and calculated muscle-tendon lengths during locomotion in the cat. They showed that the four ankle extensor muscles (medial gastrocnemius, lateral gastrocnemius, soleus, and plantaris) generated their maximum forces at different ankle angles, ranging from 40° to 100°, and that the angle corresponding to maximum muscle moment arm was 120°! Again, we have a situation where muscle force and moment arm are *not* optimized at the same angle. We arrive at the conclusion that the musculoskeletal system is not solely designed to produce a high force at a particular angle. Similarly, Gregor *et al.* (1988) showed that during the stance phase of gait when the soleus (SOL) first actively lengthens and then actively shortens, the total length excursion is only about 10% of its maximum length (and is probably even less when one considers the fact that the Achilles tendon can actually absorb some of the length change).

As a second means of emphasizing the relative importance of muscle length and velocity in producing force, let us consider the changes that occur in the frog ST during hopping. Obviously, it is impossible to measure sarcomeres directly during hopping. However, it is possible to measure sarcomere length-joint angle relationships, measure joint angle changes during frog hopping, and to infer the sarcomere behavior (Mai and Lieber, 1990). Note that, as the frog hop progresses, sarcomere length begins at about 2.0 μm and ends at about 2.6 μm (Figure 3.23**A**). Thus sarcomere length increases down the descending limb of the length-tension relationship, which, under isometric conditions, would decrease muscle force about 25% due to decreased filament overlap. However, note in the force record (Figure 3.23**C**) corresponding to this length change that force dramatically *increases* during the

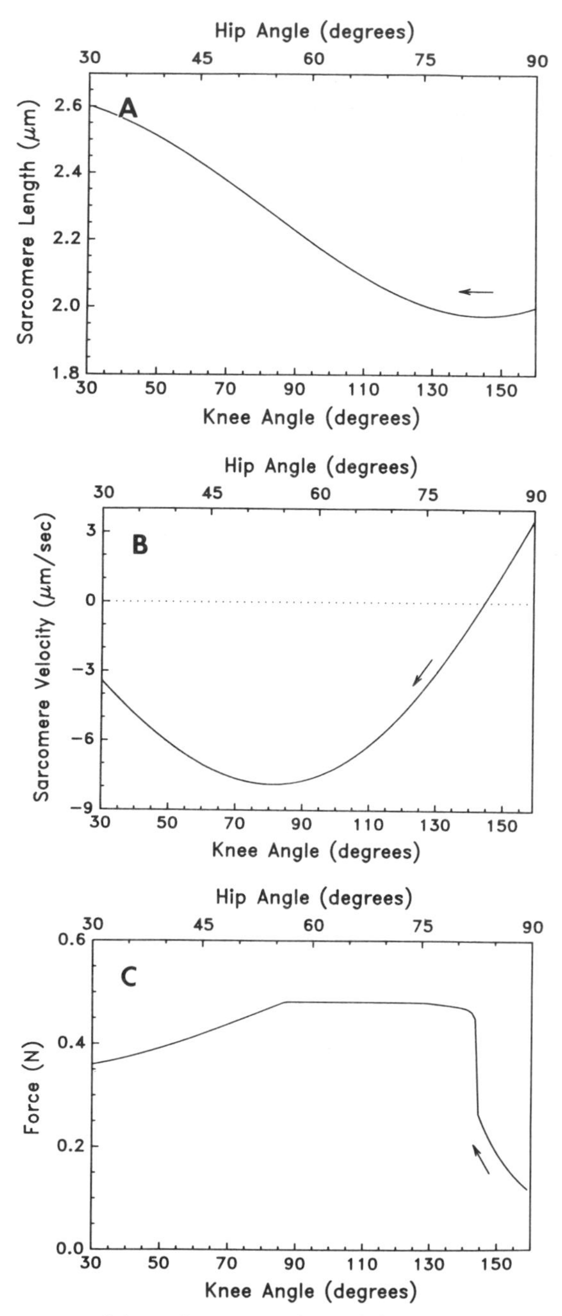

Figure 3.23. Time course of change in sarcomere length (**A**), sarcomere velocity (**B**), and muscle force (**C**) during a model frog hop, assuming full muscle activation. (From Mai MT, Lieber RL. Interaction between semitendinosus muscle and knee and hip joints during torque production in the frog hindlimb. J Biomech 1990;23:271–279.)

hop. Why? The answer lies in the muscle velocity (Figure 3.23**B**). Note that in the first portion of the hop, the muscle is shortening slightly, but shortening velocity is decreasing. Due to the force-velocity relationship, decreased muscle velocity results in increased force. For a very brief moment, the muscle is actually isometric! Next, presumably due to the action of other muscles, the ST is forced to lengthen. This forced lengthening results in very high muscle forces due to the discontinuity in the force-velocity relationship as we proceed from shortening to lengthening. We thus see that the muscle force record (Figure 3.23**C**) is not influenced as much by the muscle *length* as it is by the muscle *velocity*.

BIOMECHANICS OF BIARTICULAR MUSCLES

The sarcomere length record of Figure 3.23**A** also raises interesting points regarding the physiologic function of a biarticular muscle such as the ST. Note first that the ST undergoes shortening and then lengthening during the single coordinated hopping movement. This results directly from the relative hip and knee moment arm magnitudes, hip and knee joint angular velocities, and the relative timing of the actions at the two joints. Therefore, the view that a skeletal muscle simply contracts to cause joint rotation is not well founded. In this case we see that the velocity of the muscle itself actually reverses direction due to the sophisticated balance between muscle and joint properties.

The second observation we make is that the ST is obviously not shortening during the entire movement. In fact, we must conclude that the ST is not "hopping" the frog, it is actually being "hopped"! The ST is being actively stretched by other muscles that are extending the knee. Therefore, although we would conclude, based on anatomic considerations, that the ST is a hip extensor and a knee flexor, it is not flexing the knee at all in this motion! Physiologic function may not, therefore, be obvious based only on anatomic configuration.

How is the ST caused to lengthen? Our best guess to date is that the ST is being stretched by other muscles, resulting in its active lengthening. The most likely candidates for the other muscles are the knee extensors. We have thus hypothesized that the ST acts as an "activatable sling" at the knee joint, providing a route of force transmission from the monoarticular knee extensors to the hip joint (Figure 3.24**A**). Through the ST muscle, the monoarticular knee extensors (equivalent to humans' vasti muscles) are thus able to extend the hip joint during jumping (Lieber, 1990). Such a mechanism may operate in the mammalian ST as well as other biarticular muscles (van Ingen Schenau, 1989). For example, a similar relationship could occur between the gastrocnemius muscle (which crosses the knee and ankle) and the TA (which crosses only the ankle) whereby the dorsiflexion moment

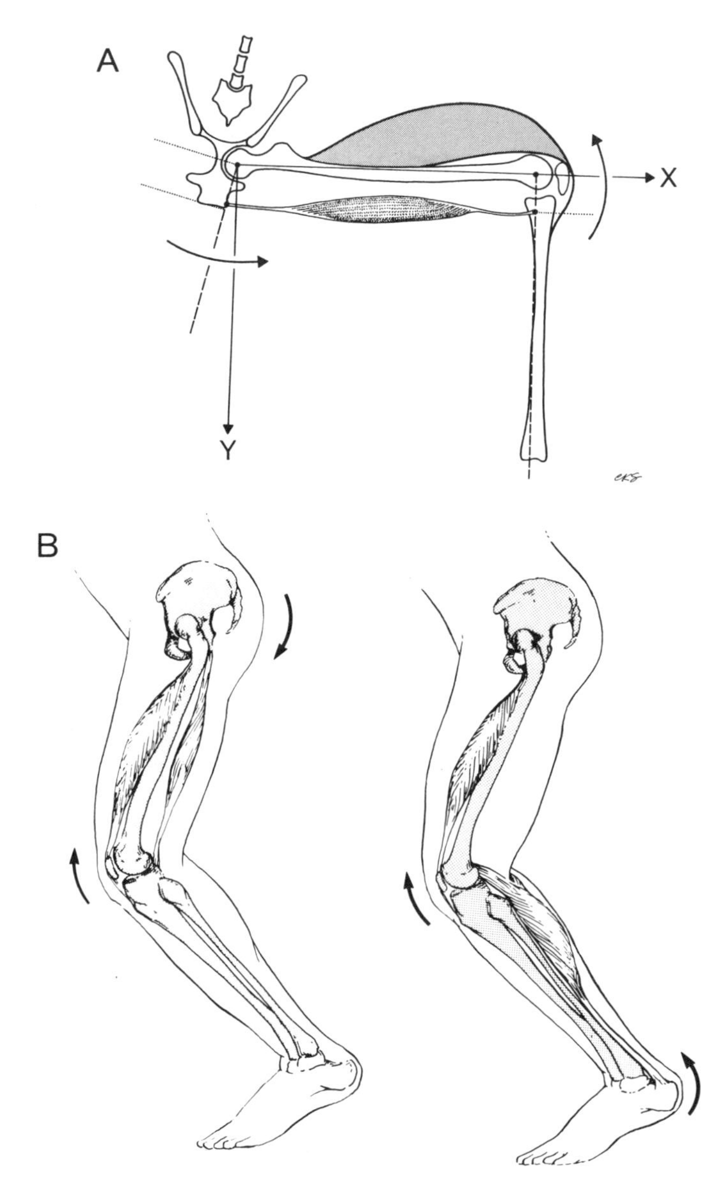

Figure 3.24. A, Schematic representation of transfer of frog knee extension into hip extension by the biarticular semitendinosus muscle. In this scenario, knee extension is converted into hip extension by the biarticular semitendinosus. (From Mai MT, Lieber RL. Interactions between semitendinosus muscle and knee and hip joints during torque production in the frog hindlimb. J Biomech 1990;23:271–279; and Lieber RL. Hypothesis: biarticular muscles transfer moments between joints. Dev Med Child Neurol 1990; 32:456–458.) **B,** Joint moment transfer by biarticular human muscles. Transfer of knee extension into hip extension by the biarticular semitendinosus (**left panel**) and transfer of knee extension into plantar flexion by the biarticular gastrocnemius (**right panel**). Other examples are proposed in Table 3.5.

Table 3.5.
Examples of Interaction between Monoarticular and Biarticular Muscles

Muscle Group	Action	Biarticular Muscle Link	Resultant Action
Quadriceps	Knee extension	Hamstrings	Hip extension
Dorsiflexors	Dorsiflexion	Gastrocnemius	Knee flexion
Gluteals	Hip extension	Rectus femoris	Knee extension
Gluteals	Hip extension	Iliopsoas	Spine flexion
Gluteals	Hip extension	Rectus femoris and gastrocnemius	Plantar flexion

generated by the TA could be transferred, via the gastrocnemius into knee flexion (Figure 3.24**B**; Table 3.5). Should such a mechanism operate physiologically, it would have a profound impact on the interpretation of studies that ascribe externally measured or calculated moments to specific muscle groups. It would also dramatically affect the interpretation of joint weakness that is observed at a particular site. For example, suppose that the ST muscle discussed above was used during standing to extend the hip, but also to transmit vasti moments to the hip. We might observe difficulty in standing and, incorrectly, conclude that the hip extensors were weak. In fact, it may be that the vasti are actually the weak muscles, unable to provide the extra moment needed to stand. This idea that muscles at remote sites might affect joints that they do not actually cross (*e.g.*, the vasti affecting the hip joint) might require new thinking regarding the basis of joint weakness or movement abnormalities.

Gait Cycle Timing at Increased Velocities

Given this relatively standardized description of the gait cycle, let us address the modifications of gait that occur as speed is changed. As gait speed increases, what happens to the relative time of each gait cycle portion? Clearly, the time in each phase must decrease, but does it decrease to the same extent? The answer is no! Goslow *et al.* (1973) showed that as gait speed increases, the proportion of the gait cycle devoted to stance decreases and is mirrored by a dramatic increase in the proportion of the gait cycle devoted to swing. During low-speed walking, stance phase occupied about 60% of the cycle while swing occupied about 40%. However, at high-speed locomotion, in spite of the decreased absolute time of each phase, the swing phase occupied over 75% of the total cycle time! Within the swing phase, both F and E_1 increased their relative proportion, although F increases to a greater extent. In stance, the relative proportion of E_2 and E_3 both decreased in proportion, but the greatest decrease was seen in E_3, which occupied nearly 60% of the gait cycle at slow speeds and only 15% of the cycle at high speeds. It is obvious that the relative amount of time spent in contact with the ground decreases as gait speed increases.

DIRECT MUSCLE FORCE AND MUSCLE LENGTH MEASUREMENTS DURING GAIT

Several animal studies measured either muscle force or muscle length or both during normal gait. The most widely studied muscles are the cat ankle extensors, which have been studied during the stance phase. This information has been used to try to understand how different muscles are used during activities of different intensities. Typical comparisons are made between the SOL and the MG muscles. Because the SOL is composed only of slow fibers (in cats) and the MG is composed primarily of fast fibers (also in cats), we also gain insights as to how different muscle fiber types are recruited during various tasks (Chapter 2, page 95).

The experimental approach used to investigate this problem involved measurement of muscle electrical activity via electromyograms (EMGs), muscle length (usually using analysis of film of walking cats), and muscle force (which recently has been measured by implanting force transducers on the individual tendons of the various muscles; Walmsley *et al.,* 1978; Whiting *et al.,* 1984; Gregor *et al.,* 1988; Hoffer *et al.,* 1989). These investigators generally agree that, at relatively low speeds, the SOL muscle is first activated to a significant extent, while only at higher speeds (requiring higher joint moments) is the faster contracting MG recruited. At still higher speeds, the SOL is probably nearly maximally activated, and any increases in extension moment are produced by recruiting the MG to higher and higher levels. How high do you think that the MG and SOL are recruited during normal walking compared to, say, maximum tetanic tension? This question was recently addressed by performing *in vivo* force measurements of muscle force during walking and then measuring maximum tetanic tension *in the same muscles* and a terminal *in situ* physiologic experiment (Fowler *et al.,* 1989). The results were, to say the least, astounding. The most surprising result to date is that muscle tensions that occur during walking often *exceed* maximum force (P_o)! This could obviously be accomplished for the ankle extensors simply by the eccentric contractions that normally occur during E_2. However, the explanation is not quite this simple. It is possible to measure muscle velocity and muscle force *in vivo* during gait and compare that force and velocity to those achieved during terminal *in situ* experiments (Gregor *et al.,* 1988). A schematic of the results for the SOL is shown in Figure 3.25. The typical force-velocity curve often lies *below* the *in vivo* force-velocity relationship. In other words, during normal gait, at a given velocity, the muscle generates more force than would be expected. How can this be? Take some time to think about this before reading on.

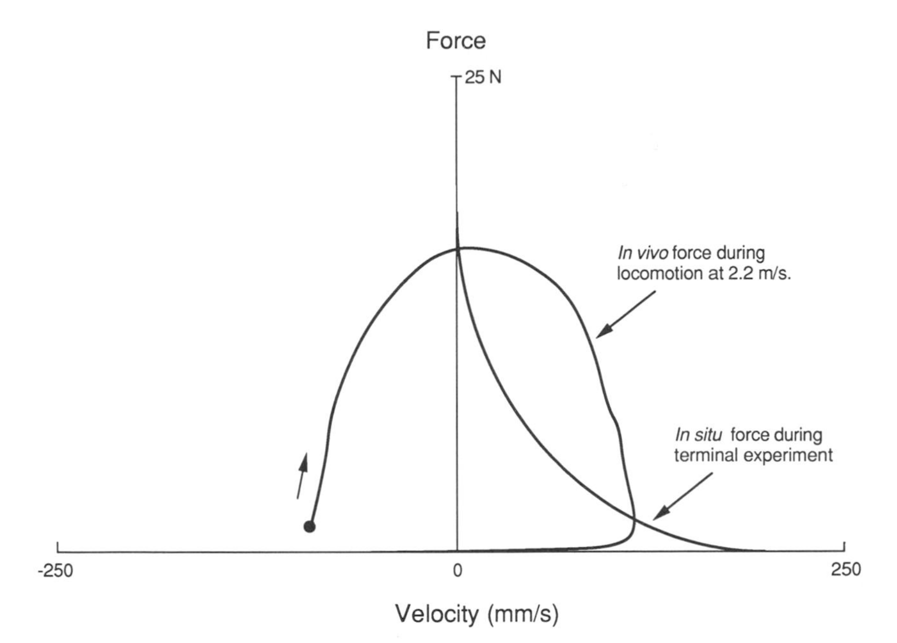

Figure 3.25. Force-velocity relationship obtained from an *in situ* isolated cat soleus muscle compared to actual *in vivo* force-velocity relationship measured during locomotion. Note that forces of the muscle-tendon unit lie above the *in situ* force-velocity curve. (From Gregor RJ, Roy RR, Whiting WC, Lovely RG, Hodgson JA, Edgerton VR. Mechanical output of the cat soleus during treadmill locomotion: in vivo vs. in situ characteristics. J Biomech 1988;21:721–732.)

Muscle-Tendon Interaction during Gait

The apparent answer to the dilemma lies in the method of measuring muscle velocity *in vivo*. Usually, such measurements are made from external movies taken of the animal during gait. Then, based on external bony landmarks, muscle origin-to-insertion distances are calculated, from which *in vivo* "muscle" velocity is obtained. However, in reality, this length represents the *in vivo* muscle-*tendon* length. It is now thought that the tendon length change during gait (at least in the ankle extensors) has a dramatic impact on this relationship. The current explanation goes something like this: During E_2, the MG is actively lengthened at a very high tension, which also lengthens the Achilles tendon. Then, as the muscle-tendon unit progresses from E_2 to E_3, the tendon begins to recoil so that it is actually shortening faster than the muscle. Therefore, energy that was stored in the tendon is now released as the tendon shortens, and the muscle is not required to shorten as fast. This could explain why the *in vivo* force-velocity relationship is above the *in situ* curve—the muscle *fiber* velocity was underestimated in the *in vivo* situation since tendon compliance was not considered. Complete resolution of this dilemma will await separate measurement of muscle length and tendon length during gait. Preliminary measurements by Hoffer *et al.* (1989) have tended to confirm this explanation, but the jury is still out. We are presented here with a new function for the tendon—that of elastic energy storage. Energy storage by tendinous structures has already been demonstrated for a number of animals, especially those with long tendons in series with large muscles (Alexander, 1988). If elastic energy storage were to be used during gait, we would expect that the perfect time would be during the E_2–E_3 transition as the muscle goes from lengthening to shortening at very high tensions. Several authors have pointed out that this stretch-shorten cycle can dramatically increase locomotion efficiency (Alexander, 1988).

Physiologic Functions of Tendons

We have spoken of the mechanical and physiologic properties of tendons. To summarize several of their roles, we reiterate those four major functions here:

1. The most obvious function of tendons is that they connect muscles to bones.
2. By virtue of their compliance, tendons can absorb length changes during high-impact motions (shock absorption), which allows muscles to lengthen at lower eccentric velocities. This in turn lowers the muscle tension achieved, which might prevent muscular injury (Chapter 6). Tendons probably function physiologically in the high compliance region of their stress-strain curves.
3. Due to their significant but limited compliance, tendons also increase the functional range of the muscle-tendon unit. This is because they allow sarcomeres to shorten during contraction, and they lengthen to absorb some of the passive tension.
4. During stretch-shorten cycles, high forces elongate tendons. The subsequent muscle shortening is accompanied by release of stored tendon elastic strain energy. This increases locomotion efficiency. In addition, since tendons recoil and take up some of the length of the muscle-tendon unit, this keeps muscle *fiber* velocity low, which keeps potential muscle tension higher.

IMPLICATIONS FOR PHYSICAL THERAPY

What are some of the implications of these concepts on current physical therapy practice? The most obvious implication is that, to fully understand the nature of *strength,* one must understand *torque.* We now know that when we measure strength we are measuring the interaction between muscle and joint properties. At the joint angle where torque is maximum, muscle force and joint moment arm are involved to varying degrees. So the first thing we must choose in strength testing is the appropriate joint angle. At what angle do we obtain the most "muscle" information? At what angle do we obtain the most "joint" information? This question must be answered on a case-by-case basis, and unfortunately, we do not have enough data on human muscles and joints to fully answer it. However, on theoretical grounds we can still make a number of statements. We must always assume that we are neither at the angle at which muscle force is maximum nor the angle at which joint moment arm is maximum (refer to Figure 3.13**B**).

In order to determine the basis for strength changes, we must consider all of the facts we have learned so far. Strength (torque) can be altered due to changes in muscle force or muscle moment arm. In situations such as simple immobilization, when muscles can become weak due to fiber atrophy (Chapter 5) and muscle force decreases, we may suspect that muscle force is the main reason that strength has decreased. Since torque is simply the product of muscle force and joint moment arm, a decrease in muscle force due to atrophy will simply be represented as a decrease in the joint torque. However, it is not necessary to have muscle atrophy to have weakness. Recall that muscle force can also dramatically change in response to the level of excitation, which recruits a varying number of muscle fibers. Changes in muscle force can, therefore, simply be a result of decreased electrical drive to the muscle. This could be detected as a change in EMG. Thus the various pieces of the puzzle—EMG, muscle force, and moment arm—must be on hand to make an unambiguous diagnosis of the basis for a strength decrease.

Finally, a surgical procedure that moves a muscle from one position to another can cause weakness. What is the basis for the weakness? Again, due to the surgical procedure itself, the muscle may have atrophied (note that I say *may have* atrophied, since surgical procedures can result in muscle *hypertrophy,* as outlined in Chapter 4!). Muscle atrophy might be the reason for the weakness. However, suppose that the muscle *moment arm* was altered by the surgical procedure itself. Could this not also produced weakness? Of course. This option is shown in Figure 3.26. Increased moment arm (and, therefore, decreased fiber length/moment arm ratio) causes the muscle to change force faster as the joint rotates. Now, if the muscle passes through its entire length-tension range, joint torque will be extremely low in spite of the fact that muscle force is exactly the same (Figure 3.26). In addition to a change in moment arm, a surgical procedure can change the muscle fiber length (Chapter 4, page 172). The functional effect would be similar: Sarcomere

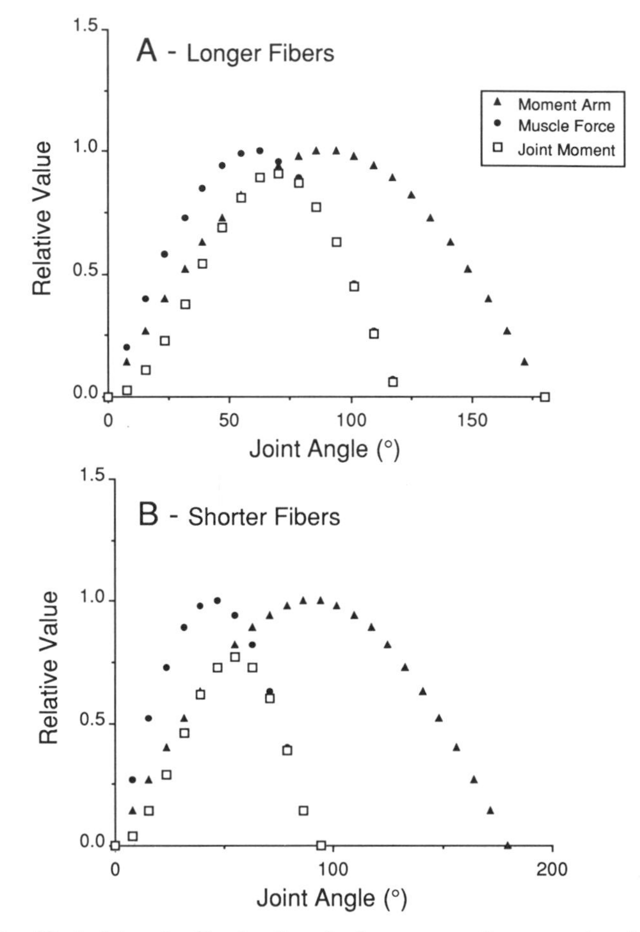

Figure 3.26. Effect of changing fiber length on the torque-generating properties of a hypothetical muscle-joint system. In this example, the moment arm (*filled triangles*) is shown for that of a simple hinge joint (*c.f.*, Figure 3.11**B**). Muscle force (*filled circles*) was calculated for two different cases. In **A**, muscle fibers were "long" and thus could generate active force over the range 0°–120°. In **B**, muscle fibers were "short" and thus could generate active force only over the range 0°–90°. Note that as muscle fiber length decreases, the joint angle at which muscle force is maximum also shifts to smaller angles. The product of muscle force and moment arm is joint moment (*open squares*) and should be compared between the two cases. Joint moment (strength) is lower for the case of shorter fibers (**B**) since the relative relationship between muscle force and moment arm has changed. Thus strength decreased in spite of the fact that the muscle generated the identical force. This example reinforces the idea that muscle force and joint strength are not necessarily uniquely related.

length would change at a different rate as the joint rotates. This could change joint torque as shown in Figure 3.26.

If we consider the muscle's ability to adapt, the situation becomes even more interesting. In spite of the relatively predictable structure and function of muscle tissue, it is one of the most plastic tissues in the body. The nature of muscle adaptation not only provides insights in a muscle's history but can also be exploited in rehabilitation to return a muscle's properties to normal. It is the subject of muscle adaptation that brings us to the next chapter. See you there!

SUMMARY

In this chapter, we painted the picture of the way in which muscles, tendons, and bones interact to produce movement. A major conclusion was that torque production results from interaction between muscles and joints and not from either muscles or joints alone. We saw that muscle architecture has a profound effect on muscle force, excursion, and range of motion and may be considered one design parameter of the musculoskeletal system. Finally, we investigated the beautifully orchestrated sequence of events known as the gait cycle. As mentioned, these concepts are being applied to sports medicine and therapy. Further advances in treatment will undoubtedly follow as more practitioners become aware of these concepts.

REFERENCES

Alexander, R. Mcn. (1988). Elastic Mechanisms in Animal Movement. Cambridge: Cambridge University Press.

Basmajian, J.V., and DeLuca, C.J. (1985). Muscles Alive. Their Functions Revealed by Electromyography. 5th ed. Baltimore: Williams & Wilkins.

Butler, D.L., Grood, E.S., Noyes, F.R., and Zernicke, R.F. (1978). Biomechanics of ligaments and tendons. Exerc. Sports Sci. Rev. 6:125–182.

Engberg, I., and Lundberg, A. (1969). An electromyographic analysis of muscular activity in the hindlimb of the cat during unrestrained locomotion. Acta Physiol. Scand. 75:614–630.

Fowler, E.G., Gregor, R.J., Hodgson, J.A., and Roy, R.R. (1989). The contribution of individual muscles to the ankle moment produced in the cat hindlimb. J. Biomech. 22:1101.

Gans, C., and De Vree, F. (1987). Functional bases of fiber length and angulation in muscle. J. Morphol. 192:63–85.

Goslow, G., Jr., Reinking, R., and Stuart, D. (1973). The cat step cycle: hindlimb joint angles and muscle lengths during unrestrained locomotion. J. Morphol. 141:1–42.

Gregor, R.J., Roy, R.R., Whiting, W.C., Lovely, R.G., Hodgson, J.A., and Edgerton, V.R. (1988). Mechanical output of the cat soleus during treadmill locomotion: in vivo vs. in situ characteristics. J. Biomech. 21:721–732.

Hoffer, J.A., Caputi, A.A., Pose, I.E., and Griffiths, R.I. (1989). Roles of muscle activity and load on the relationship between muscle spindle length and whole muscle length in the freely walking cat. Progr. Brain Res. 80:75–85.

Horii, K.N., An, W.P., Cooney, and Linscheid, R.L. (1991). Kinematics and tendon excursion of wrist movers. J. Hand Surg. Submitted.

Hoy, M.G., Zajac, F.E., Gordon, M.E. (1990). A musculoskeletal model of the human lower extremity: the effect of muscle, tendon, and moment arm on the moment-angle relationship of musculotendon actuators at the hip, knee, and ankle. J. Biomech. 23:157–169.

Lieber, R.L. (1990). Hypothesis: biarticular muscles transfer moments between joints. Dev. Med. Child Neurol. 32:456–458.

Lieber, R.L., and Boakes, J.L. (1988). Sarcomere length and joint kinematics during torque production in the frog hindlimb. Am. J. Physiol. 254:C759–C768.

Lieber, R.L., Brown, C.G., and Trestik, C.L. (1991a). Model of muscle-tendon interaction during frog semitendinosus fixed-end contractions. J. Biomech. In press.

Lieber, R.L., Fazeli, B.M., and Botte, M.J. (1990). Architecture of selected wrist flexor and extensor muscles. J. Hand Surg. 15:244–250.

Lieber, R.L., Leonard, M.E., Brown, C.G., and Trestik, C.L. (1991b). Frog semitendinosus tendon load-strain and stress-strain properties during passive loading. Am. J. Physiol. 261:C86–C92.

Mai, M.T., and Lieber, R.L. (1990). Interaction between semitendinosus muscle and knee and hip joints during torque production in the frog hindlimb. J. Biomech. 23:271–279.

Perrine, J.J., and Edgerton, V.R. (1978). Muscle force-velocity and power-velocity relationships under isokinetic loading. Med. Sci. Sports Exerc. 10:159–166.

Philipson, M. (1905). L'autonomie et la centralisation dans le systeme nerveux des animaux. Trav. Lab. Physiol. Inst. Solvay (Bruxelles). 7:1–208.

van Ingen Schenau, G. (1989). From rotation to translation. Constraints on multijoint movement and the unique role of biarticular muscles. J. Hum. Move. Sci. 8:301–337.

Walmsley, B., Hodgson, J.A., and Burke, R.E. (1978). Forces produced by medial gastrocnemius and soleus muscles during locomotion in freely moving cats. J. Neurophysiol. 41:1203-1216.

Wickiewicz, T.L., Roy, R.R., Powell, P.J., and V.R. Edgerton. (1983). Muscle architecture of the human lower limb. Clin. Orthop. Rel. Res. 179:317-325.

Whiting, W.C., Gregor, R.J., Roy, R.R., and Edgerton, V.R. (1984). A technique for estimating mechanical work of individual muscles in the cat during treadmill locomotion. J. Biomech. 17:685-694.

Zajac, F.E. (1989). Muscle and tendon: properties, models, scaling and application to biomechanics and motor control. CRC Crit. Rev. Biomed. Eng. 17:359-411.

Chapter

4

Skeletal Muscle Adaptation to Increased Use

OVERVIEW

In this chapter we change gears. Instead of discussing the design and function of the neuromusculoskeletal system, we discuss its ability to adapt—its plasticity. We will consider specific cases where the level of muscle use is increased. In order to understand adaptation to increased use, we focus on five diverse experimental models. In spite of the fact that different models are used to present increased levels of use, the adaptations are fairly consistent. We begin with the "cleanest" model of adaptation—chronic electrical stimulation—and end with the most physiologic—voluntary exercise. Patterns develop as each model adds its part to our understanding of muscle adaptation. Our goal is a complete understanding of how a muscle adapts in order to be able to exploit this plasticity to the benefit of the patient.

INTRODUCTION

In spite of the relatively dogmatic presentation of skeletal muscle structure and function in Chapters 1 to 3, I must state emphatically that skeletal muscle is one of the most adaptable (plastic) tissues in the body. Virtually every structural aspect of muscle that we have discussed can change given the proper stimulus. This includes architecture, fiber type distribution, tendon length, fiber diameter, myosin profile, mitochondrial distribution, capillary density, fiber length, *etc.* What determines each skeletal muscle property? One factor is the level of muscle activity relative to "normal." Interestingly, "normal" is different for different muscles and different fibers within the muscles. It is difficult to make sweeping generalizations regarding adaptation; however, the unifying concepts presented will allow you to "think" the way a muscle does.

ADAPTATION TO CHRONIC ELECTRICAL STIMULATION

Chronic electrical stimulation provides one of the cleanest views of muscle adaptation to increased use. It has long been used by basic scientists to study skeletal muscle adaptation since it induces a repeatable, quantifiable amount of exercise. In this setting, a well-defined progression of muscular changes is observed, which enables investigation of the mechanism and time course of muscle adaptation. In addition, observations of muscular changes following electrical stimulation provide insights into other forms of muscle adaptation, such as those that occur following immobilization, disease, surgery, or exercise.

Experimental Method for Chronic Muscle Stimulation

Normally, skeletal muscles that play a postural role and thus have a high proportion of slow fibers are physiologically frequently activated at low frequencies. Conversely, muscles with a high proportion of fast fibers are activated intermittently with high frequency bursts. The best documented effects of electrical stimulation on skeletal muscle are those occurring after chronic, low-frequency stimulation (similar to the activity of a "slow" muscle) is imposed upon a predominantly "fast" muscle. Experimentally, a cuff electrode is wrapped around a peripheral nerve, and the wire leads from the electrode are routed to a neuromuscular stimulator. The stimulator may be small enough to implant within animal, or the leads may exit and externally connect to a stimulator (Figure 4.1).

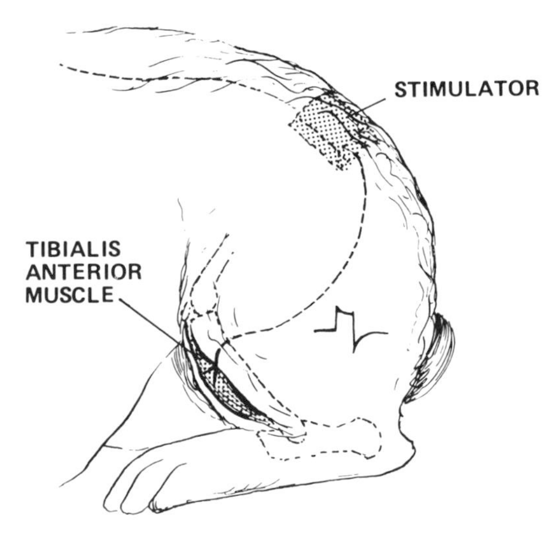

Figure 4.1. Experimental method for chronic activation of skeletal muscle. Stimulating electrodes are placed around a motor nerve and routed to either a subcutaneous stimulator or leads that exit and connect to an external stimulator (not shown). Stimulator activation pattern is then chronically imposed on the muscle.

If the stimulator is activated at a nominal frequency of about 10 Hz and allowed to operate 24 hours a day, a well-defined progression of changes is observed whereby the fast muscle first changes its metabolic and then its contractile properties to completely transform into a slow muscle. Below is a description of the details of the transformation process, which is based on experimental data from a number of laboratories. For the sake of brevity, I have combined the results, primarily representing the work of Brenda Russell Eisenberg (Chicago), Dirk Pette (Konstanz), and Stanley Salmons (Liverpool). Before describing the details of the transformation process, let's make a few general statements. Transformation has been documented in a number of different muscles and species, so that the effects observed are probably not species or muscle specific. The fast-to-slow transformation that occurs is detectable using all of our modern techniques such as measurement of muscle contractile, ultrastructural, histochemical, biochemical, and morphologic properties. In all cases, following transformation, the new slow fibers are almost completely indistinguishable from normal slow skeletal muscle fibers (Eisenberg and Salmons, 1981; Eisenberg *et al.,* 1984; Pette, 1980 and 1990; Salmons and Henriksson, 1981). It is also generally accepted, based on time-series studies and single fiber biochemistry, that the changes that occur result from *transformation* of a single fast fiber into a slow fiber and not from loss of fast fibers with subsequent slow fiber regeneration or proliferation. The fast fibers actually *become* slow fibers.

How does the transformation occur? Remember that this muscle is being completely rearranged *while it is still physiologically active!* This is analogous to fixing a car while it is moving! Let's look at the details of the various muscle cellular components as the adaptation occurs.

Time Course of Muscle Fiber Transformation

If low-frequency stimulation is applied 24 hours a day, the total transformation process requires only about 8 weeks. The earliest observed changes occur within a few hours after the onset of stimulation, where the sarcoplasmic reticulum (SR) begins to swell (Figure 4.2**A**). Within the next 2–12 days, increases are measured in the volume percent of mitochondria, oxidative enzyme activity, the number of capillaries per square millimeter, total blood flow, and total oxygen consumption, reflecting the increased metabolic activity of the muscle (Figures 4.2**B, C**). Histochemically, this is reflected in an increased percentage of fast oxidative glycolytic (FOG) fibers at the expense of fast glycolytic (FG) fibers (Figure 4.2**C**). The increase in oxidative enzymes and capillary density is manifested as a decrease in muscle fatigability. At this point, the width of the Z-band begins to increase toward the wider value

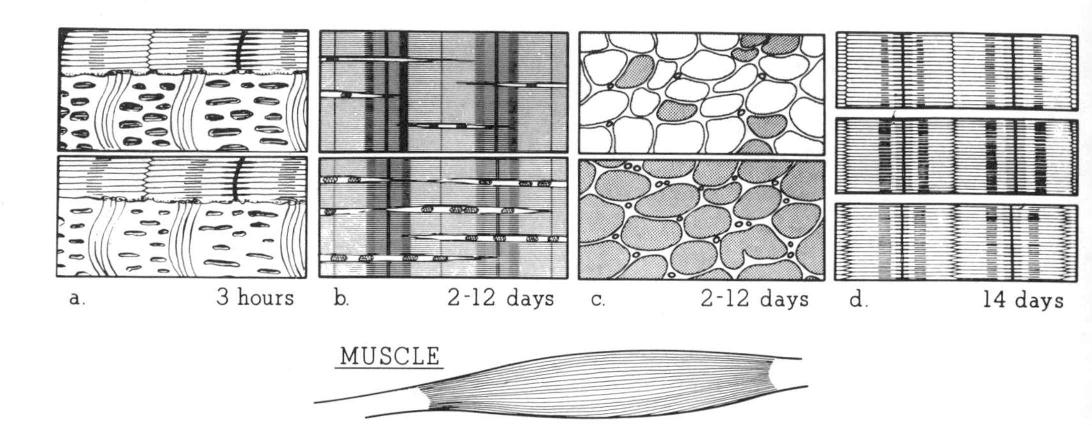

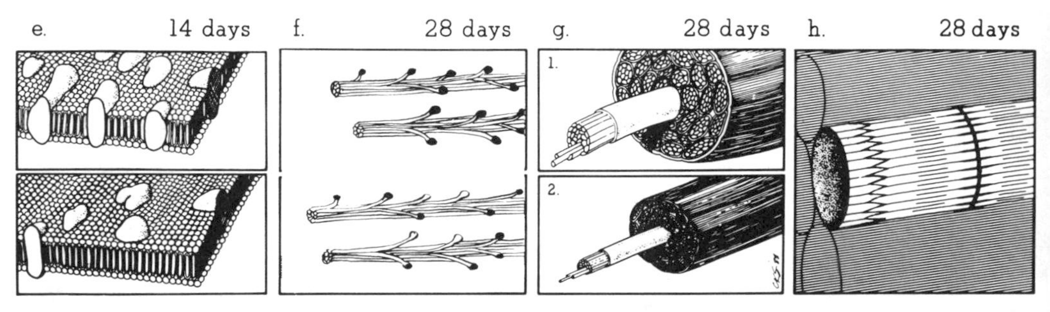

Figure 4.2. Schematic representation of the time course of muscular adaptation to chronic stimulation transforming a fast to a slow fiber. In each panel the normal fiber is shown in the upper panel and the stimulated fiber in the lower panel. **A,** SR begins to swell after 3 hours of stimulation. **B,** After 2–12 days of chronic stimulation, an increase in the volume percent of mitochondria is observed. **C,** After 2–12 days of chronic stimulation, an increase in capillary density and in type FOG (represented as darkly staining in this simulated SDH stain) fibers is observed. **D,** After 14 days, the Z-band begins to increase in width. **E,** After 14 days, a decrease in the amount and activity of calcium ATPase is observed. **F,** After 28 days, the myosin profile is altered with different myosin monomers incorporating into single filaments (this figure is schematic, and actual structural changes associated with myosin incorporation are not known). **G,** After 28 days, muscle mass and fiber area are decreased. **H,** After 28 days, the Z-band is the full width of a normal slow-contracting muscle, and the density of the T-system has decreased. At this point, the transformed fast-contracting muscle is indistinguishable from a normal slow-contracting muscle. (From Lieber RL. Time course and cellular control of muscle fiber type transformation following chronic stimulation. ISI Atlas Animal Plant Sci 1988;1:189–194.)

observed for normal slow muscle (Figure 4.2**D**). The amount and activity of the calcium transport adenosine triphosphatase (ATPase) decreases and changes its particle distribution within the SR bilayer (Figure 4.2**E**). This decrease in the amount and activity of the calcium ATPase can be detected physiologically as a prolonged time-to-peak twitch tension and a prolonged relaxation time of a muscle twitch, or as a decrease in the fusion frequency (Figure 4.3). Note that at every frequency shown in Figure 4.3, the stimulated leg demonstrates partial tetanic fusion due to the decrease in activity and/or amount of calcium transport proteins within the SR, while the control leg completely relaxes between stimuli. As a result, the stimulated leg tension record completely fuses at a lower frequency than the control leg. Finally, after about 4 weeks of continuous stimulation, an alteration in the myosin light chain profile is observed whereby the normally fast muscle, containing only light chains LC1f, LC2f, and LC3f, now contains light chains characteristic of slow fibers, *i.e.,* LC1s and LC2s (Figure 4.2**F**, Figure 4.4; see Chapter 2, page 25). In fact, the muscle cell is now synthesizing new myosin heavy chain molecules, which are being incorporated into the myosin filaments! Thus during this transition time, myosin filaments contain different myosin types (Figure 4.2**F**). By this time, muscle fiber cross-sectional area, maximum

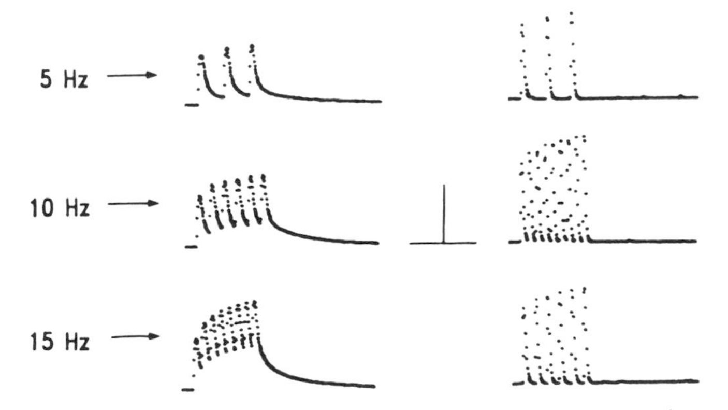

Figure 4.3. Comparison of the form of unfused tetani between a stimulated leg (left panel) and control leg (right panel) of the rabbit tibialis anterior. Note that at every frequency, the stimulated leg demonstrates partial tetanic fusion while the control leg completely relaxes between stimuli due to the decrease in activity and/or amount of calcium transport proteins within the SR. As a result, the stimulated leg tension record completely fuses at a lower frequency than the control leg. Calibration bar in center of figure. Vertical bar = 500 g, horizontal bar = 500 msec. Muscle temperature = 37.2°C. This muscle was transcutaneously stimulated at 10 Hz 1 hr/day for 5 days/week for 4 weeks. (From Lieber, 1988.)

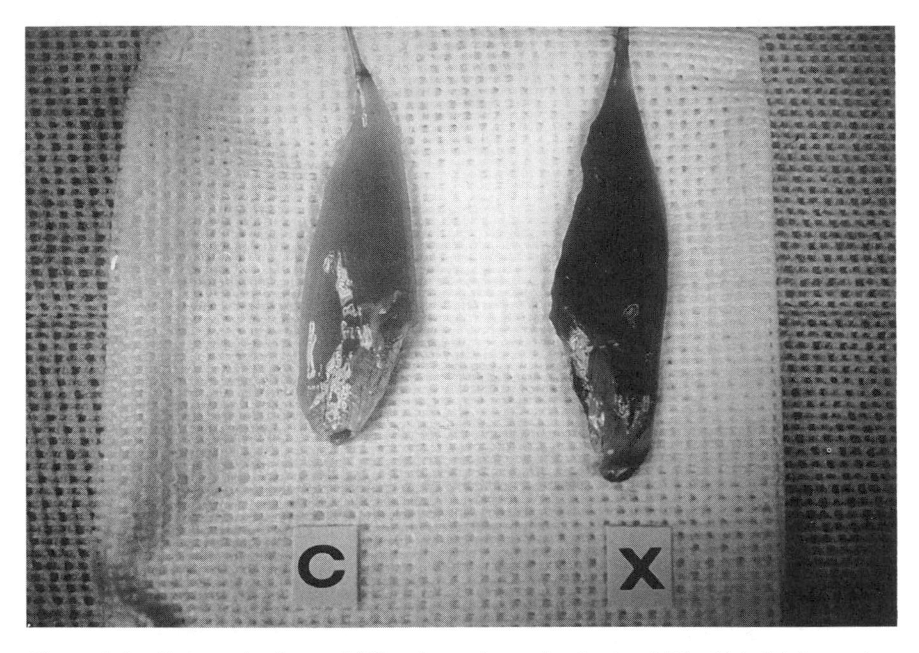

Figure 4.4. Photograph of control (*C*) and experimentally stimulated (*X*) rabbit tibialis anterior muscles, whose contractile properties were shown in Figure 4.3. Note the dark (actually deep red) appearance of the stimulated muscle, representing the dramatic capillary proliferation and increased myoglobin content.

tetanic tension, and muscle weight have decreased significantly (Figure 4.2**G**). The Z-band is now the full width of that normally observed in a slow fiber, and the density of the T-system is greatly decreased (Figure 4.2**H**). The muscle is now indistinguishable from a normal slow skeletal muscle in every respect. We conclude two things from this typical time-course of transformation:

1. Muscle metabolic enzymes, capillaries, SR, and T-system are much more easily changed than contractile proteins, and
2. while chronic stimulation does increase muscle endurance capacity it is not an effective means for increasing fiber size in normal muscle.

The decrease in skeletal muscle mass and fiber area should not be viewed as an atrophic or degenerative response. Rather, it appears that fiber atrophy represents a *deliberate* adaptive response of the muscle fiber to chronic stimulation—perhaps to decrease diffusion distances from the muscle fiber to the interstitial spaces that contain the capillaries.

Cellular Regulation of Transformation

Chronic stimulation experiments have provided insights into the cellular control of muscle fiber types. The elegantly coordinated transformation sequence suggests that chronic stimulation induces a shift in the protein synthesis and degradation machinery within the muscle cell. Transformation occurs at slightly different rates along the length of individual fibers, suggesting that the nuclei of transforming fibers do not act in a completely uniform manner (remember that we also saw that, in development, muscle fiber nuclei retained some independence, with nuclei under the neuromuscular junction preferentially expressing the acetylcholine receptor gene). For example, if a single fast fiber is stained histochemically for fiber type during the transformation, some regions will contain slow myosin and others will contain fast myosin. In fact, during transformation, many regions show *both* fast and slow myosins! In other words, individual sarcomeres are composed of different myosin types. This presents some interesting contractile possibilities that are only currently being investigated.

It is known that alterations in the amount and type of proteins present in the cell can result from numerous different mechanisms, including alterations in DNA replication (duplication of the DNA template coding for a protein), change in transcription rates (the rate at which mRNA is made from the DNA template), alteration in translation rates (the rate at which protein is translated from mRNA codes), and changes in protein degradation rates. Which of these several mechanisms operates within the transforming muscle cell? Do changes in different proteins occur by the same mechanism?

In an effort to address this question, Williams *et al.* (1986a) measured the concentration of mRNA coding for a glycolytic enzyme (aldolase) and mRNA coding for an enzyme involved in oxidative phosphorylation (cytochrome b) in muscles chronically stimulated at 10 Hz for 5 or 21 days. They documented an asynchronous change in mRNA levels coding for the two proteins. After 21 days, aldolase mRNA fell to one-fourth of control levels, during which cytochrome b mRNA increased fivefold, paralleling the observed decrease and increase of glycolytic and oxidative enzymes, respectively. However, after only 5 days of stimulation, aldolase mRNA concentration had decreased significantly but cytochrome b mRNA concentration remained unchanged. These data suggested that chronic stimulation results in reciprocal changes in the expression of the aldolase and cytochrome b genes at the level of transcription since changes in the amount of mRNA coding for the different enzymes were found. However, because of the different time courses, the transcriptional changes might have occurred by different regulatory mechanisms. This leaves open the intriguing possibility that different proteins might be controlled by different factors. As more data are accumulated, we should be able to make more general statements (see Chapter 5, page 241).

ADAPTATION TO CHRONIC STRETCH

The next model of increased use is one that is still causing great interest in the scientific community. This is the model whereby the length of a muscle is chronically changed by immobilizing a joint in a fixed position or by surgical intervention. In spite of the fact that only muscle length is changed, without concomitant muscle stimulation, this type of treatment represents an *extremely intense* increased use model since the imposed treatment is "experienced" by the muscle 24 hours a day for the duration of the experiment.

Experimental Method for Fiber Length Alteration

To chronically stretch a muscle requires joint fixation in an extreme position. For example, to chronically stretch the soleus muscle, the ankle is immobilized in a fully dorsiflexed position. Conversely, to stretch the tibialis anterior, the ankle is immobilized in full plantarflexion.

The seminal experiments on chronic stretch were performed in England by the muscle cell biologists Pamela Williams and Geoffrey Goldspink (1973; 1978). They immobilized cat ankles fully dorsiflexed to stretch the soleus muscles (Figure 4.5). The results were impressive. After only 4 weeks of immobilization, the number of sarcomeres in the soleus fibers had increased dramatically by about 20%, such that the resting sarcomere length of the stretched soleus was nearly that of the control soleus. Consider the sequence of events that occurred: The soleus muscle-tendon unit was stretched, thereby creating an increased sarcomere length. Next, the muscle, sensing this increase, *synthesized new sarcomeres* to return sarcomere length to normal. In a similar experiment, these investigators immobilized the muscles with the ankle fully plantarflexed, thus shortening the soleus and, after 4 weeks, the number of sarcomeres *decreased* by about 40%. They measured the length-tension properties of the immobilized muscles and found that they generated their maximum tetanic tension at a length corresponding to the length at which they were immobilized. It seemed that the muscle adjusted sarcomere number to reset optimum length (L_0) to the immobilization length!

The muscles immobilized in the shortened position became much stiffer, much more resistant to passive stretch. Williams and Goldspink interpreted this latter result as the muscle protecting itself from overstretch of the sarcomere. With so few sarcomeres, they felt that the increased stiffness would prevent sarcomere length from increasing to the length beyond which force

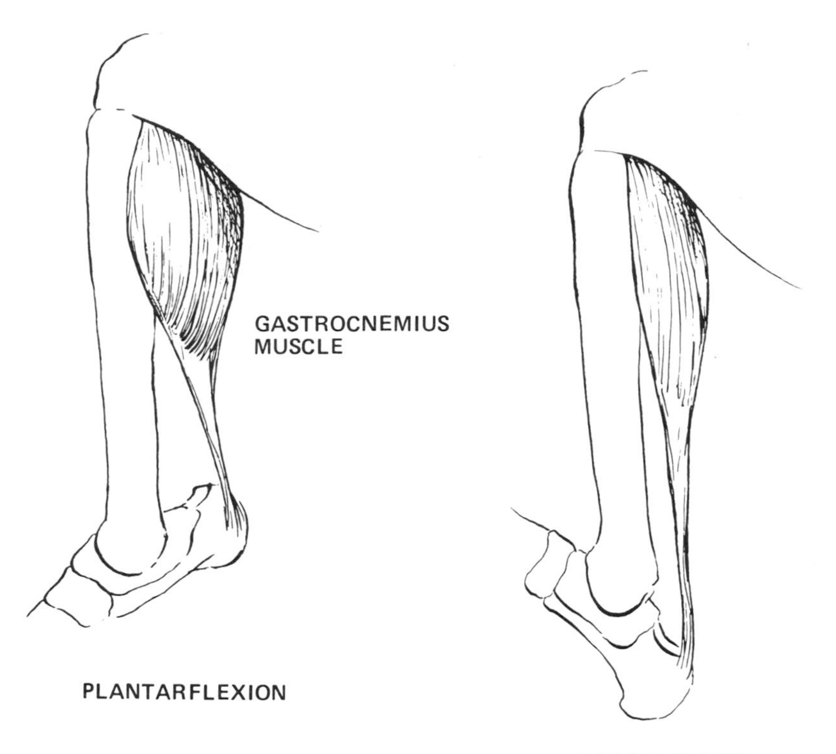

Figure 4.5. Experimental method for chronic length change of the cat gastrocnemius muscle. Ankles are immobilized either in full dorsiflexion (right) to stretch the gastrocnemius or in full plantarflexion (left) to shorten the gastrocnemius muscle. Following the immobilization period muscle, morphologic and contractile properties are determined.

could be generated. They also demonstrated that the entire process occurred even in the absence of the nerve since identical results were obtained when the nerve was severed prior to the experiment. The control center for the adaptive process was in the muscle itself.

Age-Specific Adaptation to Chronic Length Change

Another factor influencing the extent of sarcomere number adaptation was the age of the animal at the time of immobilization. Tardieu and her colleagues immobilized the ankles of young and old rabbits fully dorsiflexed and examined soleus muscle fiber and tendon lengths after 18 days in this lengthened position (Tabary *et al.,* 1972). They found, as expected, that the length of the muscle-tendon unit increased dramatically in both old and young animals. However, young animals increased muscle-tendon length by *shortening* muscle fiber length and *lengthening* tendon length while older animals increased muscle-tendon length only by lengthening their muscle fibers.

Location of Sarcomere Adaptation

A second surprising result was demonstrated in follow-up experiments by Tabary *et al.* to determine the location of the new sarcomeres added during stretch. The investigators fed the animals radioactive adenosine so that any new sarcomeres synthesized would be radioactive (adenosine is incorporated into the structural adenosine diphosphate [ADP] of the actin monomers) and thus visible using modern autoradiographic methods (a procedure whereby a film emulsion is placed over the tissue and the radioactive areas expose the film grains, causing dark spots on the tissue). Where so you suppose the newly synthesized sarcomeres were located? Throughout the tissue? At specific locations? The answer was that all of the newly synthesized sarcomeres were located at the fiber ends, near the muscle-tendon junction (Figure 4.6). Control experiments demonstrated that this change only occurred if the muscle was stretched, since immobilization with the ankle in the neutral position produced no such change (Tabary *et al.,* 1972).

In a similar type of experiment, Edgerton and his colleagues investigated the differential sensitivity of various muscles to immobilization-induced sarcomere number alteration (Spector *et al.,* 1982; Figure 4.7). Using a rat immobilization model, they compared the response of the soleus (SOL), the medial gastrocnemius (MG), and the tibialis anterior (TA) to chronic length change. Recall that the SOL is a monoarticular ankle extensor composed of primarily slow fibers, while the gastrocnemius is a biarticular ankle extensor composed primarily of fast fibers. The TA is a monoarticular ankle flexor composed mainly of fast fibers.

Spector *et al.* demonstrated that while the SOL and MG adapted in much the same way as the studies mentioned above had found (adding or subtracting sarcomeres appropriately), the TA was relatively less responsive (Figure 4.7). We thus encounter a case where all muscles are not created equal—either with respect to physiologic properties or adaptation. We will see again and again that the muscles that are often used (*e.g.,* antigravity muscles such as the SOL and MG) are often the most plastic. We will encounter this differential muscle sensitivity of muscles to adaptation again in Chapter 5 as we investigate muscle's response to decreased use.

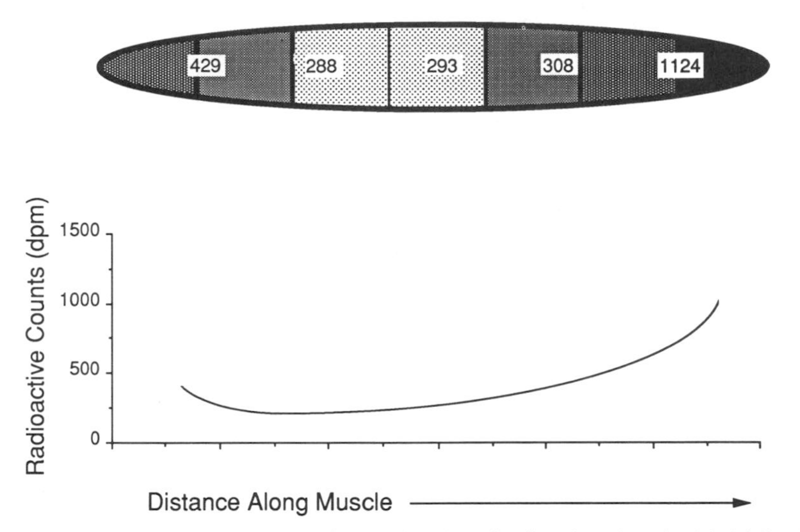

Figure 4.6. Schematic representation of spatial location of radioactive adenosine label following chronic stretch of mouse soleus muscle. Darker shading represents higher labeling (numbers superimposed on shading patterns). Most of the radioactive label was located at the fiber ends, suggesting that sarcomere addition occurred mainly at the ends of the fibers and not throughout the muscle length. Lower panel represents graphical form of the labeling density along the muscle length in counts per minute (cpm). (Adapted from Tabary JC, Tabary C, Tardieu C, Tardieu G, Goldspink G. Physiological and structural changes in the cat's soleus muscle due to immobilization at different lengths by plaster casts. J Physiol 1972;224:231–244.)

Sarcomere Type Added during Chronic Stretch

Recently, Williams and Goldspink (1986b) examined the type of sarcomeres added during chronic stretch combined with electrical stimulation. They showed that regardless of the type of myosin in the sarcomeres of the parent fiber, the sarcomeres added were always of the slow myosin type. (This was actually probably embryonic myosin which cross-reacts with the slow myosin antibody.) This observation indicates that under conditions of increased use, the muscle cell is programmed to synthesize slow myosin and create sarcomeres composed of slow myosin.

These experiments also confirmed that the "hot spot" of a muscle fiber is the muscle-tendon junction. Eisenberg and her colleagues have provided supportive data for this concept by demonstrating that muscle fiber ends have a

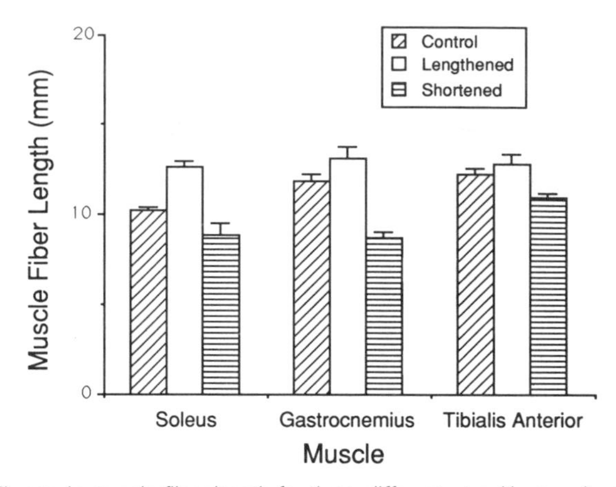

Figure 4.7. Change in muscle fiber length for three different rat ankle muscles following chronic length change. For each muscle, a group of animals was immobilized in the shortened (*striped bars*), neutral (*hatched bars*), and lengthened (*open bars*) position. Note that all fiber lengths adapted; the ankle extensors (soleus and gastrocnemius) were more adaptable than the ankle flexor (tibialis anterior). (Data from Spector SA, Simard CP, Fourier M, Sternlicht E, Edgerton VR. Architectural alterations of rat hindlimb skeletal muscles immobilized at different lengths. Exp Neurol 1982;76:94–110.)

greater concentration of mitochondria, suggesting a high metabolic activity. Numerous mitochondria provide the ATP required for protein synthesis of contractile proteins and sarcomere assembly. When muscle fiber length is chronically altered, the number of sarcomeres is adjusted to attempt to compensate for the change. The exact extent of the adaptation is not known, but it is clear, based on our extensive discussion of architecture (Chapters 2 and 3), that fiber length changes will have dramatic consequences (see below).

Eisenberg and her colleagues have also investigated the cellular control for muscle fiber elongation and hypertrophy (Dix and Eisenberg, 1990 and 1991; Kennedy *et al.*, 1988). They showed that very high concentrations of myosin mRNA were found just under the sarcolemma. Interestingly, this was the same location as many of the satellite cells (Chapter 1, page 32; Chapter 6, page 263), which were recruited into use as myonuclei to form new myotubes, myofibers, *etc.* While this type of adaption (adding new muscle fibers) may be an exceptional form of accommodating the increased load, it clearly demonstrates the impressive muscle adaptive capacity.

Effect of Intermittent Stretch on Sarcomere Adaptation

Recently, Pamela Williams addressed a question relevant to physical therapy: How much passive manipulation of the limb is needed to prevent the sarcomere number decrease observed following immobilization in a shortened position? She immobilized mouse soleus muscles in a shortened position as described above but removed the casts every day to perform passive range of motion exercises (Williams, 1988). The joints from different experimental groups were manipulated for different amounts of time ranging from 15 min/day to 3 hr/day (Figure 4.8). The joint was passed through its

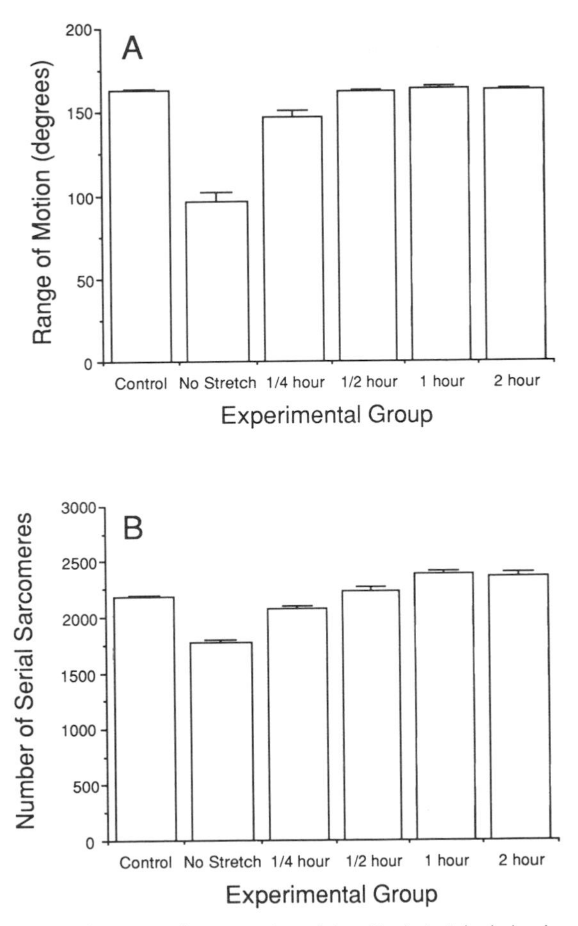

Figure 4.8. Response of mouse soleus muscle to intermittent stretch during immobilization in a shortened position. At least a half hour of stretch was required to prevent the fiber shortening seen with no immobilization or with stretch of shorter duration. **A,** Range of ankle motion following intermittent stretch. **B,** Number of serial sarcomeres following intermittent stretch for different times shown. (Data from Williams P. Effect of intermittent stretch on immobilised muscle. Ann Rheum Dis 1988;47:1014–1016.)

entire physiologic range of motion at a relatively slow rate. She found that passive stretch for 30 min/day was sufficient to prevent the dramatic decrease in range of motion (Figure 4.8**A**) and decrease in sarcomere number (Figure 4.8**B**) that was normally observed. Even 15 min/day was significantly better than nothing.

Functional Consequences of Altered Fiber Length

What if fiber length adaptation does occur in humans? What are the functional consequences? Our previous discussion of the functional significance of fiber length applies here. As we stated in Chapter 3 (page 136), if muscle fiber length increases, muscle velocity and excursion increase (assuming that excursion is not limited by the joint itself). In addition, a muscle's ability to generate force will increase at a given shortening velocity if fiber length increases (Figure 3.20). If the fiber length increase is secondary to joint immobilization, when the immobilization is removed, presumably, fiber length will return to normal. However, what if the change in fiber length is due to a surgical procedure? A fairly common procedure is the tendon transfer, whereby the insertion site of a muscle is moved to a new location to substitute for lost or impaired function. What are the consequences of such a surgical procedure?

We have simulated a tendon transfer in Figure 4.9 using a "typical" muscle with a "normal" starting fiber length. We have symbolized the muscle as a schematic sarcomere. Note that as the joint rotates from flexion to extension, sarcomere length increases and muscle force changes (Figure 4.9). The active range of joint motion is thus limited by the active muscle range, which is itself a function of sarcomere number. The total active range of motion in this example is 40°. If a surgical procedure moves the muscle insertion site such that muscle fiber length increases to 1.25, what changes would be observed? Immediately postoperatively, the muscle would obviously be at a longer length with the increased moment arm. Now, as the joint rotates, the amount of sarcomere length change per joint angle rotation increases. Thus muscle force changes over a narrower range of motion, which is decreased to only 25° in this example (Figure 4.10). Therefore, as a result of increased moment arm, active range decreases, and the joint angle at which muscle force is maximum changes (Figure 4.11). Since the joint angle corresponding to maximum muscle force is now even farther from the angle at which the maximum moment arm occurs (assumed in this case to be 90°), a torque (strength) decrease would occur simply due to the fact that the angles at which optimal muscle and joint properties occur at different angles. This is a very important point. *Weakness can be observed because of a change in*

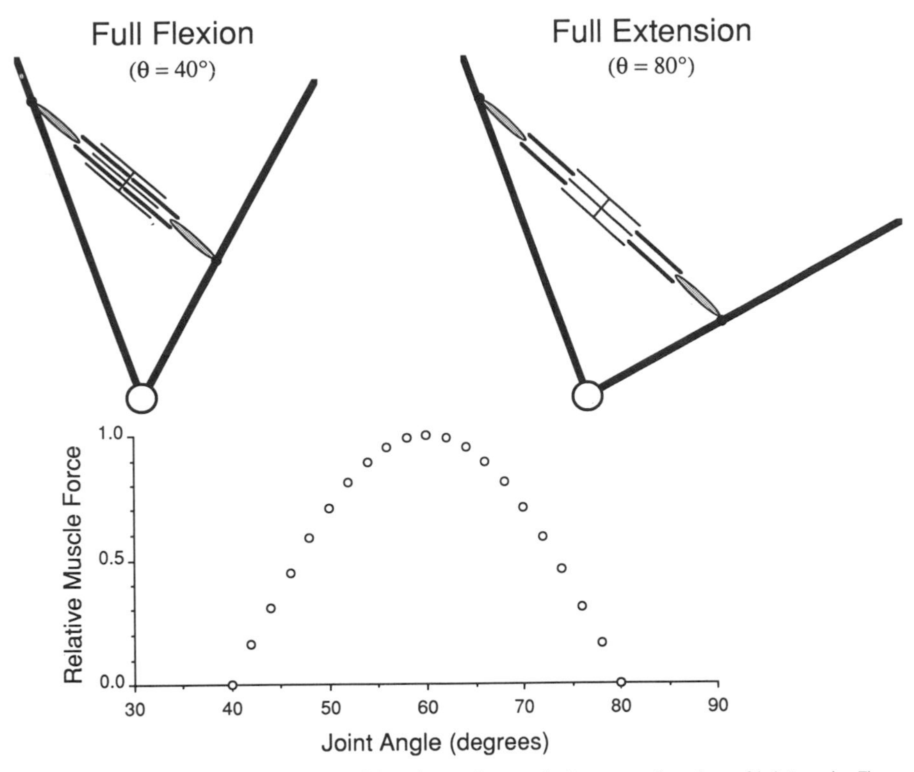

Figure 4.9. Schematic illustration of the change in muscle force as a function of joint angle. The muscle (illustrated as a sarcomere) changes from fully shortened to fully lengthened as the joint rotates from extreme flexion to extreme extension. Because of the length-tension relationship, muscle force changes as a function of joint angle. This is simulated by the sine function over a 40° range of motion.

muscle fiber length, not just a change in the muscle's ability to generate tension. This is an extension of our previous discussion which noted that strength results from the *interaction* between muscle and joint properties and not either property alone. In this case, weakness resulted from a change in the angle at which muscle generated its maximum muscle force.

Let us continue to follow the patient postoperatively. Suppose that stretch stimulates the muscle to add sarcomeres and compensate for the increased moment arm. As more sarcomeres are added, three things will happen: The subject's active range of motion will increase, torque (strength) will increase, and the joint angle at which maximum strength occurs will shift. At the end of the experiment the increase in sarcomere number could compensate for the

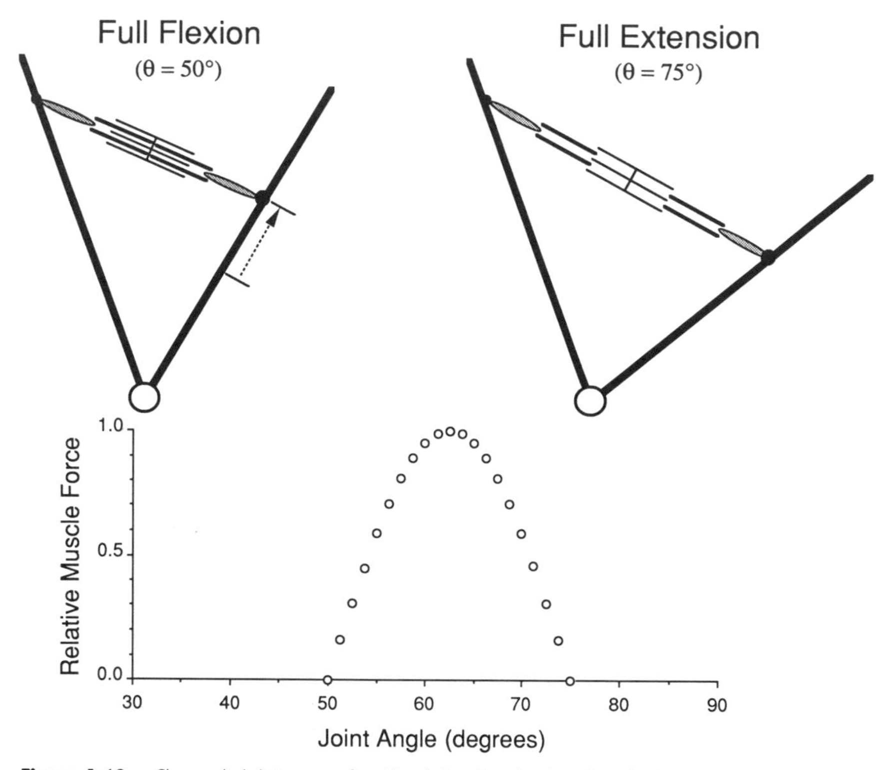

Figure 4.10. Change in joint range of motion following simulated surgical tendon transfer. Moment arm is increased (*arrow*), which causes muscle length to change more for a given rotation. Thus, muscle active range is used up over a smaller joint range, which decreases active range from 40° (Figure 4.9) to only 25°.

increase in moment arm, and the muscle-joint interaction could be normal. This scenario is good news for the surgeon—if it is true. There are still not sufficient data to demonstrate that this process occurs exactly as described here, but it is a reasonable guess based on the experimental data available. This scenario makes several specific predictions that can be experimentally tested.

If complete adaptation does not occur, it is obviously important to transfer donor muscles to new sites that are architecturally similar to the muscles whose function they are replacing. Inspection of the surgical literature reveals a relatively poor appreciation of this point. The major exception is the excellent treatise by Paul Brand *et al.* (1981), which describes the muscles and joints of the arm and hand, their interaction, and the appropriate surgical procedures for restoration of hand function following disease or injury.

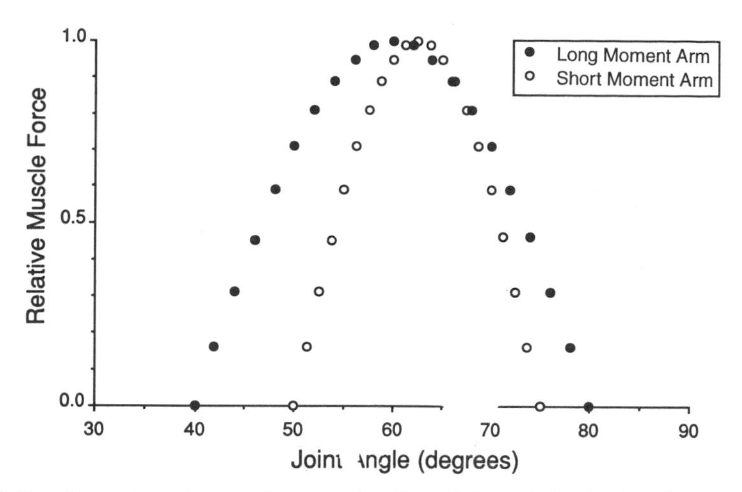

Figure 4.11. Comparison of muscle force as a function of joint angle for the short (*open circles*) and long (*filled circles*) moment arms. The increased moment arm results in decreased range of motion, as illustrated in Figures 4.9 and 4.10.

ADAPTATION TO COMPENSATORY HYPERTROPHY

Yet another means of increasing the level of skeletal muscle use is to force a muscle to work harder than normal using surgical means. Ken Baldwin and Roland Roy performed a series of experiments that defined muscle and fiber responses to compensatory hypertrophy. Their experimental model was the laboratory rat.

Rat Ankle Extensor Muscle Anatomy

By way of anatomic background, recall that the rat soleus, gastrocnemius, and plantaris muscles are the major plantarflexors in rats. (Whereas the human plantaris represents a vestigial muscle, the rat plantaris is quite sizeable.) The relative proportions of the various muscles that make up the rat ankle extensors are shown in Table 4.1. Note that the medial and lateral gastrocnemius together make up about 80% of the total ankle extensor mass, while the plantaris and soleus make up only about 15% and 5%, respectively. These muscles also vary with respect to fiber type distribution, as shown in Table 4.1. The soleus is composed of about 80% type slow oxidative (SO) fibers and 20% type FOG fibers. The medial gastrocnemius (MG) has two distinct regions, identifiable based on gross color. The red MG is highly oxidative, composed of about 50% FOG fibers and 50% SO fibers. The white MG has no SO fibers and is composed of almost all (95%) FG fibers. It is easy to appreciate, therefore, that the MG has many more fast fibers than the soleus. The plantaris

Table 4.1.
Relative Mass and Fiber Type Distribution of Rat Ankle Extensors[a]

Muscle	Relative Mass of Dorsiflexors (%)	Percent SO Fibers (%)	Percent FOG Fibers (%)	Percent FG Fibers (%)
Lateral gastrocnemius	36	—	—	—
"White" medial gastrocnemius	38	—	5	95
"Red" medial gastrocnemius		50	50	—
Plantaris	16	10	45	45
Soleus	8	80	20	

[a]Data from Roy *et al.*, 1982.

muscle is primarily composed of fast fibers (90% FG + FOG). Thus it is possible to overload any one of the ankle extensors by surgically removing one (or more) of its synergists. Following surgery, the remaining muscle(s) are forced to work harder (for 12–14 weeks) because they replace the lost function of those muscles surgically removed. Based on the variety of fiber types, architecture, and muscle sizes available in the rat, it was possible to design experiments that determined the factors important in causing the hypertrophy. Another nice feature of this model was that the remaining muscles were subjected to activation by physiologic means while the rats were performing their ordinary tasks.

Effects of Compensatory Hypertrophy on Rat Plantaris

In a series of experiments by Ken Baldwin, Reggie Edgerton, and Roland Roy, specific ankle extensor muscles were overloaded and compared to muscles from age-matched animals on which no procedures (or only surgical sham procedures) had been performed. The results provided insights into the process and basis for the adaptation observed.

In the first experiment, the distal gastrocnemius and soleus were removed bilaterally, leaving only the plantaris muscle to perform the plantarflexion function (Figure 4.12). Only about 15% of the ankle extensor musculature was initially available to the rat after surgery! What adaptations were observed in the plantaris muscle following such a dramatic surgical intervention? Amazingly, the overloaded plantaris muscles nearly doubled in size. They generated a much greater maximum tetanic tension (P_o), with P_o increasing from about 5 to 8 Newtons (Figure 4.13**A**). Maximum velocity (V_{max}) decreased significantly from about 34 mm/sec to only 18 mm/sec, with no change in fiber length, and myosin ATPase activity decreased from about 0.8 μmoles/mg/min to about 0.6 μmoles/mg/min (Figure 4.13**B**). How could that be? Biochemical analysis demonstrated that although the muscles were nearly twice their

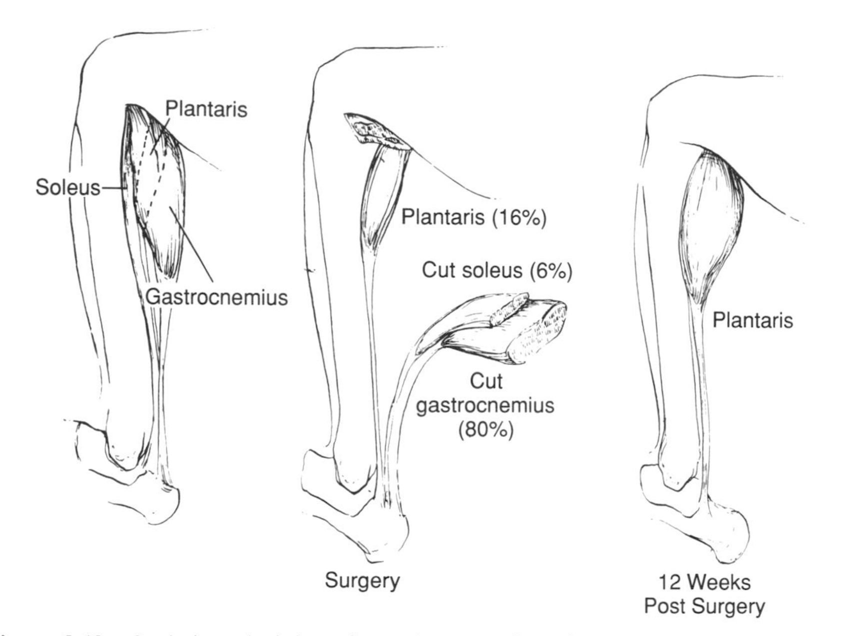

Figure 4.12. Surgical manipulation of muscles to perform the compensatory hypertrophy experiment. Muscle overload results from surgical removal or denervation of synergists. Percentage of each muscle contributing to the plantarflexors is shown in parentheses.

original size, the concentration of protein in the two groups of muscles (overloaded *vs.* control) was the same. This indicated that the muscle grew proportional to the increase in protein, which meant that the muscles were probably synthesizing new contractile filaments and not merely filling up with inflammatory cells or connective tissue. In addition to the great increase in muscle mass, the calcium transport activity of the overloaded plantaris decreased significantly compared to controls. Finally, the activities of all enzymes associated with glycolysis (phosphofructokinase, phosphorylase, α-glycerophosphate dehydrogenase, and lactate dehydrogenase [Chapter 2, page 72]) decreased significantly.

What caused the change in contractile properties? Examine Figure 4.13 to see if you can determine the nature of the adaptation. The astute reader will note that all of the changes in the plantaris tended to cause its properties to become more like a slow muscle—in fact more like that rat soleus! Baldwin *et al.* (1982) and Roy *et al.* (1982) showed that the overloaded plantaris became more like the normal soleus in every respect and that the magnitude of the conversion ranged from about 10% to 70%, depending on the property

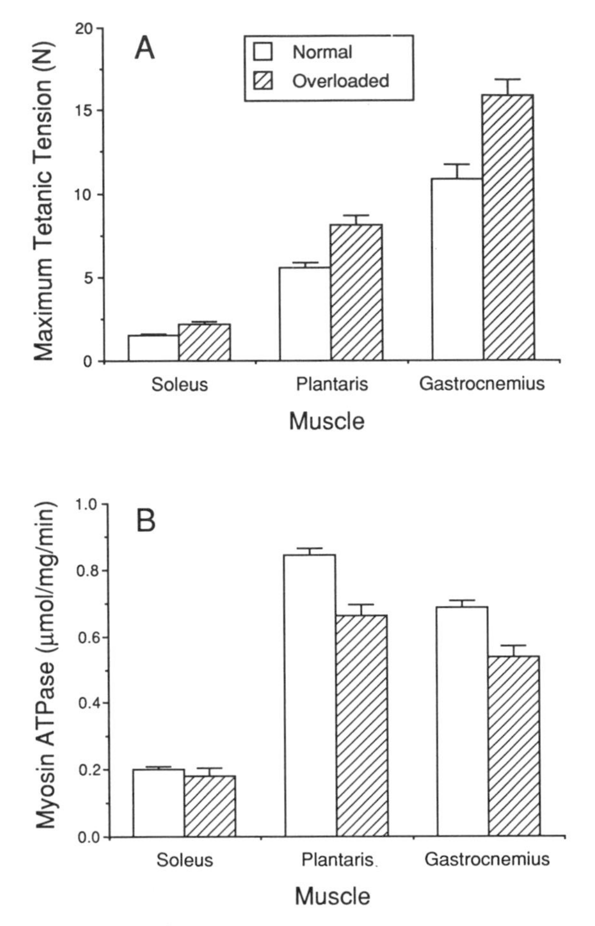

Figure 4.13. Muscle changes following compensatory hypertrophy. Properties of the soleus, medial gastrocnemius (red portion of muscle), and plantaris are shown. **A,** Maximum tetanic tension (P_O) of muscles subjected to compensatory hypertrophy. **B,** ATPase activity of muscles subjected to compensatory hypertrophy. (Data from Baldwin *et al.,* 1982 and Roy *et al.,* 1982.)

measured. In fact, the proportion of slow fibers in the overloaded plantaris increased from about 15% to about 50% in the deepest muscular region!

In some ways, then, the results of this experiment paralleled the results of the chronic stimulation experiment: Increased use tended to cause the muscle to become slower in the generic sense. In addition, at least in the overload model, the plantaris muscles became significantly stronger. The degree of the transformation (which was not complete) depended on the intensity of the adaptive stimulus. Perhaps if the overloaded plantaris were left for a year, or if a portion of the plantaris were removed, leaving only part of the muscle, it might become composed of 100% slow fibers.

Effects of Compensatory Hypertrophy on Rat Soleus and Medial Gastrocnemius

The plantaris experiments showed that 12 weeks of overloading fast muscle caused it to become much slower in every respect. What would happen following overload of a muscle that was already slow (*e.g.,* the soleus) or a fast muscle with a greater proportion of slow fibers (*e.g.,* the MG)? The same group of investigators performed analogous experiments, overloading either the soleus or the MG by surgically removing the synergists (similar to what is depicted in Figure 4.12). The results extended and helped to interpret the results from the previous experiments.

Again, the major finding was that functional overload caused the muscles to become stronger and have properties that were more like those of slow muscles (Figure 4.13). What was most interesting about the study was the relative degree of adaptation observed in these two muscles compared to the previous studies of the plantaris (Roy *et al.,* 1982; Vaughn and Goldspink, 1979). The degree of hypertrophy seemed to be related to two major factors: fiber type proportion and the relative fraction of ankle extensor complex represented by the overloaded muscle. The plantaris was the smallest muscle (15% of the ankle extensors) and hypertrophied the most (almost 100%), while the MG was the largest (60% of the ankle extensors) and hypertrophied the least (about 50%), nearly the same amount as the soleus. All fibers within the overloaded muscles hypertrophied, but the SO fibers hypertrophied the most and the FG fibers the least. Based on our discussion of muscle fiber recruitment (Chapters 2 and 3), this finding makes sense: The SO fibers were activated most often, and thus it made sense that they would experience the greatest amount of activity and tension following overload. By increasing muscle tension, we increase muscle fiber strength. Perhaps the reason the FG fibers did not hypertrophy to the same extent was that they were not activated as much due to their relatively high threshold to recruitment.

Finally, all overloaded muscles increased their proportion of SO fibers. Note that, in fact, all overloaded muscles increased their proportion of oxidative fibers, whether FOG or SO fibers, at the expense of FG fibers. A fiber type conversion in the direction FG → FOG → SO resulted from the overload. Presumably, FG fibers became FOG fibers, and given the appropriate stimulus, FOG fibers became SO fibers.

From the point of view of metabolism, all overloaded muscles demonstrated a decrease in glycolytic enzyme levels. Previously we outlined the complementary relationship between the metabolic and contractile muscle machinery. We see here that this relationship is generally maintained following muscle fiber transformation. It is not clear just why low glycolytic activity and low myofibrillar ATPase activity ought to be complementary. However, the data do suggest that there is some advantage to balancing the fiber's ability to deliver ATP under anaerobic conditions with its maximum contractile speed. We often observe that an increase in glycolytic potential parallels an increase in muscle fast fiber percentage. Similarly, when slow fiber percentage increases, glycolytic potential decreases. Overall, the balance between oxidative enzymes, glycolytic enzymes, and myofibrillar ATPase activity appears to be maintained in many different models of adaptation. New methods that allow quantitative determination of enzyme activities in single muscle fibers have been developed and provide insights into this balance. We will discuss these methods and their results as we consider several decreased use models in the next chapter.

ADAPTATION TO INTERMITTENT ELECTRICAL STIMULATION

As opposed to the chronic stimulation described above, clinical application of electrical stimulation and many experimental studies have only intermittently activated muscle. The purpose of these intermittent stimulation experiments has been to determine exactly what the muscle responds to: stimulation pulse number, stimulation frequency, total stimulation time, or a combination of one or more of these. Generally, the results are in stark contrast to those of the chronic stimulation experiments in that significant muscle *strengthening* has been documented. Why the difference?

Intermittent Stimulation with Different Activation Patterns

The Dutch neurophysiologist Dan Kernell attempted to understand the factors that influenced muscle properties by stimulating skeletal muscles using various activation patterns (Kernell *et al.,* 1987a and 1987b). He and his colleagues deafferented cat hindlimb muscles (so no pain would be felt during stimulation) and implanted electrical stimulators to activate the peroneal nerve. After 4–8 weeks, they measured the contractile properties of the peroneus longus muscle (PerL), which is normally a fast muscle. In addition, they measured dorsiflexion twitch torque noninvasively during the course of the experiment. The stimulation patterns used were designed to provide "physiologic" amounts of activity (in contrast to chronic stimulation experiments) and to mimic the activity of motor nerves normally activating fast and slow muscles (Figure 4.14). Recall that, physiologically, motor nerves

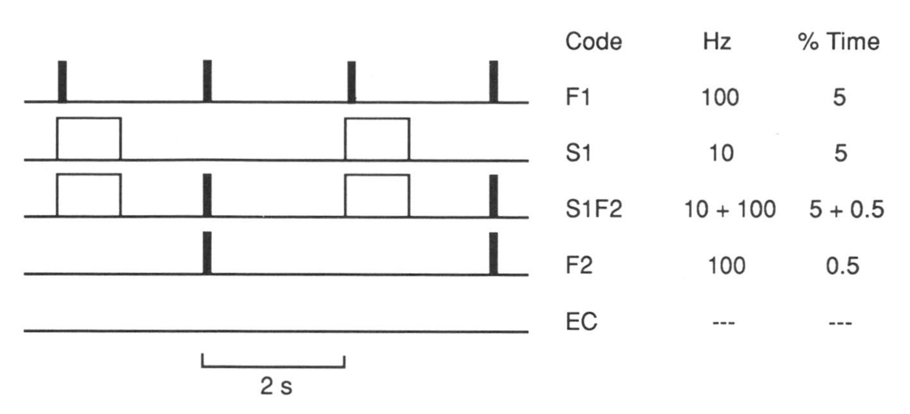

Figure 4.14. Activation patterns of muscles used to study intermittent stimulation. Muscle stimulation treatment was varied by using either 10 Hz (S1), 100 Hz 5% of the day (F1), 100 Hz 0.5% of the day (F2), or 10-Hz and 100-Hz stimulation 5.5% of the day (S1F2). EC represents experimental controls. (Adapted from Kernell *et al.*, 1987.)

innervating slow muscles tend to be activated at low frequencies relatively continuously while motor nerves innervating fast muscles tend to be activated at higher frequencies and in short bursts.

Kernell *et al.* designed the stimulation pattern to alter the total amount of electrical activity experienced by the muscles as well as the contractile tension reached during activation. Contractile tension during activation was altered by stimulating muscles at either 10 Hz or 100 Hz. Clearly, the tension reached using 100-Hz stimulation was very close to P_0 while 10-Hz stimulation would elicit only about 25% of P_0. Two experimental groups were stimulated at 100 Hz (Figure 4.14): one for 5% of the day (72 minutes of 100-msec-long trains spread evenly over 24 hours; group F1) and one for 0.5% of the day (7.2 minutes of 100-msec-long trains separated into three separate sessions spaced about 5 hours apart; group F2). Thus in both 100-Hz groups, muscles were activated to high tensions, but F1 was activated ten times more often than F2. A third group was activated at only 10 Hz for 5% of the day (72 minutes of 1-sec long trains separated into three separate sessions spaced about 5 hours apart; group S1). The final fourth group received a hybrid stimulation protocol consisting of 10-Hz and 100-Hz stimuli alternately given for a total of 5.5% of the day (superimposing the patterns of the S1 and F2 groups; group S1F2). What changes do you expect were elicited in these different experimental groups?

Maximum Tetanic Tension Resulting from Intermittent Electrical Stimulation

The P_0 of PerL muscles was measured after 8 weeks. The results demonstrated the importance of stimulation frequency *and* stimulation amount. In other words, the muscle adapted was based on the absolute tension reached during stimulation and the total number of contractions (Figure 4.15). For example, muscles from the group stimulated only 0.5% of the day at 100 Hz (group F2) generated the greatest tetanic tension (Figure 4.15**A**)! Muscles from the group stimulated 5% of the day at 100 Hz (group F1; ten times more activity but the same type of contraction as F2) generated only about half the tetanic tension compared to F2! More was not better in terms of strength! In spite of the similar *individual* contractile events between the F1 and F2 groups (*i.e.,* near-maximal contractions), the ultimate muscular adaptation was quite different. Interestingly, there was no difference in P_0 between groups F1 and S1 stimulated 5% of the day at 100 Hz or 10 Hz, respectively. However, when the F2 pattern was superimposed on S1 (group S1F2), P_0 was significantly increased relative to *either* F1 or S1! Combine these results plotted in Figure 4.15**A** with the stimulation conditions and your knowledge of muscle to see if you can sort out the way the muscle was "thinking" in response to the various stimulation protocols.

The first conclusion drawn was that high tensions were required for muscle strengthening. The only groups that demonstrated strengthening were F2 and S1F2. Note that S1 alone did not cause strengthening, whereas the addition of F2 did cause strengthening. The second conclusion drawn from the study was that increasing the amount of muscle activity tended to cause a decrease in P_0. Thus groups S1 and F1, both activated 5% of the time, showed decreased tension. However, S1F2 was activated 5.5% of the time, and P_0 still did not decrease. The fact that F1 did not cause strengthening but S1F2 did was somewhat paradoxical. It might have been that the F2 contractions, which were farther apart, caused higher tensions than the F1 patterns and thus caused the strengthening. Incidentally, wouldn't it have been interesting to have an S2 group—a group activated at 10 Hz but only 0.5% of the day—to see if decreasing total stimulation duration alone, with low tension contractions, could perhaps cause strengthening?

Twitch Kinetics and Fiber Properties Resulting from Intermittent Electrical Stimulation

Interestingly, twitch kinetics (contraction time, CT, and half-relaxation time, HRT) from all muscles in all four experimental groups were significantly increased. Thus the experimental PerL muscles all had CTs of about 30 msec

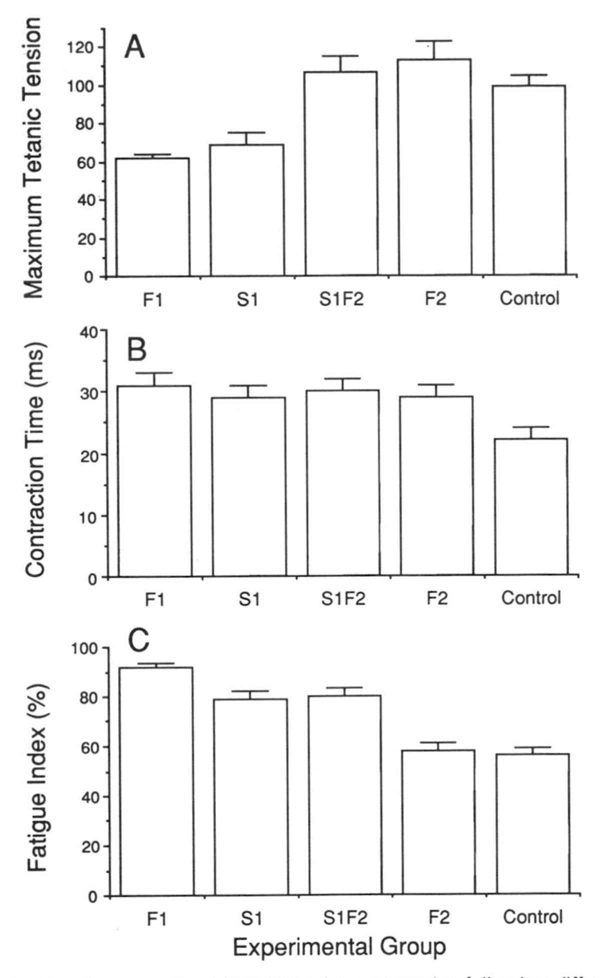

Figure 4.15. Contractile properties of peroneus longus muscles following different patterns of intermittent stimulation. The control values of the unstimulated muscle show that it was originally a fast-twitch muscle. **A,** Maximum tetanic tension (P_O) of muscles following stimulation using the parameters shown. Note that P_O was only increased if high tension contractions were elicited. **B,** Twitch contraction time (CT) of muscles following stimulation using the parameters shown. Note that CT increased for all groups studied. **C,** Fatigue index of muscles following stimulation using the parameters shown. Note that fatigue index was only increased if contractions were elicited for at least 5% of the day. (Data from Kernell *et al.,* 1987.)

compared to the normal PerL CT of 20 msec (Figure 4.15**B**). This result demonstrates that increased activity (in any "flavor," 10 Hz or 100 Hz) causes muscle slowing. The twitch response slowed after only about 10 days of stimulation and continued at a relatively constant level throughout the 8-week experiment. What was the basis for the increased CT? Since twitch kinetics were primarily determined by the rate at which the muscle can release and take up calcium (Chapter 2), the data suggested a significant loss in SR calcium transporting ability due to the loss in the amount and/or activity of the calcium transport ATPase. Again, increased use resulted in muscle slowing. Interestingly, the same observation has been made in humans following exercise training (see below).

Endurance of Muscles Subjected to Intermittent Electrical Stimulation

In order to determine the endurance level of the PerL muscles following stimulation, a fatigue test was used identical to that developed by Burke *et al.* (1973) for classifying motor units. For the fatigue test, muscles were stimulated at 40 Hz for 330 msec and allowed to rest for 670 msec. This cyclic activation pattern was continued for 2 minutes to determine fatigue index defined as the ending tension divided by the initial tension. During the fatigue test, both tension and an electromyogram (EMG) were recorded.

The results demonstrated that the muscle fatigability was dependent only on the *amount* of stimulation time and not the frequency (Figure 4.15**C**). Thus all three groups that were activated either 5% or 5.5% of the day (groups F1, S1, and S1F2) significantly increased their fatigue index to about 0.8 compared to the normal PerL, which has a fatigue index of only about 0.5! The F2 group, stimulated only 0.5% of the day, demonstrated no change in fatigue index relative to control muscles. (Again, it would have been interesting to see the results from an "S2" group, stimulated at 10 Hz only 0.5% of the time.)

The EMGs measured from muscles demonstrated that during the fatigue test performed on normal muscles, the amplitude of the EMG declined to about 50% of the initial EMG amplitude. In contrast, *all* experimental groups demonstrated no decline in peak EMG throughout the test. These results indicated that (*a*) even 0.5% daily stimulation caused significant adaptations of the nerves and/or neuromuscular junction and (*b*) that the magnitude of the EMG was not very closely related to the tension. The latter conclusion was based on the observation that the F2 and control groups had the same fatigue index but dramatically different EMG amplitudes. Even in normal muscles, an electromyogram may not be very important in determining the magnitude of the contractile tension during fatigue since force decreased with no EMG change. The neuromuscular junction or other parts of the muscle must have a built-in safety factor.

Summary of Intermittent Stimulation Experiments

We learned from these experiments that skeletal muscles respond to both the amount and type of contractile activity imposed. Greater activity is associated with muscle slowing, increased endurance, and decreased strength. High-tension contractions are necessary for increased muscle strength. This is a theme we will hear over and over as we study muscle adaptation to increased use. These data from animal studies have important implications in physical therapy, where the use of intermittent electrical stimulation is common. Such implications will be further discussed at the chapter's end.

All previous discussions apply to muscles that are innervated appropriately. This is a good time to present the results of chronic muscle stimulation under conditions where the muscles have been denervated. We will see that these results contrast sharply with those presented above. This example provides our first illustration that a denervated muscle is qualitatively different from an innervated muscle.

STIMULATION OF THE DENERVATED RAT SOLEUS MUSCLE

Several early studies of chronic stimulation were performed in order to identify the factors that controlled the expression of muscle properties. In a classic early experiment, Buller *et al.* (1960) surgically transferred the proximal portion of a motor nerve from a slow muscle to the distal stump of a motor nerve to a fast muscle. They found that the fast muscle took on the properties of a slow muscle and concluded that the nerve contained some type of chemical—a neurotrophic factor—that determined the muscle's properties. However, others contended that it was the electrical activity of the nerve, not the nerve itself, that imparted the specific properties on the muscle. To test this hypothesis, the Norwegian neurophysiologist Terje Lømo severed the nerve to the rat soleus muscle and stimulated it with various patterns (similar to the studies above).

Following severe muscle atrophy (due to the denervation), Lømo and his colleagues electrically activated the muscle using either low-frequency (10 Hz) or high-frequency (100 Hz) stimulation for a short time (1 hr/day) or long time (24 hrs/day) over a 4-week period (Lømo *et al.*, 1980). After the stimulation period, contractile properties were measured and easily interpreted (Figures 4.16 and 4.17). In all cases where the muscle was stimulated at 100 Hz, the twitch contraction times were fast and the muscle demonstrated posttetanic potentiation (independent of stimulation amount). In contrast, after 4 weeks of

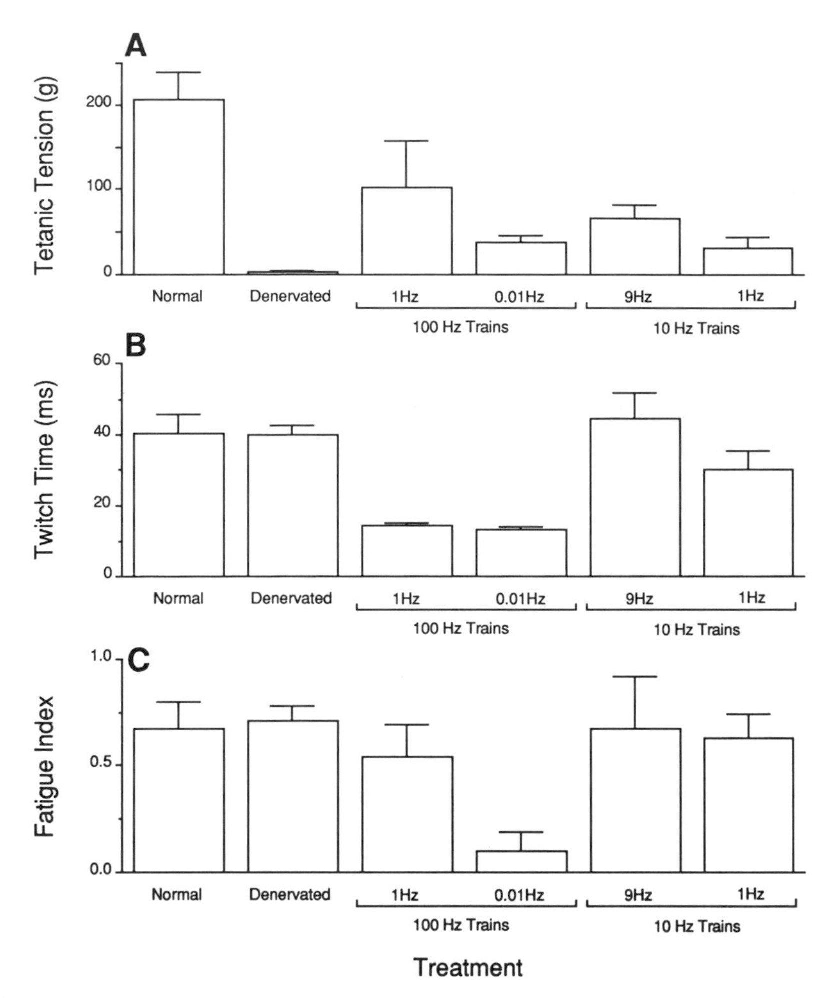

Figure 4.16. Contractile properties of denervated rat soleus muscles following chronic stimulation (Data from Lømo *et al.*, 1980.)

stimulation at 10 Hz, twitch contraction times were slow (Figure 4.16**B**), and muscles demonstrated no posttetanic potentiation (Figure 4.17). All muscles stimulated for a long time had much greater endurance than those stimulated for a short time, independent of frequency (Figure 4.16**C**). The conclusion was obvious: In the denervated soleus muscle, *speed (twitch contraction time) and endurance were independently regulated.* In other words, the muscle's speed was determined only by the frequency at which it was activated, while its endurance was determined only by the total amount of muscle activity.

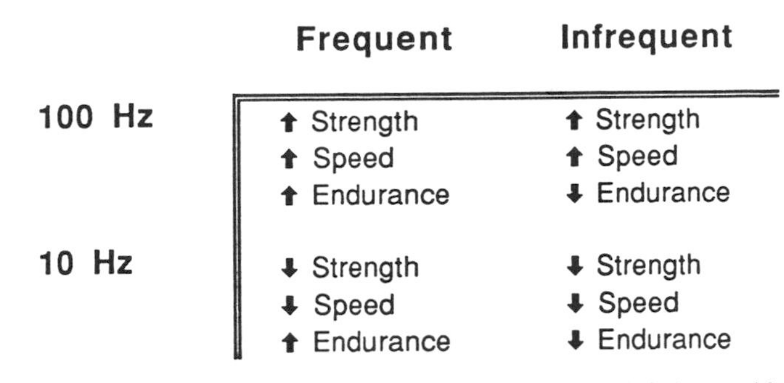

Figure 4.17. Effects of chronic stimulation on tetanic tension, twitch time, and fatigue index of the denervated rat soleus muscle. Note that if a great amount of stimulation was used (independent of stimulation frequency), endurance was high, whereas if low amounts of stimulation were used, endurance was low, suggesting that stimulation amount alone determines muscle endurance. Similarly, if high-frequency stimulation was used (independent of stimulation amount), twitch time was fast, suggesting that stimulation frequency alone determines muscle speed. (Adapted from Lømo *et al.*, 1980.)

These results were very appealing in that they seemed to provide a unifying explanation for speed and endurance regulation by the muscle cell as well as an end to the search for the elusive neurotrophic factor. Unfortunately, these same results could not be obtained for normally innervated muscle (as in the Kernell *et al.*, 1987a and 1987b experiments). This specific, predictable muscle response seemed only to hold true for the denervated soleus muscle. The data demonstrated that the control system of a denervated muscle is significantly different from that of a normally innervated muscle. It should also be noted that although the denervated muscles were able to adapt to the stimulus pattern imposed, they still generated only 10–30% of the force of the normal rat soleus (Figure 4.16**A**). These muscles should thus be considered severely atrophic. We will further consider the denervation model in Chapter 5 as a model of decreased use, but remember that denervated muscles are qualitatively different in structure and plasticity compared to normal muscles.

Comparison between Animal and Human Studies of Electrical Stimulation

Several hundred human and animal studies of electrical stimulation have been performed. It is fair to say that nearly all stimulation studies in humans or animals show increased muscle endurance following stimulation. However, while numerous human studies demonstrate muscle strengthening with electrical stimulation, almost all animal science studies show just the opposite—muscle weakening with electrical stimulation (Figure 4.2). Why the difference?

The three major differences between most animal and human studies presented to date are that (*a*) chronic animal stimulation is usually accomplished using implanted electrode systems while most human muscles are stimulated transcutaneously, (*b*) stimulation doses in animal studies are 10–1000 times greater than the doses used in human studies, and (*c*) human muscle are usually stimulated isometrically while animal limbs are allowed to move freely. Which of these, if any contribute to the disparity of results?

First, the difference is probably not related to the implantation of electrodes. Because motor nerves have a lower threshold to activation than muscle fibers (see below), during electrical activation of a whole muscle, the nerves are first depolarized, causing depolarization of the muscle fibers. Thus even with transcutaneous stimulation, the actual mechanism of muscle fiber activation is probably the same as using implanted electrodes. Second, increased stimulation causes decreased muscle force in animal studies (Kernell *et al.,* 1987a), so the fact that human stimulation doses are much lower than animal doses might be responsible for the different degrees of strengthening noted. Finally, consider the mechanical environment of stimulation (*i.e.,* isometric vs. free-moving). As we have seen, muscle strengthening depends, to a large degree, on the stress imposed on the muscle. The change in maximum force that occurs as a function of shortening velocity, given by the force-velocity equation, predicts a large difference in force developed under "isometric" vs. free-moving conditions. For example, if a muscle is stimulated isometrically at 10 Hz, the force developed is equivalent to about 30% of its P_o. Now, if the same muscle is allowed to shorten at a velocity equal to only 10% of its maximum shortening velocity (easily obtainable under physiologic conditions), the force developed is only about 18% P_o. Thus to the extent that muscle stress results in muscle strengthening, the stimulating conditions can influence the results of electrical stimulation training.

In order to reconcile the disparity of animal and human research on neuromuscular electrical stimulation (NEMS), future research on NMES application to muscle rehabilitation requires an examination of the physiologic basis for the muscular response to transcutaneous stimulation. In order to utilize the available literature on muscle plasticity, experiments must be designed such that muscle treatment force is well defined. It is crucial to understand the tension generated by human muscles during therapeutic stimulation. Numerous values have been reported for human quadriceps NMES, ranging from 5% maximum voluntary contraction (MVC) to 120% MVC. Many times the tension levels are simply not reported at all! Obviously, if

NMES is to be effective in strengthening skeletal muscle, it must be possible to activate muscles to high tensions.

Human Neuromuscular Electrical Stimulation

In light of the discussion presented above, we were interested in determining the tension generated by the human quadriceps muscles during clinically applied NMES (Lieber and Kelly, 1991). We measured knee extension torque in 150 subjects using a specially designed apparatus that fixed the distal leg to a force transducer (Figure 4.18). Carbon rubber electrodes were placed on the quadriceps musculature, and stimulation intensity was increased to the subject's maximum tolerable level. We found that it was possible, using young healthy naive subjects, to activate the muscles to about 25% of their MVC. Interestingly, this relatively low fraction of MVC *felt* like an extremely intense contraction to the subject. These patients felt like their tension levels were much higher. Why were the tensions so low? Several recent studies have suggested a number of factors that affect the tension generated during NMES.

NERVE AND MUSCLE ACTIVATION WITH NMES

To address the mechanism of human muscle activation, Hultman *et al.* (1983) stimulated the muscles from human subjects who were preparing to have surgery. Following informed consent, they first measured the maximum elicited electrically induced torque from patients under anesthesia so that pain would not limit tension levels. Then they "curarized" the patient (infused curare, a chemical agent that prevents conduction of the neuromuscular junction by binding to the muscle's acetylcholine receptor). Now, the stimulator was again activated and joint torque measured. Muscle force dropped dramatically! In fact, it was nearly impossible to electrically induce a muscle contraction. Why? The interpretation was that, with NMES, *motor nerves* were actually activated, which, in turn, activated muscles. This conjecture was reasonable in light of the fact that nerves are known to have a significantly lower threshold to electrical activation than muscle fibers. (This is one of the reasons why it is so difficult to electrically activate denervated skeletal muscle.) Since motor nerves activate muscles, perhaps the low relative forces achievable with NMES result from the fact that many motor nerves are anatomically too deep to reach with an electrical stimulator.

Factors Affecting NMES-Induced Torque Production in Humans

We attempted to further clarify the basis for the relatively low muscle activation levels achieved with NMES. The variety and number of stimulators

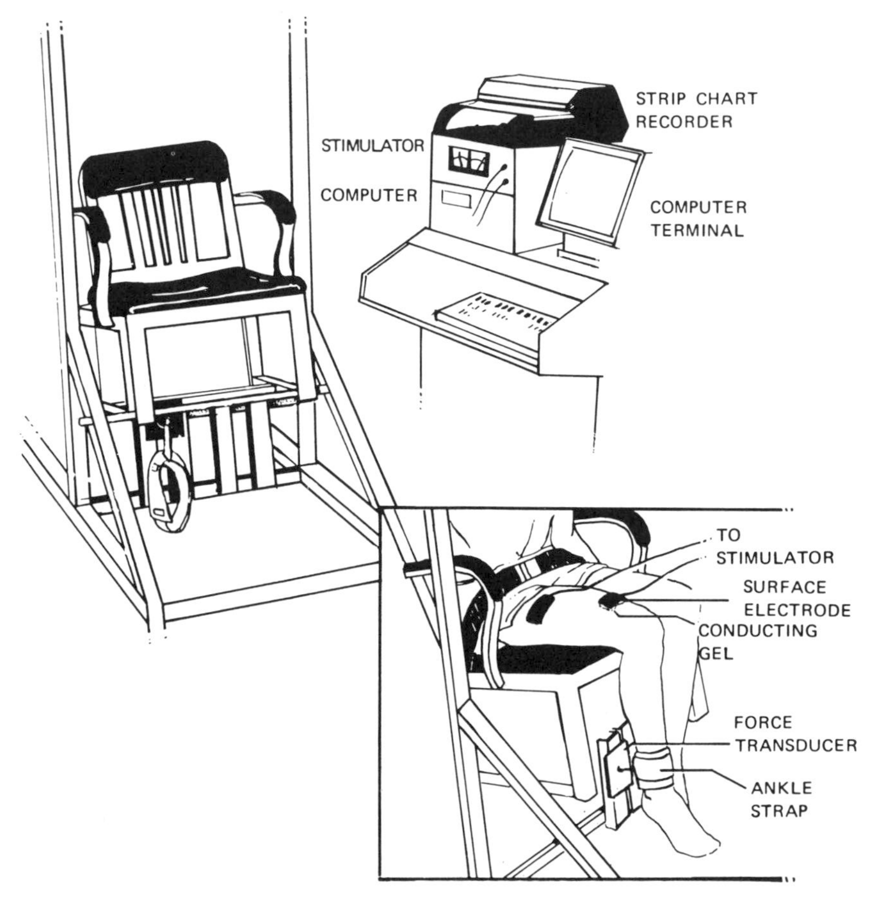

Figure 4.18. Experimental apparatus used to study NMES-induced joint torque in humans. Surface electrodes are placed over the quadriceps musculature, and the ankle is attached to a strap connected to a strain gauge (*inset*). Stimulation parameters, timing, and data acquisition are computer controlled. (From Lieber RL, Kelly J. Factors influencing quadriceps torque using transcutaneous electrical stimulation. Phys Ther 1991;71:715–721.)

and stimulating electrodes on the market suggests that there is not general agreement on the optimal method for muscle activation. What if very large electrodes are used? Would torque be higher? How about the amount of electrical current? Would that affect the maximum torque elicited by NMES? We addressed this question by using three different electrode types of different sizes and measuring stimulation current, skin resistance, and joint torque associated with the electrical activation. In 40 subjects, we randomly varied the order of electrode application, using all electrodes on all subjects.

The difficulty of this type of experiment is sorting out which of the various parameters measured is important in determining torque. This experimental design is known as a multivariate design (because of the *multi*ple *varia*bles). The appropriate analysis of this experimental design requires a stepwise linear regression. Simply stated, the experiment was designed to determine which (if any) of the many experimental variables were important in determining the magnitude of the electrically induced torque. The analysis also determined the relative importance of each parameter. We found that although torque increased with voltage, current, and electrode size (Figure 4.19), the most important factor was one we could not explicitly identify—one that was an intrinsic property of the individual (Figure 4.19**D**).

We defined a parameter called "efficiency"—the amount of torque produced by an individual for a given amount of electrical current (in units of Nm/mA, *i.e.,* units of torque/current). Some individuals (for reasons still not known to us) generated a great deal of torque for a given amount of current, while others, in spite of high current tolerance, generated relatively low torques. Taken together, all parameters measured accounted for over 90% of the experimental variability (Figure 4.20). Again, the dominant term was the subject's intrinsic activation ability. Studies are underway to clarify this issue further.

In summary, our current thoughts are that NMES can activate muscle to a relatively low percent of MVC. We do not know whether, in generating 25% MVC, 25% of the muscle fibers are generating maximum tension, if all the fibers are generating 25% of their maximum tension, or if some in-between combination is occurring. We also do not know the spatial distribution of the muscle fibers or nerves activated using NMES. Are the fibers activated in a superficial "shell" of fibers? Are fibers of a certain type activated throughout the muscle? You students and future medical researchers are responsible for addressing questions such as these.

The possibility that electrical stimulation may activate a small fraction of muscle relative to voluntary contraction raises the possibility that electrical stimulation and voluntary exercise may not be equally effective in strengthening muscle. We have seen that skeletal muscles respond to the tension imposed upon them—high tensions result in much strengthening. A review of the literature demonstrates that there is not general agreement as to whether electrical stimulation and voluntary exercise are equally effective. This may be because the tensions imposed upon the muscles are so different in the two cases.

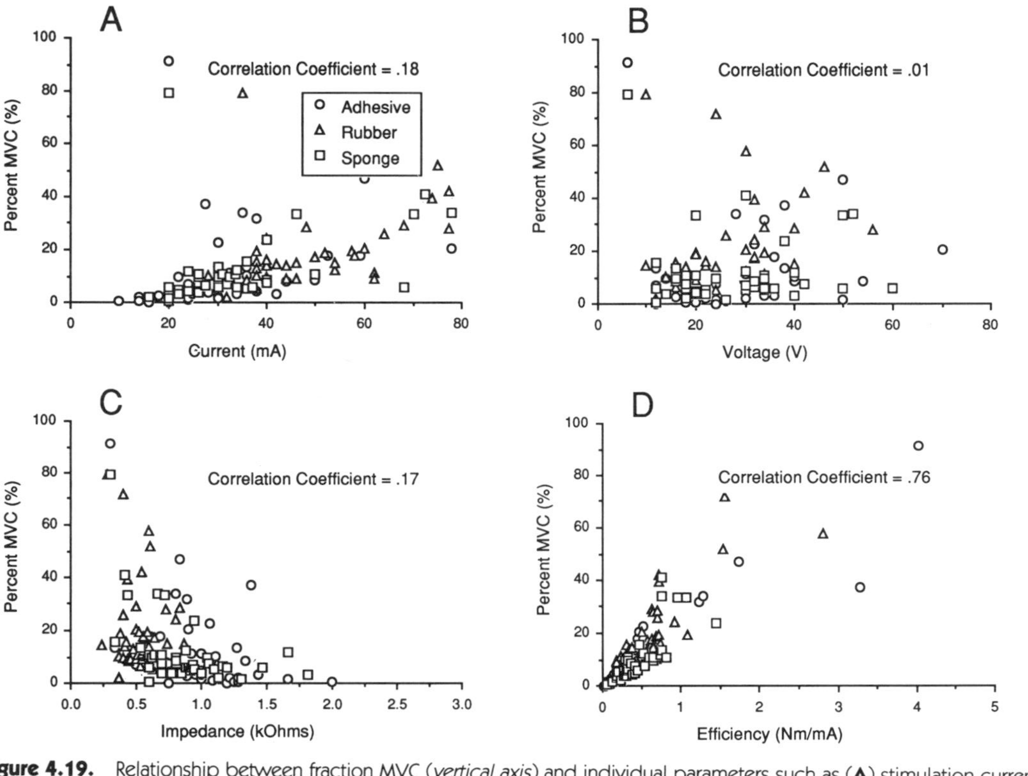

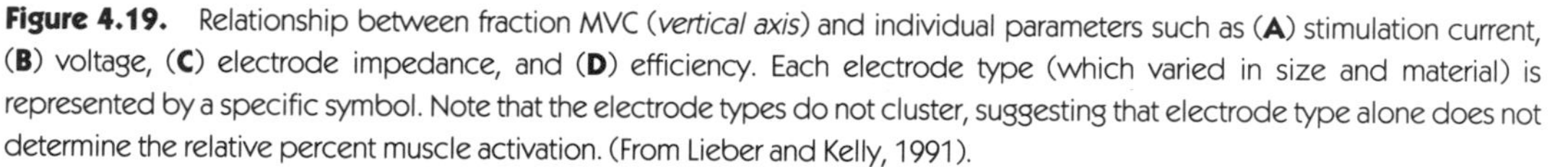
Figure 4.19. Relationship between fraction MVC (*vertical axis*) and individual parameters such as (**A**) stimulation current, (**B**) voltage, (**C**) electrode impedance, and (**D**) efficiency. Each electrode type (which varied in size and material) is represented by a specific symbol. Note that the electrode types do not cluster, suggesting that electrode type alone does not determine the relative percent muscle activation. (From Lieber and Kelly, 1991).

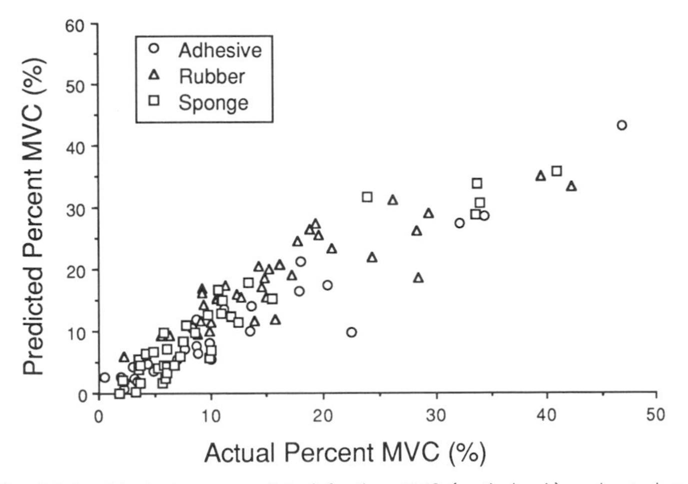

Figure 4.20. Relationship between predicted fraction MVC (*vertical axis*) and actual measured fraction MVC (*horizontal axis*) for all 120 tests. Prediction based on multivariate analysis of all parameters in Figure 4.19. Each *symbol* represents a different electrode type, as shown in the legend. Note that all three electrodes are well represented across the entire ranges of relative torque. Note also that the actual data are well approximated by the predicted data. (From Lieber and Kelly, 1991.)

In order to control for the effect of tension on skeletal muscles, we performed an experiment where muscle tension was tightly controlled (Lieber *et al.,* 1992). We randomly placed 40 patients recovering from anterior cruciate ligament reconstruction into either a voluntary exercise group or a stimulated group. While this *type* of experiment has been performed many times before, the novel aspect of this experiment was that we matched the tension imposed upon the muscles between the two groups (Figure 4.21**A**). The stimulated group was allowed to receive maximum tolerable intensity and the exercise group was matched to their tension. In spite of this careful regulation, the exercise groups still generated slightly more tension over the 4-week treatment period (Figure 4.21**A**).

Both groups of subjects dramatically increased their strength over the 4-week treatment period (Figure 4.21**B**). Interestingly, there was no significant difference in relative strengthening under these conditions where treatment tensions were tightly controlled. Of course more work is required in this area, but our provisional interpretation is that muscle strengthening with electrical stimulation is equally as effective as voluntary exercise *provided the tension levels are equivalent.*

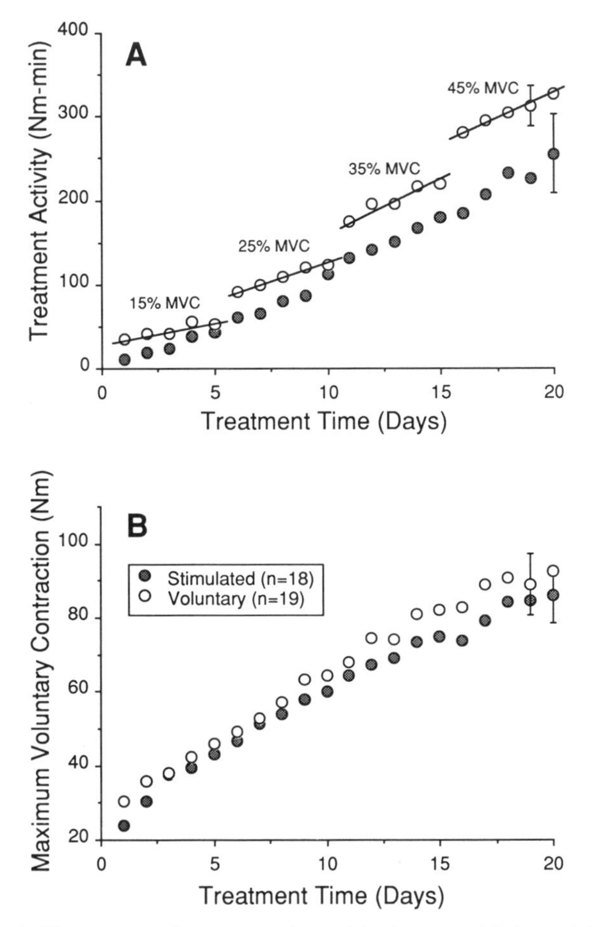

Figure 4.21. **A,** Time course of neuromuscular activity (measured in Nm-min) for two study groups. Voluntary contraction group (*open circles*) was paced at the intensity shown above the symbols. Note that exercise intensity increased during the treatment period. Stimulation group was exercised at the maximum tolerable level. At the end of the treatment period, there was no difference between groups in the total treatment activity (symbols are mean ± SEM for n = 40 subjects). **B,** Maximum voluntary contraction level (in Nm) for experimental subjects during the time course of treatment. Note that both groups gained strength at the same rate and to the same extent. (From Lieber RL, Silva PD, Daniel DM. Electrical and voluntary activation are equally effective in strengthening quadriceps muscles. Trans 38th Orthop Res Soc 1992, in press.)

ADAPTATION TO EXERCISE

All of the models presented above have as an advantage, that they are relatively well-defined. However, we know that muscle adaptations occur simply as a result of exercise. The literature (and magazines) is replete with examples of the many exercises that increase strength and stamina. Is it possible to apply the principles we have already learned to the more physiologic types of increased use normally experienced by muscle? The answer is, fortunately, yes! In fact, based on what you already know, I bet that you can predict the themes we will reiterate in the context of exercise. What are they? The answer: Muscle tension and total activity dictate the nature of the adaptation seen in exercise. In this section we will investigate the changes occurring in muscle and in performance in response to exercise. We will reinforce the concept that skeletal muscle responds to the amount and type of activity that is imposed upon it. The added nuance of training studies is that muscle fiber recruitment is an important variable. For example, in sprint training a greater fraction of the quadriceps muscles are activated compared to slow running or swimming. The response of the muscles as a *whole* might be quite different between the two training protocols. However, much of the difference can be ascribed to differences in recruitment, not to differences in an individual fiber's response to training.

Definition of Exercise Intensity

Comparisons between various exercises often distinguish between endurance exercises and strengthening exercises. However, this can be largely a matter of semantics. In most exercise studies (as opposed to the studies described above), it is very difficult to define precisely the muscular conditions that cause the adaptation. Thus we generally refer to the type of exercise in terms of the *amount* of exercise (minutes per day × days per week × total weeks) and make some reference to exercise *intensity* (percent of maximum voluntary contraction, for example, in isometric exercise). We must also make the disclaimer that we might not really know what is happening at the muscular level and think that most people perform exercise in a similar manner.

Ideally, we would like to know the force and length history of a muscle during exercise in order to attempt to provide an explanation for any adaptations seen. We showed in Chapter 2 that a knowledge of muscle length and velocity allows prediction of muscle tension. We would also like to know which of the various fiber types are activated at which particular frequency and for how much time. However, in noninvasive studies (and in relatively few invasive studies!) such information is simply not available.

Exercise Intensity in Terms of the Maximal Rate of Oxygen Consumption

One index that describes exercise intensity is the maximal rate of oxygen consumption (the chemical abbreviation for the oxygen molecule is O_2), which we abbreviate as $\dot{V}o_2$max. $\dot{V}o_2$max can be expressed in absolute units of liters of oxygen consumed per minute (liters/min or liters·min^{-1}) or in units that are normalized to the size of the person as milliliters of oxygen consumed per kilogram body mass per minute (ml/kg/min or ml·kg^{-1}min^{-1}). Because metabolic rate during exercise is primarily determined by muscle metabolism, Vo_2max represents the maximum rate at which oxygen can be used by the muscles of the individual. Clearly, it says nothing about the anaerobic capacity of an individual and therefore should not necessarily be closely linked with performances not requiring oxygen. However, the greater the aerobic capacity of the individual's muscles and the greater the ability for the cardiovascular system to deliver oxygen, the greater will be $\dot{V}o_2$max.

It is not possible to exercise near $\dot{V}o_2$max for very long. In fact, the relationship between exercise intensity (in terms of $\dot{V}o_2$max) and exercise duration is well established (Figure 4.22). For example, if an individual exercises at an intensity of about 80% $\dot{V}o_2$max (very intense exercise), he or she will only be able to exercise for about 10 minutes before exhaustion. However, if the exercise intensity is relatively low (20%–30% $\dot{V}o_2$max) the exercise can last much longer.

Exercise Intensity in Terms of Maximum Voluntary Contraction

A second index often used as a standard for exercise intensity is the percent effort compared to a MVC. The MVC is usually performed at a fixed joint angle in order to prevent complications having to do with inertia and kinetics. The important (and unlikely) assumption in using the MVC is that there actually exists such a measurable entity as MVC in humans. In fact, there is excellent evidence that the MVC, in contrast to $\dot{V}o_2$max, is not at all a measurable entity. In the first place, there is a great deal of histochemical and electrophysiologic evidence that when a person voluntarily activates his or her muscles to "maximum," that not all muscle fibers are activated (Enoka and Fuglevand, 1991). Current estimates regarding the fraction of muscle fibers activated during an MVC range from 50% to 80%. Future studies are required to determine the exact nature of physiologic muscle activation. Suffice it to say that although MVC is used as a standard by which to compare intensities between individuals, it is not an absolute standard that represents the muscle's maximum contractile ability.

Muscle Fiber Activation during Exercise at Varying Intensities

Which muscle fibers are activated under different exercise conditions? Technically, it is difficult to quantify muscle fiber activation during exercise. Even if muscle biopsies are taken immediately after exercise, only a small portion of the muscle is sampled, and it is not possible to sample repeatedly the same region of the same individual. However, using methods similar to those used in motor unit identification, exercise physiologists have demonstrated, as expected, that different muscle fiber types are activated at different exercise intensities and durations.

For example, Phil Gollnick, Bengt Saltin, and their colleagues performed a series of studies in which they obtained small muscle biopsies from different subject's vastus lateralis muscles following exercise of various intensities and durations (Figure 4.22; see summary in Saltin and Gollnick, 1983; Gollnick *et al.,* 1972). By staining the biopsies for glycogen, they identified the muscle fibers activated during exercise. (As with motor unit identification, the potential problem was that they might not be able to deplete the SO muscle fibers of glycogen since they rely primarily on aerobic metabolism.) Using a bicycle ergometer, they exercised subjects at intensities ranging from 31% to 85% $\dot{V}o_2$max for durations ranging from 12 to 180 minutes (of course, long durations could only be used at low intensities). They found that at an exercise intensity of about 30% $\dot{V}o_2$max, mostly SO fibers were used to perform the exercise. Only after about 3 hours did FOG fibers begin to be used significantly at this low intensity (Figure 4.22). However, at a higher intensity of 75% $\dot{V}o_2$max, after 20 minutes SO fibers and some FOG fibers were recruited, and after 2 hours of exercise at 75% $\dot{V}o_2$max, all three fiber types were recruited. At 85% $\dot{V}o_2$max, all three fiber types were recruited after only 12 minutes! These studies support (but do not prove) the orderly recruitment of muscle fibers during exercise and demonstrate increased fiber recruitment at longer durations and higher exercise intensities.

Recall that the spike-triggered averaging method was also used in humans to measure motor unit tension as a function of recruitment order. At low voluntary exertion levels, motor units with small twitch forces were recruited. At higher exertion levels, high-force motor units were recruited. If we combine these results with the observation that slow motor units tend to develop lower tensions than fast motor units, we arrive at the implication that even in humans, motor unit recruitment occurs in an orderly fashion, as originally proposed by Henneman *et al.* (1965), with slow motor units being recruited at low exertion levels and faster motor units being recruited at higher exertion levels.

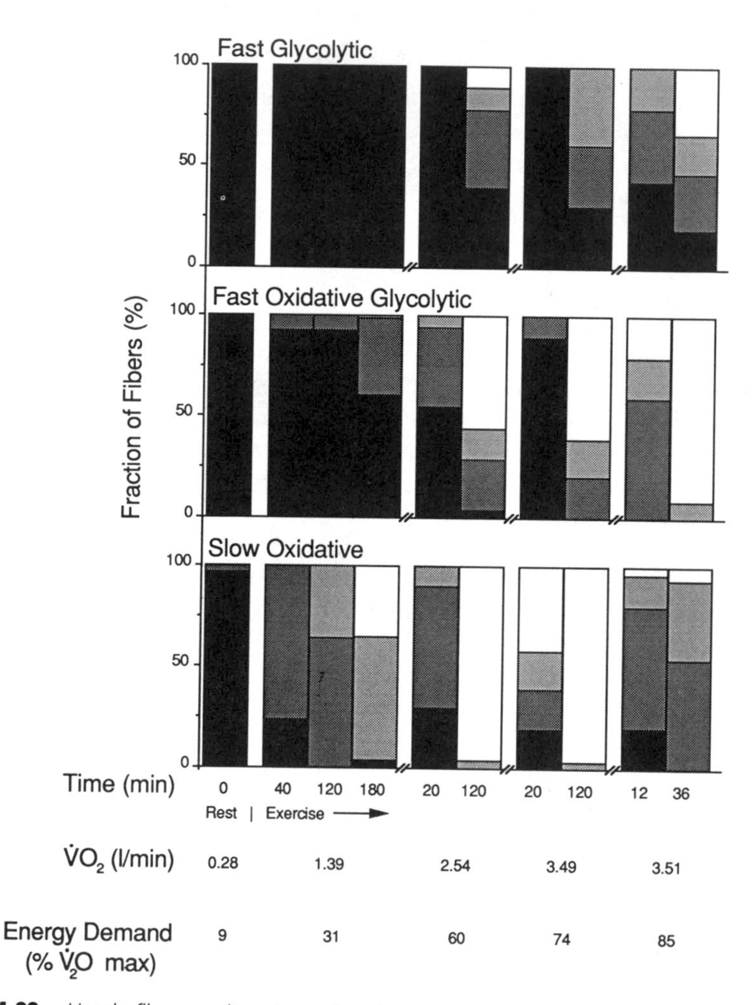

Figure 4.22. Muscle fiber recruitment as a function of exercise intensity and duration. Shading patterns represent the amount of glycogen in the muscle fibers. Filled pattern represents fibers full of glycogen, while open bars represent glycogen-depleted fibers. Other patterns represent intermediate glycogen contents. Note that as exercise intensity and duration increase, muscle fibers are recruited in the order SO → FOG → FG. (From Saltin B, Gollnick PD. Skeletal muscle adaptability: significance for metabolism and performance. In: Peachey LD, ed. Handbook of physiology. Bethesda, MD: American Physiological Society, 1983:539–554.)

Having given the disclaimers regarding the difficulties in measuring exercise intensity and muscle activation mechanism, what are some of the muscular changes that have been demonstrated during exercise? We will see that in many ways, these adaptations represent special cases of the experiments discussed above: Exercise causes muscle slowing, increased endurance, and, given the appropriate tensions, muscle strengthening.

Oxidative Enzyme Adaptation to Exercise

One lesson we learned was that typical early muscular changes are increased oxidative metabolism. Does this also occur in humans? The Swedes Jan Henriksson and Jan Reitman (1977) demonstrated that it does. They trained 15 college males in bicycle ergometry for 1 hr/day for 3 days/week at 45% $\dot{V}o_2$max for 8 weeks. They then allowed the subjects to detrain for an additional 6 weeks. At weekly intervals, muscle biopsies were taken from the subjects' vastus lateralis, and the enzyme activity of succinate dehydrogenase and cytochrome oxidase (two oxidative enzymes, Chapter 2) were measured along with the subjects' new $\dot{V}o_2$max. They found that oxidative enzyme activity increased significantly within about 3 weeks and was paralleled by increases in $\dot{V}o_2$max (Figure 4.23). For the next 5 weeks of training, muscle enzyme activity continued to increase, but $\dot{V}o_2$max increased at a much slower rate. The authors interpreted the data as saying that while $\dot{V}o_2$max and muscle oxidative enzyme activity are *correlated,* they are not *causally* related. Clearly, in spite of a great capacity for oxidative metabolism in the subjects' muscles after 3 weeks of training, the body did not utilize this entire capacity. Perhaps $\dot{V}o_2$max was limited by the ability of the cardiovascular system to deliver oxygen, or perhaps muscle oxidative enzyme activity was not simply responding to the tissue oxygen debt.

This experiment again cautions against making performance statements based only on muscle properties. The cardiovascular system, nervous system, and musculoskeletal system are all involved in performance. Coordination and complementation are generally the rule; however, it is unwise to make global performance statements based on analysis of only one of these systems.

In detraining, when the subjects stopped exercising, $\dot{V}o_2$max and oxidative enzyme levels dropped preciptiously—almost twice as fast as they had increased! This is bad news for those who attempt to stay in shape in that it is faster to get out of shape than to get in shape.

Glycolytic Enzyme Adaptation to Exercise

In contrast to increases in muscle oxidative capacity, training does not appear to increase muscle glycolytic capacity (if anything, it tends to decrease). It is

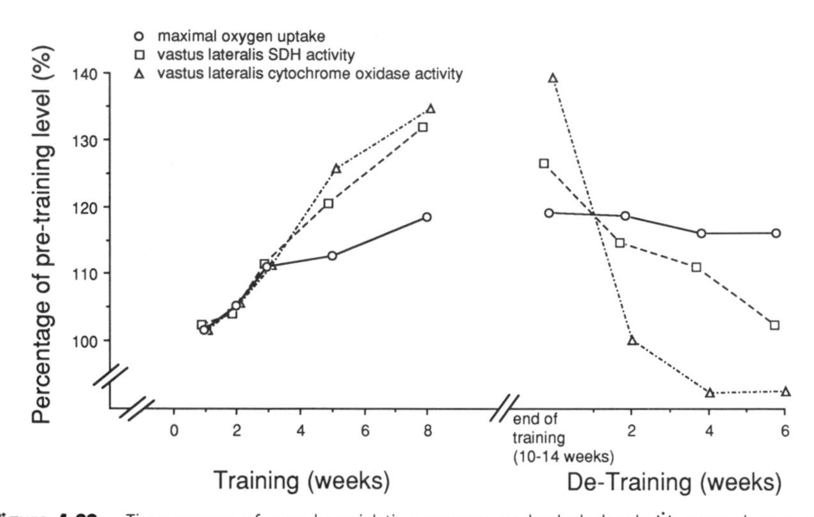

Figure 4.23. Time course of muscle oxidative enzyme and whole-body $\dot{V}O_2$max change after exercise and detraining. Note that early in the training period, $\dot{V}O_2$max and oxidative enzyme activity increase at similar rates. After about 3 weeks, enzyme activity increased faster than $\dot{V}O_2$max. Also note that detraining occurs faster than training. (From Henriksson and Reitman, 1977).

difficult to generalize across all training studies since the amount and type of exercise differs considerably between studies. It is thus not surprising that experimental results are also conflicting. In studies where the training intensity is at or close to $\dot{V}O_2$max, increases in glycolytic enzymes have been documented. For example, Gollnick *et al.* (1972) trained young males for 5 months by having them pedal a bicycle ergometer for 1 hr/day for 4 days/week at an intensity of about 75% $\dot{V}O_2$max (of course, initially, the subjects could not pedal this long at this intensity, but at the study's end, they were pedaling for *1 hour* at 85%–90% of their $\dot{V}O_2$max!).

After 5 months, the subjects increased their $\dot{V}O_2$max by 13%. As in the previous example, SDH activity increased much more—by almost 100%—while glycolytic activity (using phosphofructokinase, PFK) also increased by about 100%. Inspection of biopsies from the various subjects revealed that while both fast and slow muscle fibers appeared to increase their SDH activity, only fast fibers increased their PFK activity. This coincided with an increased muscle glycogen content.

More recent quantitative studies of muscle metabolic activity during different models of transformation have shown that muscle metabolic enzymes generally maintain a certain relationship even after adaptation. We will discuss this concept further in the next chapter while discussing muscle adaptation to spinal cord isolation.

Fiber Type Changes Following Exercise

We know already that, given the appropriate stimulus, muscle fiber type transformation can occur. Numerous exercise studies have claimed emphatically that muscle fiber type cannot change, and just as many, just as emphatically, have claimed that fiber type transformation does occur. Neither set of reports should surprise us based on our understanding of the way in which muscle fibers adapt. If the exercise stimulus is relatively low, increased oxidative capacity will occur with no change in fiber type. However, at very intense exercise levels, we would not be surprised to observe a fast-to-slow muscle fiber type transformation. Such transformations have been reported/debated in the literature, and precise definitions for the conditions under which they occur are lacking. My bias is that even if such transformation does occur, its influence on performance is relatively small compared to intrinsic architectural and kinematic musculoskeletal properties. Muscle fiber types are clearly *related* to performance but are not the cause for performance at a particular level.

Muscle Strength Changes Following Exercise

Changes in muscle strength are easily documented following many types of exercise. While changes in the nervous system can also account for performance changes (see below), changes in muscle strength are a direct result of muscle fiber hypertrophy. While selected cases of muscle fiber "splitting" have been documented in the literature, by far the main mechanism for increasing muscle strength is increasing fiber size (hypertrophy), not fiber number (hyperplasia).

Which muscle fibers increase in size following exercise? To answer this question, remember that for a fiber to increase in size, the tension imposed upon it must be high. Therefore, any time fiber tension is high, it will hypertrophy. Whether or not a given fiber type will hypertrophy will depend on whether it is recruited during the particular training regimen and to what tension.

For example, endurance training protocols are usually performed at low intensities, during which mostly SO and FOG fibers are used. Since the tensions are relatively low, fiber hypertrophy is small, and no strength increase is detected. With power lifting, on the other hand, most fibers are recruited and tension levels are very high. As a result, muscle fibers hypertrophy, and muscle strength increases. It is obvious, therefore, that exercise prescriptions ought to be considered in terms of muscle fiber recruitment and muscle fiber tension. In this way, specific adaptations can be more easily predicted.

Evidence for Exercise Specificity

Having documented muscular changes occurring during exercise, let us return for a moment to consider performance changes following exercise. A number of exercise training studies, taken together, support the idea of *exercise specificity,* which states that exercise benefits are *specific* to the method and quality of training.

Cross-sectional studies (studies performed across different populations of individuals) allude to the concept of training specificity—that measures of performance are specific to training mode. Stromme *et al.* (1977), in a study of specifically trained athletes, demonstrated that the highest $\dot{V}o_2max$ measured for each athlete was measured in the sport for which that athlete was specifically trained. For example, elite cross-country skiers yielded a ski-$\dot{V}o_2max$ that was significantly higher than their $\dot{V}o_2max$ measured during running. Similarly, rowers and cyclists attained significantly higher $\dot{V}o_2max$ values while rowing or cycling compared to treadmill running.

In addition to the information gathered from cross-sectional studies, several investigators have addressed the phenomenon of training specificity by studying the performance of a group of individuals before and after exercise training (a longitudinal study since the same group is tested before and after training). In a study involving young male swimmers, Magel *et al.* (1975) utilized interval swim-training for 1 hr/day for 3 days/week for 10 weeks. Swim- and run-$\dot{V}o_2max$ were measured before and after training. While the swim-trained individuals significantly increased their swim-$\dot{V}o_2max$, there was no significant change in run-$\dot{V}o_2max$. Similarly, in a study from the same laboratory, Magel *et al.* (1978) found that 10 weeks of arm training significantly increased $\dot{V}o_2max$ measured during arm-crank ergometry but had no effect on $\dot{V}o_2max$ measured during treadmill running. How can these results be explained based on the concept of muscular training previously presented?

Physiologic Basis of Exercise Specificity

It appears that support for the concept of training specificity is found when peripheral (muscular) adaptations occur without significant accompanying central (cardiovascular) adaptations. We must keep in mind that performance is based on a complex interaction between systems and cannot be attributed to one system alone. Differential peripheral *vs.* central adaptation arises when a relatively small muscle mass is trained and, therefore, total metabolic demand is insufficient to cause significant central cardiovascular adaptation. For example, in the study cited above, Magel *et al.* (1978) trained subjects in one-arm ergometry at about 85% of each subjects one-arm ergometry-$\dot{V}o_2$max. They then measured a significant increase in one-arm ergometry-$\dot{V}o_2$max but no significant increase in treadmill-$\dot{V}o_2$max. These results suggest that the one-arm ergometry training is insufficient to elicit a significant central cardiovascular training adaptation. Thus it is apparent that exercise training *can* result in significant adaptations that occur at the peripheral level alone. Under these conditions, testing in the mode that exercises the trained peripheral musculature will yield significant improvements. However, testing in modes that do not exercise the trained musculature will show less improvement due to the lack of central cardiovascular adaptation. Presumably, if the exercising muscle mass were to provide sufficient cardiovascular stress, adaptation would be expected to occur at both the peripheral and the central levels.

The difficulty in *not* demonstrating exercise specificity comes from the fact that in different exercise modes, oxygen consumption values differ. Thus the relative cardiovascular stresses between exercise modes (say, running *vs.* swimming) differ, resulting in specificity demonstration. If absolute exercise intensities could be made to be exactly equivalent, presumably specificity would not be required.

Support for the *lack of specificity* was recently provided by Debbie Lieber in a study comparing run-training and swim-training *at the same absolute intensity* (Lieber *et al.,* 1989). She had to overcome several technical obstacles to force runners and swimmers to exercise at the same absolute intensity (for example, it is well-known that swimming can cause heart-rate variations simply due to body angle and facial immersion in water). After correcting for these factors, and exercising 20 young men for 1 hr/day for 3 days/week for 12 weeks at an intensity equivalent to 75% of their *treadmill* $\dot{V}o_2$max, she found that both swimmers and runners increased their treadmill $\dot{V}o_2$max to the same extent, by about 25%!

This study leads to the hypothesis that if different groups of subjects are trained in different exercise modes at a identical absolute intensities and in modes (or at intensities) sufficient to cause central cardiovascular adaptation, and then tested in a mode eliciting sufficient cardiovascular demands, training specificity will not be demonstrated. This notion does not deny the existence of training specificity but, instead, merely places numerous restrictions on the conditions under which training specificity *will not* be demonstrated. In addition, it emphasizes the importance of choice of training mode, training intensity, and testing mode on the interpretation of results from training studies.

Thus specificity is not a magic aspect of exercise but a reaffirmation that performance results from interaction between numerous physiologic systems and not a single system alone.

Neural Adaptation to Exercise

It is clear that exercise training does alter cardiovascular and muscular properties. Let us now consider the final component of the performance system—the nervous system. Obviously, performance indices such as strength depend on the level of neural activation. Is it actually possible for the nervous system to adapt to exercise? If it occurs, is it significant?

An illustrative experiment was performed by Moritani and Devries (1979), who trained young men and women in elbow flexion. Elbows were maintained at a joint angle of 90°, and subjects were asked to perform the following isometric elbow flexion protocol: 10 repetitions at an intensity of two-thirds MVC twice daily three times a week for 8 weeks. Note that this was not an extraordinarily long or intense training protocol.

At 2-week intervals during the training period, subjects' MVCs and EMGs were measured. At the end of 8 weeks of training, the trained arms increased strength almost 25% (from 58 to 79 pounds). However, the *untrained* contralateral limb also increased in strength by about 15% (from 54 to 67 pounds). How could these results be explained? How could the untrained muscle, which was never exercised, increase in strength?

In order to investigate the basis for the strength change in both trained and untrained arms, Moritani and Devries developed a method to determine whether strength changes were due to muscular and/or neural factors. These methods could easily be used today in strength evaluation, so let us explain the methods before proceeding with the experimental results.

METHOD FOR EVALUATING NEURAL AND MUSCULAR COMPONENTS OF STRENGTH CHANGE

By measuring both EMG and MVC, it is possible to "decompose" strength changes into neural and muscular factors. Figure 4.24 presents a scheme used to perform the decomposition. If strength increase is due only to increased level of neural activation, this will be manifest as an increased force that is directly proportional to the increased EMG (Figure 4.24**A**). If, however, the strength increase is due only to muscle hypertrophy, this will be manifest as an increased force with no increase in EMG (Figure 4.24**B**). Finally, if the

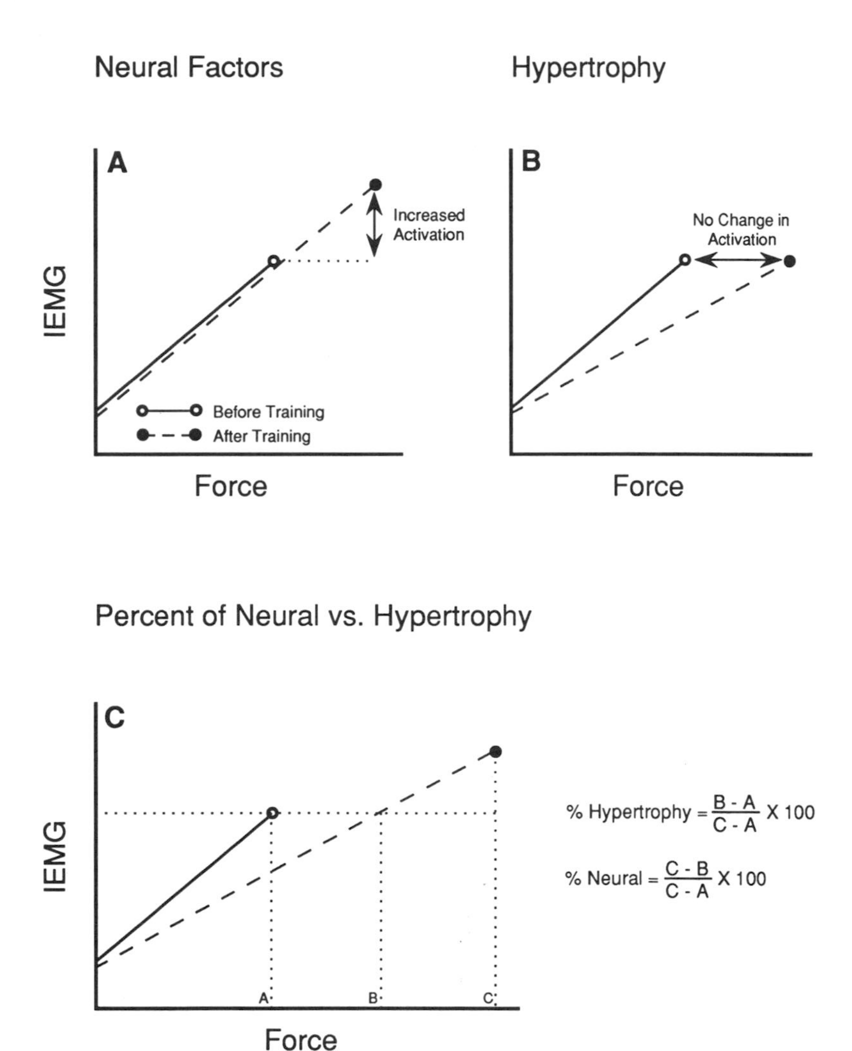

Figure 4.24. Experimental method for determination of neural and muscular factors responsible for strength increase. **A,** Situation in which all strength increase is due to neural factors since strength increase is directly proportional to peak EMG increase. **B,** Situation in which all strength increase is due to muscular factors since strength increase occurs with no change in EMG. **C,** Situation in which both neural and muscular factors are involved in strength increase, along with method for calculation of the relative component of each. (Adapted from Moritani T, Devries HA. Neural factors versus hypertrophy in the time course of muscle strength gain. Am J Phys Med 1979;58:115–130).

strength increase is due to a combination of neural and muscular factors, the relative proportion of each can be determined algebraically, as shown in Figure 4.24**C**. The beauty of this method is that strength changes can be ascribed totally or in part to muscular and/or neural changes.

This method was used in the study described above to document the time course of strength change in the study as well as the time course of the relative contributions of neural and muscular components to the strength change.

Time Course of Muscular and Neural Components to Strength Change

After 2 weeks of training, about 80% of the strength change of the trained arm was due to increased muscle activation (neural factors), and only 20% was due to changes in the muscle itself. In the remainder of the study, the relative proportion of neural contribution decreased while the relative muscular contribution increased until, after 8 weeks, over 95% of the strength change was due to muscular factors while only about 5% could be ascribed to neural factors (Figure 4.25). We thus see that early strength changes are primarily neural in basis, while later on the strength changes are almost all muscular.

In the untrained arm, the relative contributions of muscular and neural

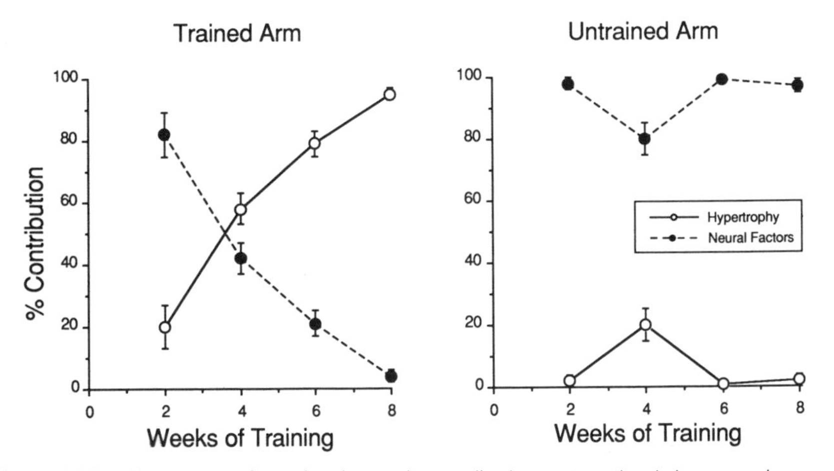

Figure 4.25. Time course of neural and muscular contribution to strength gain in men and women. Note that for the untrained arm (*right panel*) strength increase was due to neural factors, whereas for the trained arm (*left panel*) strength change was more complex. Early in training, the trained arm strength increased due to neural factors whereas later in training strength increased due to muscular factors. (Adapted from Moritani T, Devries HA. Neural factors versus hypertrophy in the time course of muscle strength gain. Am J Phys Med 1979;58:115–130.)

factors were relatively constant throughout the study (Figure 4.25). Almost all of the strength increase seen in the untrained arm at all time points was due to changes in neural activation level.

In summary, isometric joint strength was increased by about 25% in the trained arm and 15% in the untrained arm. The trained arm increased strength early on by increasing neural activation (increased muscle fiber recruitment), while later on strength increased because of muscle changes. Finally, strength increases in the untrained arm were due to increased EMG activity. The significance of these findings was the demonstration of the neural plasticity in normal subjects following training. How much more might these factors be involved in rehabilitation of disabled patients? There is pressing need to perform these types of studies in the rehabilitative setting in order to determine the physiologic basis of recovery. If recovery is due primarily to neural factors, emphasis should be placed on treatments emphasizing muscle activation. If, however, recovery is due primarily to muscular factors, emphasis should be placed on treatments that strengthen muscle.

(As a side note, let me mention a recent study that demonstrated that myosin changes can occur in contralateral limbs from rats who had one leg electrically stimulated while the other was allowed to rest. This study suggested that, indeed, muscles changes in contralateral limbs may be due to systemic effects such as altered hormone levels. Not all contralateral changes observed are necessarily due to neural changes. Future research will provide insights into the mechanism of these observed results.)

SUMMARY

We have learned the stereotypic response of muscle to increased use. The main lesson, aside from the almost unbelievable capacity of a muscle to adapt, is that muscle responds to increased use by becoming slower. The extent to which this adaptation occurs depends on the intensity of the increased use: In chronic stimulation, we observe complete fast-to-slow fiber type transformation, whereas in exercise training the specific results depend on the exercise duration and intensity. We saw that metabolic adaptation of muscle occurs more readily than change in myosin type. Muscle fibers appear to maintain a balance between oxidative, glycolytic, and contractile enzyme activities. Finally, using the examples of exercise specificity and neural adaptation, we saw that external performance indices such as strength cannot be uniquely ascribed to muscle properties alone but represent the interaction between the body's various physiologic systems. In the next chapter, we will consider the reverse stimulus—decreased use—and again make some generalized conclusions.

REFERENCES

Baldwin, K.M., Valdez, V., Herrick, R.E., MacIntosh, A.M., and Roy, R.R. (1982). Biochemical properties of overloaded fast-twitch skeletal muscle. J. Appl. Physiol. 52:467–472.

Brand, P.W., Beach, R.B., and Thompson, D.E. (1981). Relative tension and potential excursion of muscles in the forearm and hand. J. Hand Surg. 3:209–219.

Buller, A.J., Eccles, J.C., and Eccles, R.M. (1960). Interactions between motorneurons and muscles in respect to the characteristic speeds of their responses. J. Physiol. 150:417–439.

Burke, R.E., Levine, D.N., Tsairis, P., and Zajac, F.E. (1973). Physiological types and histochemical profiles in motor units of the cat gastrocnemius. J. Physiol. 234:723–748.

Dix, D.J., and Eisenberg, B.R. (1990). Myosin mRNA accumulation myofibrillogenesis at the myotendinous junction of stretched muscle fibers. J. Cell Biol. 111:885–894.

Dix, D.J., and Einsenberg, B.R. (1991). Distribution of myosin mRNA during development and regeneration of skeletal muscle fibers. Dev Biol. 143:422–426.

Eisenberg, B.R., Brown, J.M.C., and Salmons, S. (1984). Restoration of fast muscle characteristics following cessation of chronic stimulation. Cell Tissue Res. 238:221–230.

Eisenberg, B.R., and Salmons, S. (1981). The reorganization of subcellular structure in muscle undergoing fast-to-slow type transformation. A stereological study. Cell Tissue Res. 220:449–471.

Enoka, R.M., and Fuglevand, A.J. (1991). Neuromuscular basis of the maximum voluntary force capacity of muscle. In: Grabiner, M., ed. Current Frontiers in Biomechanics. Human Kinetics Press.

Gollnick, P.D., Armstrong, R., Saubert, C., Piehl, K., and Saltin, B. (1972). Enzyme activity and fiber composition in skeletal muscle of untrained and trained men. J. Appl. Physiol. 333:312–319.

Henneman, E., Somjen, G., and Carpenter, D.O. (1965). Functional significance of cell size in spinal motorneurons. J. Neurophysiol. 28:560–580.

Henriksson, J., and Reitman, J. (1977). Time course of changes in human skeletal muscle succinate dehydrogenase and cytochrome oxidase activities and maximal oxygen uptake with physical activity and inactivity. Acta Physiol. Scand. 99:91–97.

Hultman, E., Sjoholm, H., Jaderholm-Ek, I., and Krynicki, J. (1983). Evaluation of methods for electrical stimulation of human skeletal muscle in situ. Pflugers Arch. 398:139–141.

Kennedy, J.M., Eisenberg, B.R., Reid, S.K., Sweeney, L.J., and Zak, R. (1988). Nascent muscle fiber appearance in overloaded chicken slow-tonic muscle. Am. J. Anat. 181:203–215.

Kernell, D., Donselaar, Y., and Eerbeek, O. (1987a). Effect of physiological amounts of high- and low-rate chronic stimulation on fast-twitch muscle of the cat hindlimb. II. Endurance-related properties. J. Neurophysiol. 58:598–613.

Kernell, D., Eerbeek, O., Verhey, B.A., and Donselaar, Y. (1987b). Effects of physiological amounts of high- and low-rate chronic stimulation on fast-twitch muscle of the cat hindlimb. I. Speed- and force-related properties. J. Neurophysiol. 58:598–613.

Lieber, D.C., Lieber, R.L., and Adams, W.C. (1989). Maximal oxygen uptake and body composition changes resulting from run-training and swim-training at equivalent absolute intensities. Med. Sci. Sports Exerc. 21:655–661.

Lieber, R.L. Time course and cellular control of muscle fiber type transformation following chronic stimulation. ISI Atlas Animal Plant Sci 1988;1:189–194.

Lieber, R.L., and Kelly, J. (1991). Factors influencing quadriceps torque using transcutaneous electrical stimulation. Phys. Ther. 71:715–721.

Lieber, R.L., Silva, P.D., and Daniel, D.M. (1992). Electrical and voluntary activation are equally effective in strengthening quadriceps muscles. Trans. 38th Orthop. Res. Soc. 38:513.

Lømo, T., Westgaard, R.H., and Engelbretsen, L. (1980). Different stimulation patterns affect contractile properties of denervated rat soleus muscles. In: Pette, D., ed. Plasticity of Muscle. New York: Walter de Gruyter, 297–309.

Magel, J.R., Foglia, G.F., McArdle, W.D., Gutin, B., Pechar, G.S., and Katch, F.I. (1975). Specificity of swim training on maximum oxygen uptake. J. Appl. Physiol. 38:151–155.

Magel, J.R., McArdle, W.D., Toner, M., and Delio, D.J. (1978). Metabolic and cardiovascular adjustment to arm training. J. Appl. Physiol. 45:75–79.

Moritani, T., and Devries, H.A. (1979). Neural factors versus hypertrophy in the time course of muscle strength gain. Am. J. Phys. Med. 58:115–130.

Pette, D., ed. (1980). Plasticity of Muscle. New York: Walter de Gruyter.

Pette, D., ed. (1990). The Dynamic State of Muscle Fibers. Berlin: Walter de Gruyter.

Roy, R.R., Medows, I.D., Baldwin, K.M., and Edgerton, V.R. (1982). Functional significance of compensatory overloaded rat fast muscle. J. Appl. Physiol. 52:473–478.

Salmons, S., and Henriksson, J. (1981). The adaptive response of skeletal muscle to increased use. Muscle Nerve. 4:94–105.

Saltin, B., and Gollnick, P.D. (1983). Skeletal muscle adaptability: significance for metabolism and performance. In: Peachey, L.D., ed. Handbook of Physiology. Bethesda: American Physiological Society, 539–554.

Spector, S.A., Simard, C.P., Fourier, M., Sternlicht, E., and Edgerton, V.R. (1982). Architectural alterations of rat hindlimb skeletal muscles immobilized at different lengths. Exp. Neurol. 76:94–110.

Stromme, S.B., Ingjer, F., and Meen H.D. (1977). Assessment of maximal aerobic power in specifically trained athletes. J. Appl. Physiol. 42:833–837.

Tabary, J.C., Tabary, C., Tardieu, C., Tardieu, G., and Goldspink, G. (1972). Physiological and structural changes in the cat's soleus muscle due to immobilization at different lengths by plaster casts. J. Physiol. 224:231–244.

Vaughn, H., and Goldspink, G. (1979). Fibre number and fibre size in a surgically overloaded muscle. J. Anat. 129:293–304.

Williams, P. (1988). Effect of intermittent stretch on immobilized muscle. Ann. Rheum. Dis. 47:1014–1016.

Williams, P., and Goldspink, G. (1973). The effect of immobilization on the longitudinal growth of striated muscle fibers. J. Anat. 116:45–55.

Williams, P., and Goldspink, G. (1978). Changes in sarcomere length and physiological properties in immobilized muscle. J. Anat. 127:459–468.

Williams, R.S., Salmons, S., Newsholme, E.A., Kaufman, R.E., and Mellor, J. (1986a). Regulation of nuclear and mitochondrial gene expression by contractile activity in skeletal muscle. M. Biol. Chem. 261:376–380.

Williams, P., Watt, P., Bicik, V., and Goldspink, G. (1986b). Effect of stretch combined with electrical stimulation on the type of sarcomeres produced at the ends of muscle fibers. Exp. Neurol. 93:500–509.

Chapter

5

Skeletal Muscle Adaptation to Decreased Use

OVERVIEW

Having laid the foundation in the previous chapter, we now consider muscle's response to decreased use: what happens to muscle when it is *not* used. Again, the details of adaptation are investigated by considering various models of decreased use. We begin with the simplest model, immobilization; then we consider various spinal cord interruption models, followed by simulated weightlessness (such as occurs in spaceflight). Finally, the denervation model is presented in order to demonstrate its uniqueness. Again, patterns develop as each model adds to our understanding of muscle adaptation. The description of these abnormal states also clarifies the normal structure-function relationships in neuromuscular units.

INTRODUCTION

We continue our investigation of muscle adaptation. In contrast to increased use, where the "clean" model of chronic stimulation set the stage for later models, there is no clean model demonstrating decreased use. In other words, there is no chronic decreased use model because a dramatic decrease in use is sometimes difficult to elicit. Some muscle fibers are rarely used, so that immobilization, for example, may not have much influence on them. As a result, investigators have developed various models that cause a decrease in muscle use, although in different ways. The model that is easiest to understand is immobilization. We'll use it as our starting point.

ADAPTATION TO IMMOBILIZATION

Limb immobilization has been used since the turn of the century to protect fractured bones and injured tissues from repeated injury. The most common complication of immobilization is the muscle wasting that occurs due to decreased muscle use. However, as you saw in the previous chapter, immobilization *per se* does not cause atrophy. Strictly speaking, atrophy-inducing models using immobilization implement it in a shortened position. In this and other models, we will first present the model, followed by experimental data that describe the muscle changes resulting from the treatment.

The Immobilization Model

Immobilization models have long been used to study muscle adaptation. In addition to their obvious clinical relevance, immobilization models are relatively noninvasive—no surgery is necessary. We saw in Chapter 4 that a muscle responds to the tension level imposed upon it. (To date, muscle tension during immobilization has not been experimentally measured directly). Immobilization should not be viewed as placing a muscle in a state of suspended animation since electrical activity, tension, and motion occur within the cast (Figure 5.1). We should not consider immobilization as a "disuse" model, in the sense that the muscle is completely unused. Rather, it is more appropriate to consider immobilization a "reduced-use" model.

Electromyographic Changes during Immobilization

Since direct measurement of muscle tension within the cast during immobilization is technically difficult, an indirect "measure" of muscle activity during immobilization has been made by quantifying the electromyographic (EMG) activity. Experimentally, this is performed by implanting fine wire electrodes into a muscle and routing the leads to an external connector. Then, at various intervals after electrode implantation and immobilization, muscle EMG

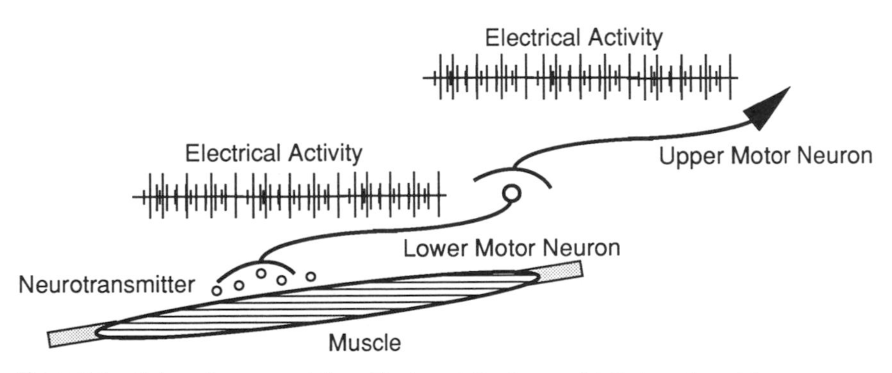

Figure 5.1. Schematic representation of the immobilization model. Each portion of the neuromuscular system is shown. Note that muscle electrical activity is present, and therefore, it is not appropriate to term this model "disuse" but rather decreased use. (Compare to Figures 5.8 and 5.20.)

activity is recorded to determine the long-term activity changes following immobilization.

Mario Fournier performed such an experiment in Reggie Edgerton's laboratory. These investigators implanted electrodes in both the fast-contracting medial gastrocnemius (MG) muscle and the slow-contracting soleus (SOL) muscle of the rat (Fournier *et al.,* 1983). Their purpose was to determine the extent to which immobilization resulted in muscle "disuse." EMG activity was first measured after electrode implantation for several days, in order to determine whether the implantation procedure itself affected muscle activity (fortunately, it did not!) Next, EMG activity was measured continuously for 15 minutes every hour, for 24 hours, on days 7, 17, and 28 postimmobilization. In order to produce varying degrees of atrophy, joints were immobilized such that the SOL and MG muscles were either shortened, at neutral length, or lengthened. At these same time intervals, animals were sacrificed and muscle mass measured as an index of "atrophy." (We will see below that wet muscle mass was probably not the best index of atrophy, but for the purposes of this study, it was probably adequate. Since skeletal muscle may contain over 80% protein, protein content is actually the most common index of muscle atrophy.)

Fournier *et al.* demonstrated that the total EMG activity of both the SOL and MG muscles decreased markedly after only 1 week of immobilization with the muscle in a shortened position. The SOL decreased to a greater extent than the MG. EMG activity continued to remain low throughout the remainder of the experiment, such that after 28 days, SOL EMG activity had decreased by 77% while MG EMG activity had decreased by 50% (Figure 5.2**A**). With the muscle immobilization in the neutral position, no change was seen in MG EMG activity, while only a 50% decrease in SOL EMG activity was measured.

The interesting aspect of the study was that the atrophic muscle response was not closely related to the magnitude of the change in EMG. For example, the shortened MG atrophied by 30% while EMG activity decreased by about 50% and the neutral MG did not decrease in EMG activity at all, yet the atrophy was the same as the shortened MG. (Figure 5.2**B**). In spite of the fact that the EMG response of the MG was markedly different, depending on muscle length, the magnitude of the atrophic response was identical. The immobilization model decreased the level of MG "use" (by decreasing EMG and thus the electrical activity of the muscle fibers), but this change in use was not related to the magnitude of the resulting muscle atrophy. Similarly, for the SOL muscle, EMG activity decreased to the same extent, whether the muscle was immobilized in the neutral or the fully lengthened position (decreasing to about 50% of control values). However, the SOL muscle immobilized in the neutral position decreased in mass by about 50%, while the SOL muscle

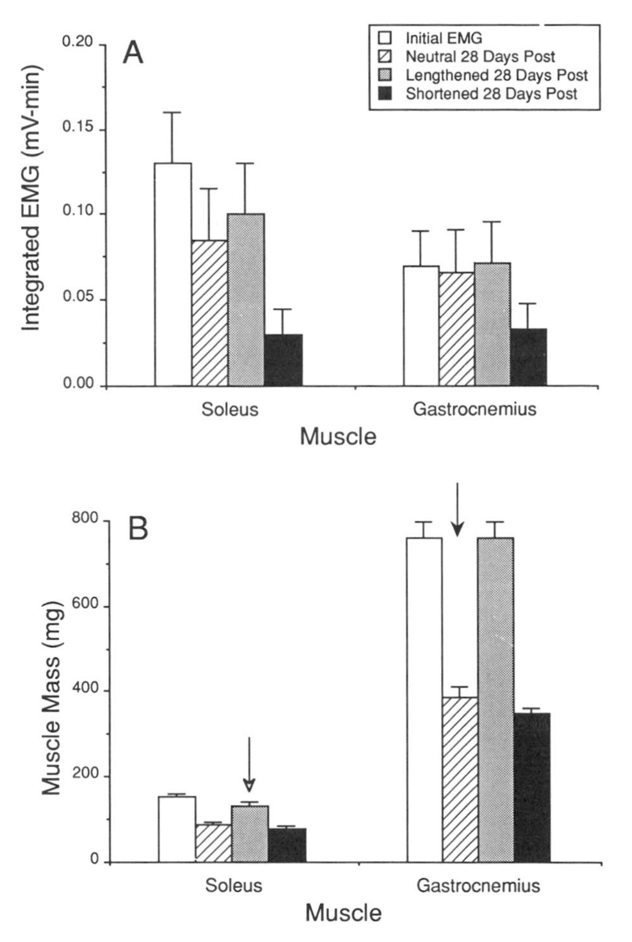

Figure 5.2. **A,** Change in integrated EMG activity from the rat soleus and medial gastrocnemius immobilized in different positions. **B,** Change in muscle mass from the rat soleus and medial gastrocnemius immobilized in different positions. Note that EMG changes are not paralleled by mass changes (*arrows*). (Data from Fournier *et al.,* 1983.)

immobilized in the lengthened position showed no decrease at all! Again, the change in level of use as measured by EMG activity was not related to the atrophic response.

The take-home lesson of this study is that the change in muscle electrical activity is not the cause of muscle atrophy. It is also obviously inappropriate to refer to atrophy that occurs secondary to immobilization as disuse atrophy since neural activity to muscles remained for the entire immobilization period.

Dog Quadriceps Model Rationale

In order to determine the nature of muscle's response to immobilization, literally thousands of experiments have been performed using every conceivable muscle group and animal or human model. There is general agreement that muscles composed mainly of slow fibers atrophy to a greater extent than muscles composed mainly of fast fibers (as we saw above). It also appears that antigravity muscles atrophy to a greater extent than their antagonists. Thus the "fast" gastrocnemius atrophies to a greater extent than its "fast" tibialis anterior antagonist. However, there are numerous exceptions to these generalizations. One reason for this lack of agreement is the lack of control of muscle length during immobilization. If, for example, the ankle joint is immobilized with the ankle plantarflexed, the soleus will dramatically atrophy (due to the lack of tension), while the tibialis anterior may actually hypertrophy (due to stretch). Do we therefore conclude that muscles composed primarily of slow fibers (such as the soleus) atrophy after immobilization, while muscles composed primarily of fast fibers (such as the tibialis anterior) actually hypertrophy? Obviously not. It is important, therefore, to control muscle length and other factors in order to properly generalize regarding the effects of immobilization on fast and slow skeletal muscles.

We recently performed a study of muscle fiber area, using as our experimental model three heads of the dog quadriceps muscles: the rectus femoris (RF), vastus lateralis (VL), and vastus medialis (VM) (Lieber *et al.,* 1988). These three muscles contain nearly identical architectures and fiber lengths but differ in fiber type percentage and number of joints crossed. For example, the RF acts both as a knee extensor and hip flexor and is composed of about 50% slow fibers. The VM and VL both function only as knee extensors, but the VL contains only about 20% slow fibers while the VM contains about 50% slow fibers. This model thus allows comparison between the VM and VL, which can be immobilized at precisely the same length but contain different percentages of slow and fast fibers. Similarly, comparisons between the RF and VM can be made, which have similar fiber type percentages but cross different joints.

Note that I have simply referred to the dog muscle fibers as "fast" and "slow" in spite of our relatively lengthy muscle fiber types discussion presented in Chapter 2. Why? Fortunately, dog muscles contain no type fast glycolytic (FG) fibers. Therefore, all fast fibers are of the fast oxidative glycolytic (FOG) type, and all slow fibers are of the slow oxidative (SO) type. As a result, unequivocal fiber type identification can be made from a single histochemical stain for myofibrillar adenosine triphosphatase (ATPase) activity (remember that this is only valid having experimentally determined that dog muscle contains no FG fibers).

Dog Quadriceps Immobilization Method

In order to ensure that muscle lengths were held constant during the immobilization period, an external skeletal fixator was used to fix the knee joint angle at 90° (Figure 5.3). One leg was immobilized, while the other leg was left as a control. (However, note that contralateral legs from experimental animals are not truly "normal" for a variety of reasons. First, they probably bear more weight since the immobilization leg is raised. Second, systemic effects might affect all muscles differentially. In this particular study, control experiments were performed that demonstrated no differences between truly normal muscles, taken from untreated animals, and contralateral muscles from those used in this study.)

Following 10 weeks of immobilization, small biopsy specimens were taken from the VL, VM, and RF and prepared for histochemical analysis as described in Chapter 2. An added point that should be mentioned here is that the specimens were always kept under some tension since fiber shortening will cause an apparent increase in fiber cross-sectional area. After staining muscles for myofibrillar ATPase activity and classifying each fiber as fast or slow, muscle fiber areas were measured. As a result, we obtained the following data for each muscle:

1. Slow fiber area (μm^2)
2. Fast fiber area (μm^2)
3. Percentage of each fiber type
4. Area fraction of endomysial/perimysial connective tissue (%).

Let's discuss what each parameter indicates. Clearly, fiber area relates to the force-generating capacity of the fiber. The greater the fiber area, the greater the number of myofibrils arranged in parallel within the fiber. Thus when we discuss muscle fiber atrophy, strictly speaking, we must consider myofibrillar number or a parameter closely related to it (such as fiber area). As myofibrillar number decreases, the force generated by the fiber also decreases. We will compare fiber area (as an index of myofibrillar number) from control and immobilized muscles in order to determine their responses to immobilization. We saw in the increased-use models, that fiber type percentage provides insights into the degree of use experienced by the muscle. Recall in the chronic stimulation model that the fast-to-slow fiber type conversion resulted from the dramatically increased level of fiber activity compared to normal. In the same way, a muscle's normal fiber type

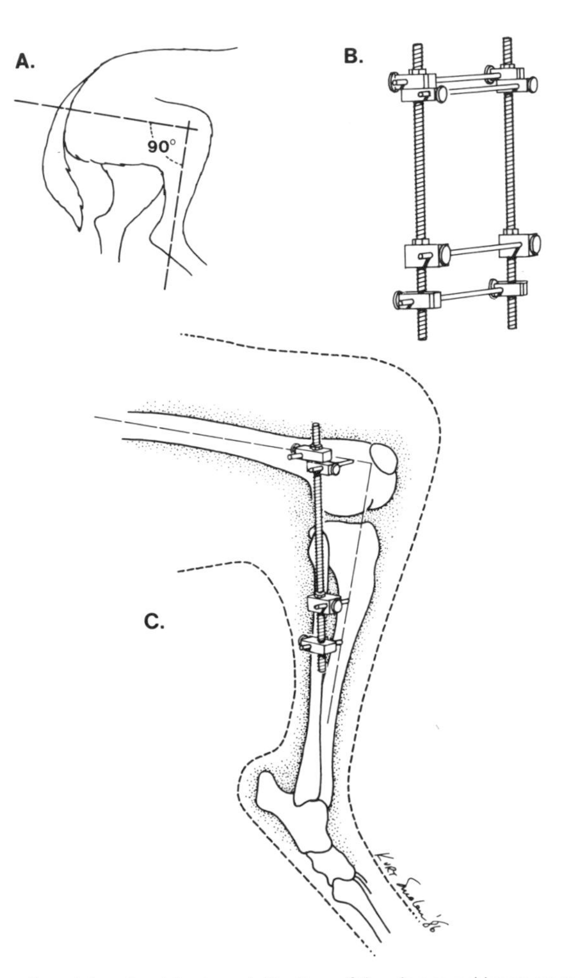

Figure 5.3. Experimental method for immobilization of the dog quadriceps muscles. An external skeletal fixator was used to maintain the knee angle at 90° of flexion. In this way, muscle length could be carefully controlled. (From Lieber RL, Fridén JO, Hargens AR, Danzig LA, Gershuni DH. Differential response of the dog quadriceps muscle to external skeletal fixation of the knee. Muscle Nerve 1988; 11:193–201.)

distribution is probably related to the amount of normal activity experienced by the muscle. We saw in Chapter 1 that, during development, differentiation past the secondary myotube required innervation. We infer from this that during maturation, muscle fibers that receive a great deal of neural activity have a greater likelihood of becoming slow fibers. The take-home lesson is that a muscle's normal fiber type distribution provides insights into its normal level of use and, as we shall see, its response to decreased use.

Muscle Fiber Response to Immobilization

The most obvious muscle response was fast and slow fiber atrophy (Figure 5.4). However, although all micrographs were taken at the same magnification, it was obvious that the magnitude of the atrophy was different for the three different muscles. Using stereometric techniques developed in Switzerland by Weibel (Weibel, 1980), we found that there was no difference in fast or slow muscle fiber area between any of the muscles on the control side (Figures 5.5 and 5.6). However, 10 weeks of immobilization caused significant decrease in both fast (Figure 5.5) and slow (Figure 5.6) muscle fiber areas. Clearly this would result in decreased muscle force-generating capacity. However, the most interesting result was not that muscle fiber areas decreased in response to immobilization (which had already been shown by numerous scientists); it was that the amount of muscle fiber atrophy was *different* for each of the three muscles. Specifically, the fast fiber area of the immobilized VM and VL was less than the fast fiber area of the immobilized RF (Figure 5.5). Similarly, in an even more dramatic differential response, slow fiber area of the VM muscles was much less than that of the VL, which was much less than that of the RF! The atrophic response for slow fibers was thus, in order from most to least atrophied, VM > VL > RF, while for fast fibers the corresponding order was VM = VL > RF. This dramatic difference was even more dramatic on inspection of the actual micrographs. Compare, for example, the fiber sizes in the micrographs of Figure 5.4, photographed at the same magnification! In addition to these changes in fiber area, a significant increase in fast fiber percentage following immobilization was observed in the VM.

Immobilization also caused proliferation of endomysial and perimysial connective tissue relative to control legs, with a significantly greater increase in the immobilized VM and VL muscles compared to immobilized RF muscles.

Proposed Explanation for Differential Muscle Fiber Response

How can the differences between these muscles be accounted for based on our understanding of muscle plasticity? Why would slow fibers in one muscle

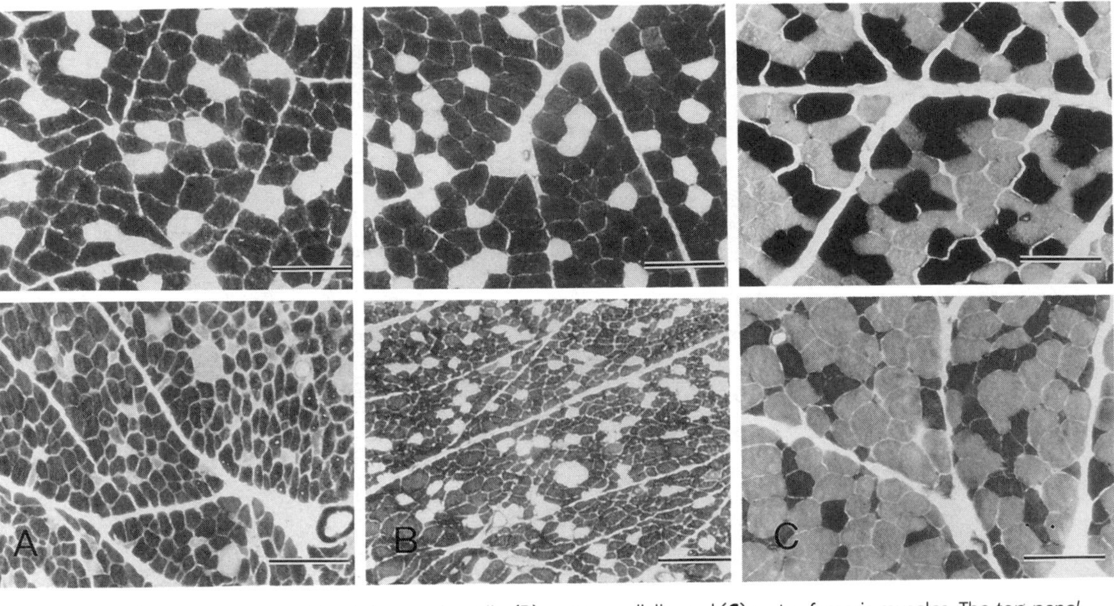

Figure 5.4. Light micrographs of the (**A**) vastus lateralis, (**B**) vastus medialis, and (**C**) rectus femoris muscles. The *top panel* represents micrographs from normal muscles, and the *bottom panel* represents micrographs from immobilized muscles. All micrographs were taken at the same magnification. Fast fibers appear dark, and slow fibers appear light. Calibration bar represents 100 μm. (From Lieber RL, Fridén JO, Hargens AR, Danzig LA, Gershuni DH. Differential response of the dog quadriceps muscle to external skeletal fixation of the knee. Muscle Nerve 1988;11:193–201.)

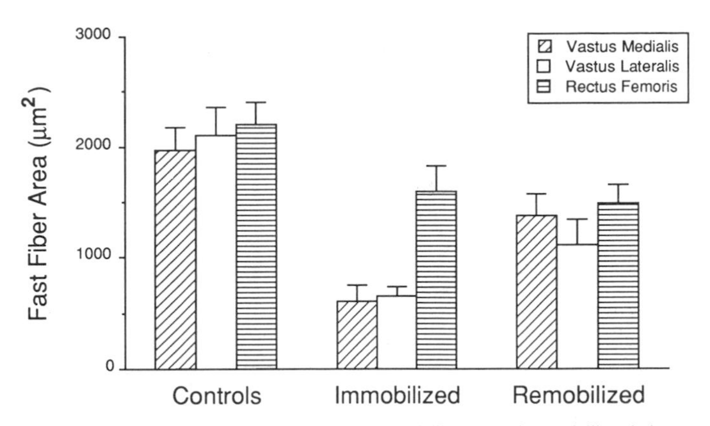

Figure 5.5. Graph of fast fiber area from control, immobilized, and remobilized dog quadriceps muscles. Note that the magnitude of atrophy was muscle specific. Upon remobilization, muscles returned to parity. (From Lieber RL, Fridén JO, Hargens AR, Danzig LA, and Gershuni DH. Differential response of the dog quadriceps muscle to external skeletal fixation of the knee. Muscle Nerve. 1988;11:193–201; Lieber RL, McKee-Woodburn T, Fridén J, and Gershuni DH. Recovery of the dog quadriceps after ten weeks of immobilization followed by four weeks of remobilization. J Orthop Res 1989;7:408–412.)

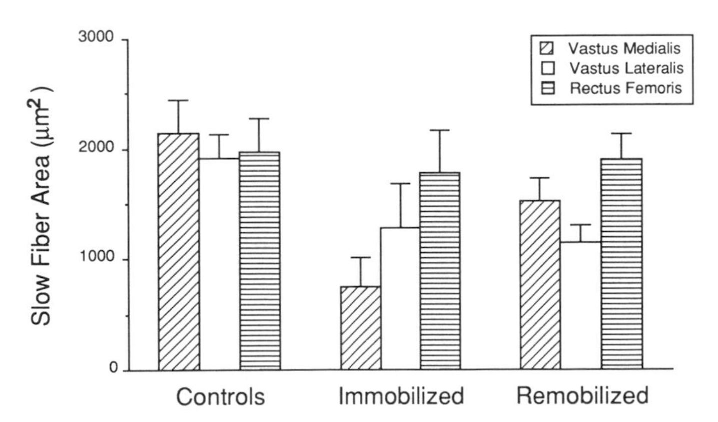

Figure 5.6. Graph of slow fiber area from control and immobilized dog quadriceps muscles. (From Lieber RL, Fridén JO, Hargens AR, Danzig LA, Gershuni DH. Differential response of the dog quadriceps muscle to external skeletal fixation of the knee. Muscle Nerve 1988;11:193–201; Lieber RL, McKee-Woodburn T, Fridén J, Gershuni DH. Recovery of the dog quadriceps after ten weeks of immobilization followed by four weeks of remobilization. J Orthop Res 1989;7:408–412.)

dramatically atrophy, while slow fibers in another muscle atrophy only slightly? Let's consider each muscle sequentially.

The differences observed between the VM and VL could not simply be explained by differences in immobilization length. We know that immobilization length strongly influences the atrophic response, but the VL and VM were fixed at identical lengths. Using a similar argument, differences observed between the VM and RF could not be explained by differences in fiber type distribution since they both began with about 50% fast and 50% slow fibers. The architecture of all three muscles is very similar. The ratio of fiber length/muscle length has been studied in humans and guinea pigs and approximates 0.2 for all three muscles, and thus, architecture differences could not account for the differences observed. In order to explain the differential atrophy of the VM, VL, and RF, several factors must be considered.

The RF demonstrated the smallest degree of atrophy of the three muscles studied. Conversely, the most severe atrophy was observed in the VM. This is interesting in light of the fact that the two muscles initially contained nearly identical fiber type distributions. We thus see that initial fiber type distribution alone does not dictate the magnitude of the atrophic response. However, the RF crosses both the hip and knee, functioning both as a knee extensor and hip flexor, while the VM and VL cross only the knee functioning as knee extensors. The RF is therefore less rigidly immobilized than either of the vasti, probably explaining the small atrophic response.

The slow fibers of the VM atrophied to a greater extent than those of the VL. Both the VM and VL were immobilized at the same length since they both arise from the proximal femur and insert together with the rectus tendon onto the patella. It seems unlikely, then, that the small difference in anatomic location could account for the markedly different response. The VM initially contained a much larger proportion of slow fibers than the VL, which indicates that the VM was probably used more since muscle fiber type distributions provide insights into muscle activation history. Therefore, following immobilization, the *change* in the amount of VM activation was probably greater than the change for the VL, even though the absolute levels following immobilization may have been similar. This provides support for the idea that immobilization represents a model of decreased use.

Fiber Type Transformation following Immobilization

Perhaps this large change in VM activation level could also account for the slow-to-fast transformation. This is our first exposure to the idea that, in contrast to increased-use models, which cause fast-to-slow transformations, decreased-use models cause the opposite, slow-to-fast, transformation. This type of transformation has been observed clinically. For example, Haggmark and his coworkers (1976) observed a significant increase in vastus lateralis fast fiber percentage following surgical reconstruction of the anterior cruciate ligament. Interestingly, the magnitude of the transformation seemed to be correlated with the change in use since elite athletes (whose muscles were most "used" to high activity levels) demonstrated the greatest degree of slow-to-fast transformation. Similarly, Gunnar Grimby observed dramatic slow-to-fast fiber type transformations in patients recovering from traumatic spinal cord lesions (Grimby *et al.,* 1976). Interesting, eh?

Generalizations of Muscle Fiber Atrophy

This study established the relative influence of two factors that contribute to immobilization-induced atrophy. The most significant factor is the degree of immobilization (number of joints crossed), and next is the change in use relative to normal function. The initial percentage of slow muscles fibers is a fair indicator of the normal muscle use level and a good predictor of the relative degree of atrophy.

These data indicate that a blanket concept of "slow fiber atrophy" cannot apply to all muscles. Rather, a combination of factors determines the muscular response to decreased use. Given the structure and fiber type distributions of the various human muscles, it is possible to predict those most vulnerable to immobilization-induced atrophy, *i.e.,* those that function as antigravity muscles, cross a single joint, and contain a relatively large proportion of slow fibers. This description fits the soleus, vastus medialis, and vastus intermedius. The next class of muscles susceptible to immobilization-induced atrophy would be antigravity muscles, predominantly slow, that cross multiple joints, namely the longissimus and transversospinalis (erector spinae), gastrocnemius, and rectus femoris. These muscle groups should be immobilized conservatively to avoid severe strength loss. Conversely, phasically activated, predominantly fast muscles (*e.g.,* tibialis anterior, extensor digitorum longus, biceps) can be immobilized with less loss of strength.

This hierarchy of susceptibility to immobilization is supported by the data of Reggie Edgerton and colleagues (Edgerton *et al.,* 1975), who measured morphologic, biochemical, and physiologic properties of immobilized hindlimb muscles from *Galego senegalensis* (a small primate commonly known as the bushbaby) and found that muscles atrophied in the order (most to least atrophy) soleus > plantaris > vastus intermedius = vastus lateralis > gastrocnemius > tibialis anterior = rectus femoris, agreeing well with the principles stated above. Remember, increased atrophy is observed upon immobilization in muscles normally used a great deal. These tend to be muscles with a relatively high percentage of slow fibers in the general population.

Descriptions of Muscle Atrophy

Numerous immobilizations studies have documented differing degrees of atrophy following different types of immobilization treatment. How should the magnitude of the atrophic response be calculated? Is the atrophy directly related to the change in muscle force? Muscle mass? In order to describe appropriate measures of atrophy, let us return to the immobilization study performed in Reggie Edgerton's laboratory, which we discussed in the Chapter 4.

Recall that muscles were immobilized at different lengths, after which time architectural (Spector *et al.,* 1982) and contractile (Simard *et al.,* 1982) properties were measured. Let's compare the conclusions we would reach using muscle mass as a measure of atrophy compared to maximum tetanic tension (P_O) as a measure of atrophy. We will consider both the SOL and MG. The normal rat MG weighs about 800 mg. Following immobilization in the lengthened position, mass dropped to 635 mg, while in the shortened position, mass decreased dramatically to 350 mg (Figure 5.7**A**). In terms of *mass,* therefore, the stretched MG atrophied by only 15%, while the shortened MG atrophied by over 50%. These represent impressive, and quite different, changes. Unfortunately, these changes in mass had almost no relation to the actual muscle performance following the immobilization period. While the shortened MG decreased in mass by 50%, MG P_O decreased by over 75% (Figure 5.7**B**)! The lengthened MG decreased in mass by 15%, while MG P_O decreased by over 40%! Quite a disparity! Using a similar argument for the SOL, the results were quite different. Following immobilization in the lengthened position, SOL mass dropped from 150 mg to 132 mg, while in the shortened position, mass decreased dramatically to 79 mg (Figure 5.7**A**). In terms of *mass,* therefore, the stretched SOL atrophied by only 12%, while the shortened SOL atrophied by 50%! In terms of contractile tension, the stretched SOL atrophied by 50%, and the shortened SOL atrophied by 80%! Again, quite different results—12% mass versus 50% P_O and 50% mass versus 80% P_O. More "atrophy" is measured if we consider P_O instead of mass.

The reason mass and performance were not closely related was that performance was a direct function of *architecture,* while mass was simply proportional to the total amount of contractile material in the muscle. To summarize, a knowledge of muscle *mass* change following immobilization provides very little information as to the functional consequence without some knowledge of the concomitant architectural changes. The explanation

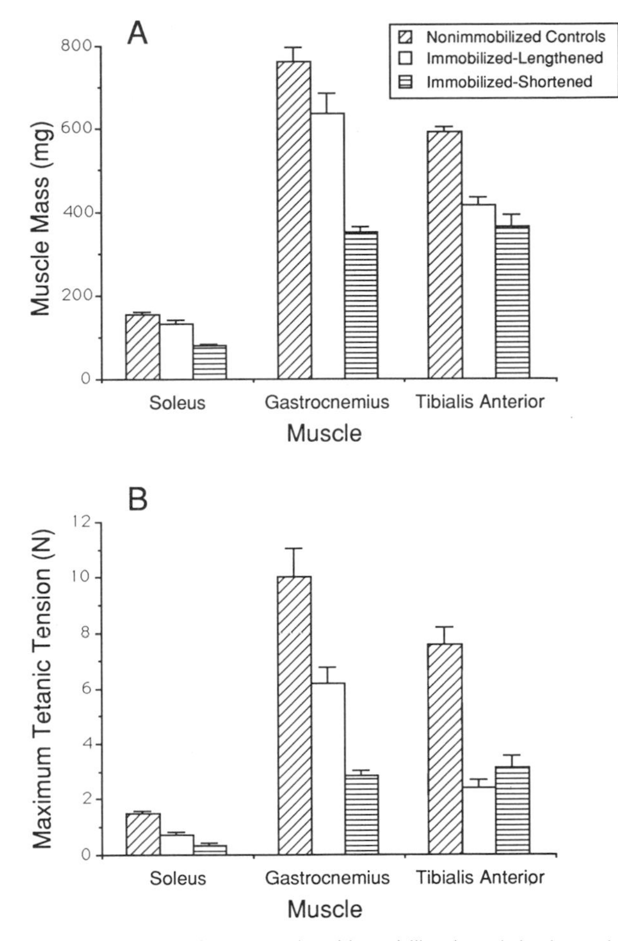

Figure 5.7. **A,** Muscle mass. **B,** P_O from normal and immobilized rat skeletal muscles. Note that mass and P_O changes are not uniquely correlated. Mass is thus a poor predictor of muscle tension. Note also that immobilization in a lengthened position spares the muscle of mass and tension losses that accompany immobilization in the shortened position. (Adapted from Simard *et al.*, 1982.)

for the disparity between the results using mass or P_O is based on the changes in fiber length occurring following the immobilization. Clearly, fiber length alterations do not necessarily affect contractile tension. Therefore, the only conditions under which mass, or protein content, provides a reliable index of tension decrease are those in which no architectural adaptations are seen.

Incidentally, Edgerton and his colleagues also demonstrated that the decrease in P_O was nearly proportional to the decrease in muscle fiber area. It thus appears that atrophy of muscle fiber cross-section truly is the primary cause for decreased muscle force—the muscle fibers themselves generate less force.

REMOBILIZATION FOLLOWING IMMOBILIZATION

How long does it take to recover from immobilization-induced atrophy? Interestingly, very few remobilization studies have been performed. The few that have been performed suggest that it takes longer to recover from immobilization than to elicit the initial response.

We essentially repeated the dog study described above by remobilizing canine quadriceps for 4 weeks following 10 weeks of immobilization (Lieber *et al.,* 1989). During this 4-week remobilization period, normal activity was permitted, and daily 1-hour walking/running outings were encouraged. Normal weight bearing resumed spontaneously within about 1 week.

As in the initial study, no difference in slow or fast fiber area was observed between any control muscles. However, 10 weeks of immobilization followed by 4 weeks of remobilization still resulted in about a 30% decrease (or not full recovery) of both slow and fast fiber areas (Figures 5.5 and 5.6). However, in contrast to the study of immobilization, no difference between immobilized muscle fiber areas was seen. While immobilization-induced atrophy was muscle- and fiber-type specific, recovery following immobilization was neither a function of muscle nor of fiber type. The fiber type transformation that had occurred with immobilization had also returned to normal following remobilization.

A second difference between the immobilization and remobilization studies was that while a large increase in the amount of extracellular connective tissue was seen following 4 weeks of immobilization, no difference was seen between control and remobilized muscles in the amount of extracellular connective tissue. Thus the previously elevated area fraction of connective tissue (about 20%) that was observed following immobilization returned to control levels (about 10%) following the remobilization period. Since connective tissue is associated with passive muscle stiffness, these data suggest that muscle stiffness, which had increased due to the immobilization process, returned to normal after remobilization.

These results can be explained based on the level of use experienced by each muscle following remobilization. Recall that we explained the differential immobilization-induced atrophy based on the *change* in level of use relative to normal. Now, during remobilization, we presume that the normal use levels returned and brought fiber back to parity. Detailed confirmation of this hypothesis will await future experiments.

ADAPTATION TO SPINAL CORD TRANSECTION

Decreased-use models have also been developed that induce either an upper or lower motor neuron lesion. As we will see, lower motor neuron disruption (denervation) induces numerous changes in muscle that are quite distinct from other decreased-use models. (We saw, for example, that a denervated muscle responds to chronic stimulation in a manner qualitatively different from that of a normal muscle.) This difficulty has been overcome by developing a model that interrupts the *upper* motor neuron pathway by transecting the spinal cord (*i.e.,* cordotomy; Figure 5.8). This procedure has been performed in several animal models. Again, we will discuss muscle contractile, histochemical, and biochemical alterations following cordotomy.

Experimental Method of Rat Cordotomy

In a previous study, we were interested in the extent of muscle adaptation that could occur following long-term spinal cordotomy (Lieber *et al.,* 1986a and 1986b). The rat was chosen as the experimental model since they live only 2–3 years, and 1 year of cordotomy would represent about one-half a lifetime of chronic disuse.

Two groups of rats were studied. Control rats were permitted normal growth for 1 year after entry into the study at age 6 weeks. At age 6 weeks (about 2 weeks after the muscle fiber types were differentiated) experimental rats were anesthetized, and following laminectomy, the spinal cord and its covering were completely transected.

Postoperative care of the cordotomized rats required special cage bedding

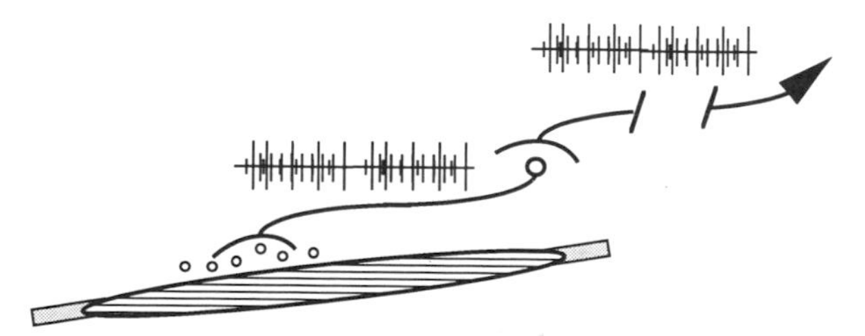

Figure 5.8. Schematic representation of the cordotomy model. The upper motor neuron pathway has been interrupted, but muscle EMG activity remains due to segmental influences. Muscle tension is low since the limbs are not used for locomotion. (Compare to Figures 5.1 and 5.20.)

to prevent pressure sores. Initially, there was a flaccid paraplegia with the limbs dragging behind the rats as they crawled about in the cage. They were able to move using their forelimbs and had no difficulty reaching food and water. At approximately 3 to 4 weeks, the paralyzed hind limbs of the animals changed from flaccid to spastic. After spasticity developed, the limbs were almost always held in extension, and no recovery of voluntary activity was ever observed.

Contractile Properties of Muscles after Long-Term Cordotomy

Two muscles were chosen for study—the soleus and the extensor digitorum longus (EDL). The two muscles differ in fiber type percentage (80% slow fibers in the soleus and 5% in the EDL). In this way, it would be possible to determine whether any observed effects of cordotomy were fiber type- or muscle-specific. The contractile properties of the soleus and EDL muscles were tested in both groups of rats at age 58 weeks. The distal muscle insertion was carefully dissected free and attached to a force transducer. The peripheral nerve innervating the muscle was also carefully isolated and electrically activated using an artificial stimulator. Note that in spite of a year of cordotomy, the lower motor neuron was completely intact and activatable! Paralyzed is not synonymous with dead.

The responses of the soleus from normal and transected rats stimulated at 5 and 10 Hz are shown in the upper and lower panels, respectively, of Figure 5.9. Note that at 10 Hz, the transected soleus developed a greater force and was less fused than the soleus, implying faster contraction and/or relaxation. Unfused tetani of the EDL stimulated at 10, 20, and 30 Hz are shown in Figure 5.10. Note that the differences between the normal EDL (upper panel) and transected EDL (lower panel) were much less dramatic than those observed for the soleus.

Analysis of the contractile responses showed that no differences were observed between normal and transected EDLs (Figure 5.11). In contrast, for the soleus muscle, dramatic changes were observed in all contractile properties measured. For example, time to peak tension decreased by about 50% (Figure 5.11**A**), suggesting a change in the properties of the sarcoplasmic reticulum (SR). The change was probably due to an *increase* in the calcium transporting ability of the SR. Supporting this hypothesis, the soleus fusion frequency increased by 100% (Figure 5.11**B**). Absolute maximum tetanic tension (in N) did not change significantly following transection. However, since the soleus muscle cross-sectional area decreased by about 50% (Figure 5.11**C**), specific tension (in N/cm^2) of the soleus significantly *increased* by over 100% 1 year following transection (Figure 5.11**D**). Thus the soleus

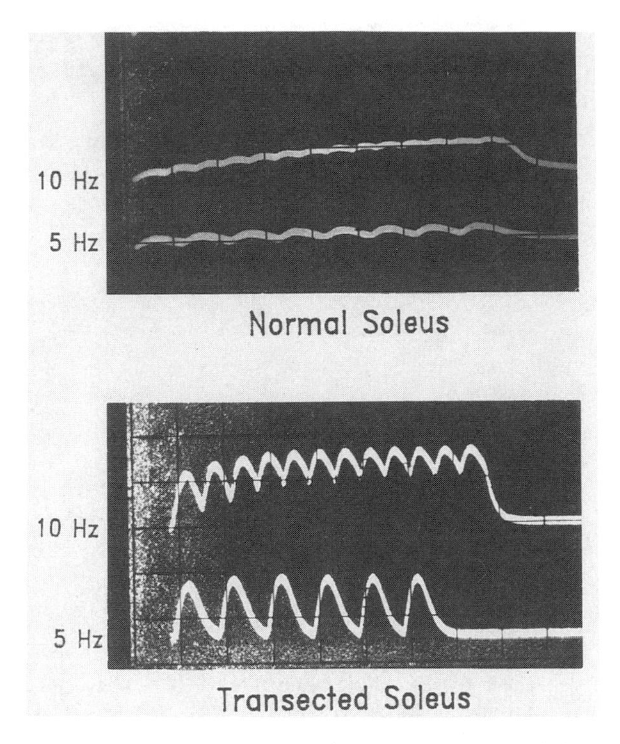

Figure 5.9. Unfused tetani from rat soleus muscles. Upper panel: Records from control animals. Lower panel: Records from transected animals. Note that the upper records demonstrate greater fusion than the lower records, implying increased calcium transport kinetics following cordotomy. (From Lieber RL, Johansson CB, Vahlsing HL, *et al.* Long-term effects of spinal cord transection on fast and slow rat skeletal muscle. I: Contractile properties. Exp Neurol 1986;91:423–434.)

muscle was half the cross-sectional area but still generated the same tension—how? (See below, page 230.)

To summarize, following transection, the soleus muscle increased in contractile speed and specific tension, while no significant contractile changes were seen in the EDL.

Morphometric Properties of Muscles after Long-Term Cordotomy

To try to understand the contractile results, we measured the fiber type and fiber size distributions within the soleus and EDL, using methods similar to those previously described.

Low-magnification cross-sections of normal soleus and EDL muscles stained for myofibrillar ATPase are presented in Figures 5.12**A** and 5.13**B,** respec-

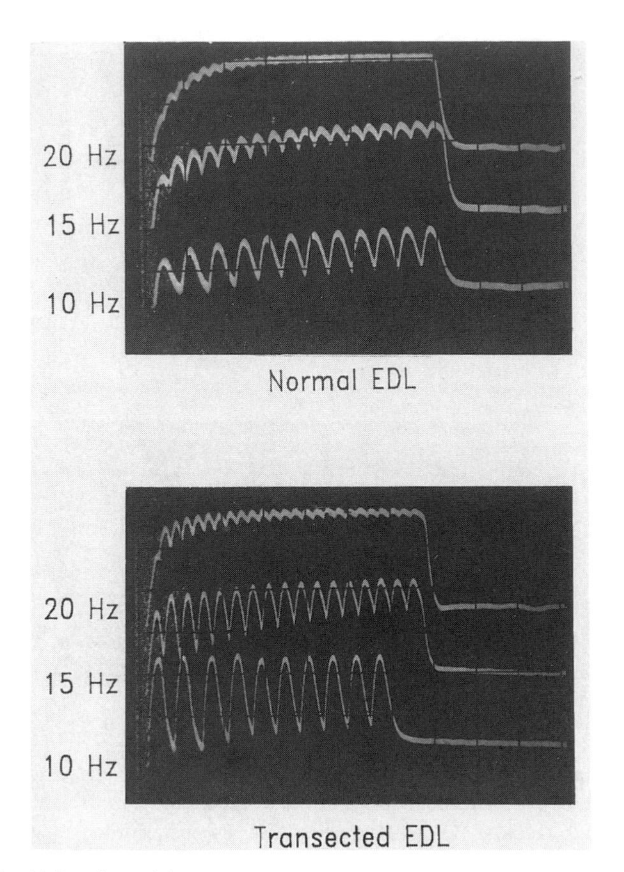

Figure 5.10. Unfused tetani from rat EDL muscles. **Upper panel:** Records from control animals. **Lower panel:** Records from transected animals. Very little difference is observed between records. (From Lieber RL, Johansson CB, Vahlsing HL, *et al.* Long-term effects of spinal cord transection on fast and slow rat skeletal muscle. I: Contractile properties. Exp Neurol 1986;91:423–434.)

tively. Note that in the normal soleus, fast fibers are scattered throughout the muscle, while the transected soleus (lower panel of Figure 5.12) is composed almost *entirely* of fast fibers. This represents a dramatic slow-to-fast muscle fiber type conversion (the opposite type of conversion that we observed with chronic stimulation). The soleus at higher magnification (Figure 5.14), showed a decreased size of both the slow and fast fibers. While fiber type transformation occurred in the EDL also (lower panel of Figure 5.13), the magnitude of the effect was small. The normal EDL contains only a few slow fibers (light staining) in the anterior superficial region of the muscle. Following transection, even fewer slow fibers were visible.

Quantitative changes observed following transection were similar for both

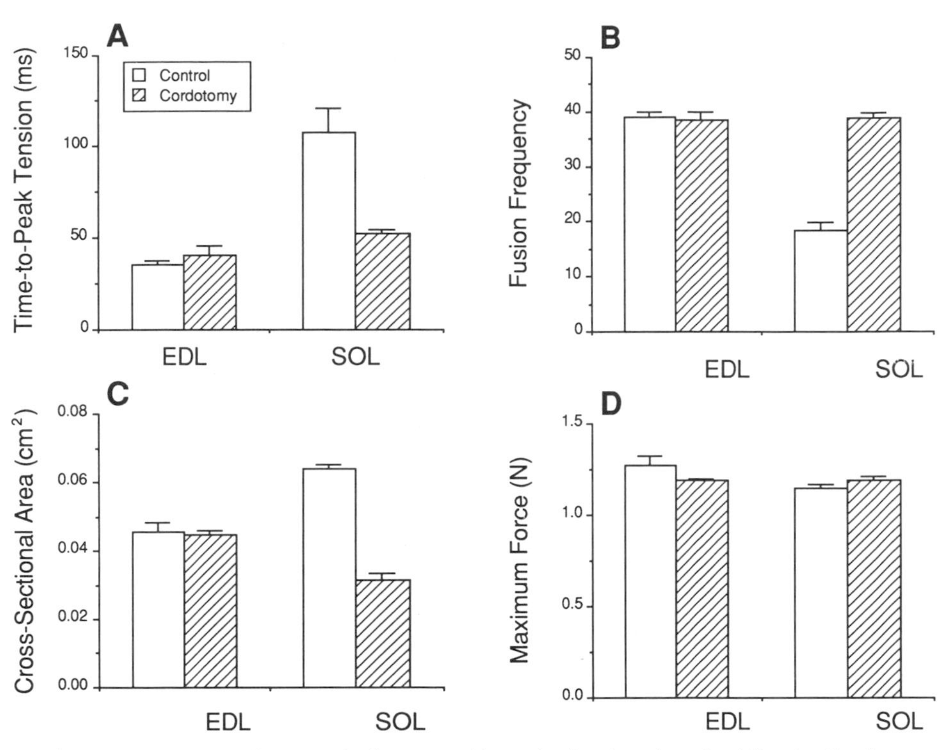

Figure 5.11. Contractile properties from normal (*open bars*) and cordotomized (*hatched bars*) rat skeletal muscles. Note that the soleus contractile properties changed dramatically, while EDL properties show little change. (Data from Lieber *et al.*, 1986b.)

muscles, although the magnitude of the changes was greater for the soleus compared to the EDL. As can be appreciated from the micrographs, the percentage of slow muscle fibers decreased significantly for both muscles. The average slow fiber area decreased significantly by about 50% for the soleus, but not for the EDL. The percentage of fast fibers increased significantly for both muscles. Again, the magnitude of the increase was greater for the soleus. Fast fiber area decreased by about 25% in both muscles. The percentage of extracellular connective tissue increased significantly for both muscles by about the same amount. Thus both muscles demonstrated fiber atrophy, a slow-to-fast fiber type transformation, and a significant increase in connective tissue. However, since the EDL is normally composed of about 95% fast fibers, these changes had no effect on contractile properties. Clearly such a transformation had much more profound consequences for the normally 80% slow soleus muscle. Finally, since the soleus muscle generated

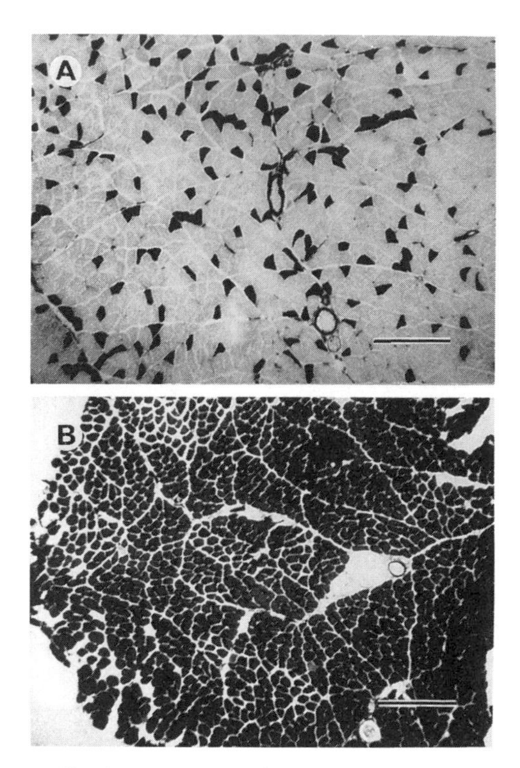

Figure 5.12. Low-magnification micrographs from soleus muscle (**A**) before and (**B**) after cordotomy. Muscle sections stained for myofibrillar ATPase activity. Fast fibers are dark. Calibration bar represents 500 μm. Note the increase in fast fiber percentage following cordotomy. (From Lieber RL, Fridén JO, Hargens AR, Feringa ER. Long-term effects of spinal cord transection on fast and slow rat skeletal muscle. II: Morphometric properties. Exp Neurol 1986;91:435–448.)

the same absolute contractile force in spite of its smaller muscle fibers, the force per unit area of muscle fiber (*i.e.,* the specific tension) must have increased. Since most of the fibers in the soleus following cordotomy are fast fibers, we conclude that the fast fibers of the rat have a higher specific tension than the slow fibers.

The Importance of Tension in Determining Muscle Properties after Cordotomy

What factors contribute to the dramatic changes observed following cordotomy? What can be done to prevent these changes or to ameliorate recovery from such changes?

In yet another interesting plasticity study, Edgerton and his colleagues

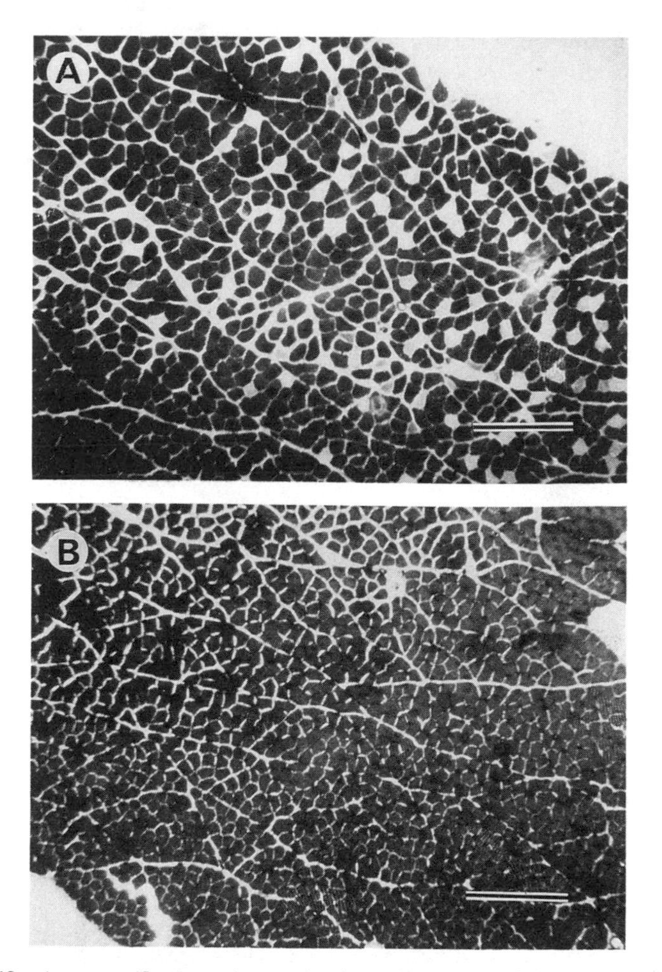

Figure 5.13. Low-magnification micrographs from EDL muscle (**A**) before and (**B**) after cordotomy. Muscle sections stained for myofibrillar ATPase activity. Fast fibers stain darkly. Calibration bar represents 500 μm. Note the increase in fast fiber percentage following cordotomy, which is more modest than that seen for the soleus. (From Lieber RL, Fridén JO, Hargens AR, Feringa ER. Long-term effects of spinal cord transection on fast and slow rat skeletal muscle. II: Morphometric properties. Exp Neurol 1986;91:435–448.)

exercised adult cats that had been spinalized (*i.e.,* the spinal cord in transected, Jiang *et al.,* 1990) and compared them to animals that had been spinalized but did not receive exercise. After transection at the midthoracic level and 1 month of postoperative recovery, animals were exercised on a treadmill for 30 minutes/day for 6 months! Measurement of EMG activity

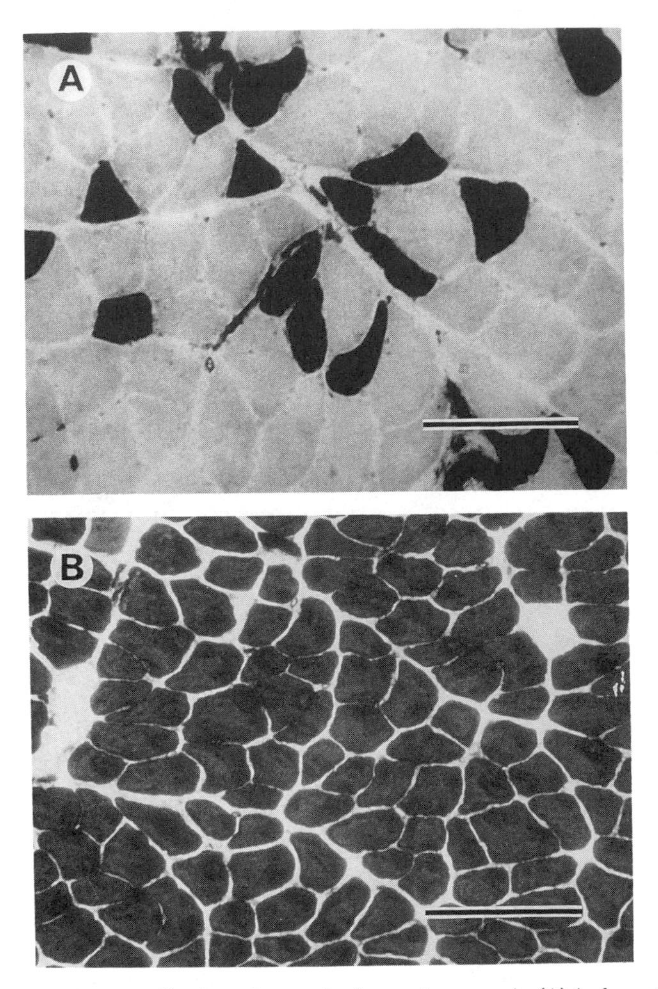

Figure 5.14. Higher magnification micrographs from soleus muscle (**A**) before and (**B**) after cordotomy. Muscle sections were stained for myofibrillar ATPase activity. Fast fiber stain darkly. Note the modest fiber atrophy. Calibration bars represent 100 μm. (From Lieber RL, Fridén JO, Hargens AR, Feringa ER. Long-term effects of spinal cord transection on fast and slow rat skeletal muscle. II: Morphometric properties. Exp Neurol 1986;91:435–448.)

following spinalization had shown a reduction similar to that seen in the immobilization models. In addition, direct measurement of muscle force during locomotion confirmed that significant weight bearing was induced by the exercise training procedure. Following the 6-month training period, muscle fiber histochemical properties were measured. Not only were the usual qualitative histochemical stains implemented, but *quantitative* measurement of myosin ATPase activity (MATPase), succinate dehydrogenase activity (SDH), and α-glycerophosphate dehydrogenase activity (αGP) were also measured, which enabled detailed description of the fiber metabolic properties themselves.

Six months of exercise tended to cause muscle fiber size to be larger than spinalization without exercise. For example, while the normal soleus muscle fiber area was about 4200 μm^2, the spinalized slow fibers were about 2700 μm^2 and the spinalized + exercised slow fibers were about 3000 μm^2. The prevention of fiber atrophy by exercise was not complete, but the data suggest that the added muscle tension due to exercise significantly improved muscle strength.

In terms of metabolic properties, muscle oxidative capacity, as indicated by SDH activity, did not change following spinalization, and even *increased* with exercise. Because the soleus muscles also demonstrated a significant slow-to-fast fiber type conversion, this increase in oxidative capacity probably reflects the increased SDH activity of the FOG fibers relative to SO fibers. This study illustrates that while *increases* in SDH activity can occur relatively rapidly and to a large extent following increased use, the converse does not occur following decreased use. It appears that skeletal muscle fiber "baseline" oxidative capacity is not very easily changed. Also, the tension of exercise significantly affected muscle fiber area and metabolic properties in spite of the fact that the upper motor neuron was cut.

Clinical Relevance

The main result of these studies is that following spinal cord injury, a slow-to-fast muscle fiber type transformation can occur. The consequences are that a dramatic change occurs in the contractile properties of muscles that have a large proportion of slow fibers. As a result of this increased contraction and relaxation speed, these "slow" muscles become less able to generate low-level, prolonged contractions, as is required of predominantly slow muscles (*e.g.,* soleus and vastus intermedius). If the specific tension of fast and slow muscle fibers differs, slow-to-fast transformation may result in increased (or at least not decreased) muscle strength, causing an imbalance of muscle forces about the joint. The joints would thus tend to remain extended, as most of the "antigravity" muscles contain a larger proportion of slow fibers. This concept may explain the observation that, following transection, the rat hindlimbs remained fully extended.

This study suggests that paraplegia is not necessarily associated with muscular weakness, especially after a relatively long period. While *neuromuscular* weakness follows injury, muscle strength is not necessarily compromised. This is not to say that muscle strength *cannot* decrease, only that it is not *necessarily* decreased.

The data also show that modalities that increase muscle tension are effective in reducing the atrophy seen with spinalization alone. Thus far, it has not been possible to completely reverse the effects of spinalization, but perhaps future treatments might include a greater exercise duration or intense muscle contractions (eccentric contractions?).

ADAPTATION TO HINDLIMB UNLOADING

Recently, a new model for studying muscle fiber atrophy was developed. In cooperation with the National Aeronautics and Space Administration (NASA), Emily Morey developed the hindlimb unloading model for simulating the weightless environment (Morey, 1979). It is well known that spaceflight results in muscle atrophy and loss of bone mineral content. Astronauts return from space weaker and more vulnerable to bone fracture. Such changes are similar to those observed with other decreased-use models, and thus there is great interest in discovering the factors that cause them. Unfortunately, it is extremely costly and difficult to perform scientific research in space. Therefore, a ground-based experimental model, the so-called hindlimb unloading model, was developed.

The Hindlimb Unloading Model

To mimic the effects of spaceflight, Morey and her colleagues removed the weight bearing function of rats' hindlimbs. Connectors were secured to the rats' tails and attached to a revolving gimbal mounted at the top of the cage. The rats could easily navigate about the cage using their forelimbs. Early experimental results were promising: Following hindlimb unloading, rats demonstrated many of the changes documented in spaceflight: muscle atrophy, bone mineral loss, interstitial fluid shifts, and decreased growth. From the point of view of studying muscle plasticity, the hindlimb unloading model also provided a unique opportunity to study a decreased-use model in which the lower motor neuron was intact and muscle tension was extremely low. (Incidentally, it was shown that the model itself caused a relatively small degree of transient stress in the rats, based on measurement of a small increase in adrenal gland mass and transiently increased plasma corticosterone levels. Both factors are known to increase with stress.)

Skeletal Muscle Activity during Hindlimb Unloading

In a manner similar to that described for the immobilization studies, Alford *et al.* (1987) measured the chronic EMG activity of both the tibialis anterior

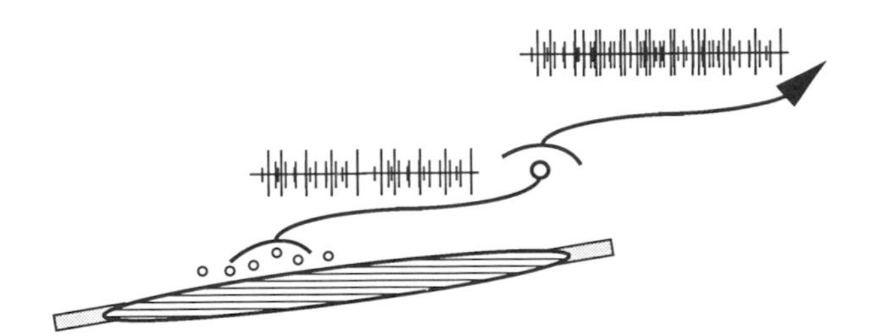

Figure 5.15. Schematic view of neuromuscular EMG activity following hindlimb unloading. Note that EMG activity remains relatively constant or even slightly increases following unloading. (Adapted from Alford *et al.*, 1987.)

ankle flexor and the soleus and medial gastrocnemius ankle extensors in order to better understand muscle use during hindlimb unloading. These investigators showed that the extensor activity of the soleus and gastrocnemius muscles decreased initially and then returned to control levels after about 4 weeks (Figure 5.15). However, the tibialis anterior, which was slightly loaded because of ankle extension, increased electrical activity by two to four times. Thus the hindlimb unloading model really does not cause complete disuse (since the muscles have some EMG activity).

Changes in Muscle Contractile Properties following Hindlimb Unloading

Interestingly, in spite of the unremarkable changes in muscle EMG activity after hindlimb unloading, dramatic changes in muscle mass and muscle contractile forces have been measured by numerous investigators. The nature of these changes was as expected based on our previous immobilization studies. For example, Don Thomason, working in Ken Baldwin's laboratory, showed that the mass of both the plantaris and soleus muscles decreased continually throughout the unloading period (Figure 5.16). Soon after unloading began, both muscle masses decreased rapidly. For the next 30 days, mass decreased at a slower and slower rate, until after about 30 days mass stayed relatively constant.

Other muscle properties have also been measured in other laboratories. Consistent with the change in mass, contractile tension of both muscle types decreased by about 50% during the same time period. Again, consistent with other decreased-use models, the soleus muscle began to demonstrate increased muscle speed, as manifest by a decrease in both contraction and half-relaxation times. Finally, significant increases in muscle maximum contraction velocity (V_{max}) (and even in single-fiber V_{max}) were observed.

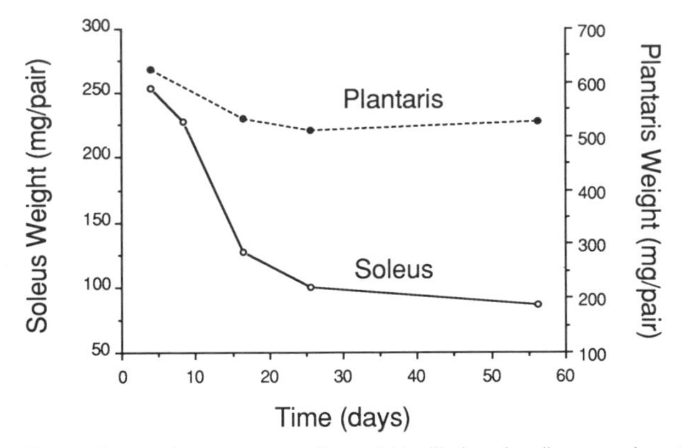

Figure 5.16. Change in muscle mass over an 8-week hindlimb unloading experiment. Muscle mass decreases quickly and then "levels off." (Adapted from Thomason *et al.*, 1987.)

Changes in Fast and Slow Muscle Fibers following Hindlimb Unloading

Using quantitative analysis of muscle fiber area and muscle fiber enzyme activity, Reggie Edgerton and Roland Roy demonstrated that the flexor and extensor muscles adapted differently to the unloading (Roy *et al.*, 1987; Hauschka *et al.*, 1987). Whereas the fast and slow fibers within the gastrocnemius muscle atrophied by about 25%, the tibialis anterior muscle fibers showed no atrophy whatsoever (Figure 5.17). Part of the tibialis anterior response was probably due to the slight stretch placed on it during unloading, but most of the change was probably attributable to the low responsiveness of the ankle flexor to decreased use. This is the same story we saw with the immobilization and cordotomy models.

In terms of metabolic activity, the single fibers of both the medial gastrocnemius and tibialis anterior dramatically decreased in SDH activity, while only the medial gastrocnemius dramatically increased in αGP activity (Figure 5.18). This actually represented a departure from the norm, in which decreased-use models usually showed no change in muscle oxidative capacity. The increased αGP activity was interpreted as before: Glycolytic capacity appears to be matched to muscle fiber type. Thus muscles that increase their proportion of fast fibers also increase their glycolytic capacity. (Incidentally, the same investigators demonstrated that 1.5 hours of exercise per day on a 30% grade at relatively high speeds could partially ameliorate these deleterious effects—further evidence that muscle tension can modulate the atrophic response.)

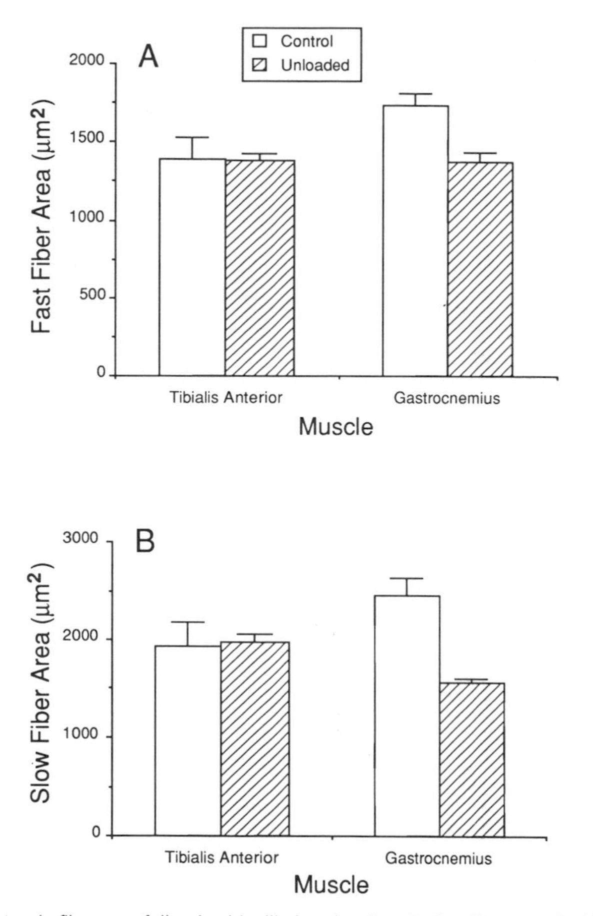

Figure 5.17. Muscle fiber area following hindlimb unloading. **A,** Fast fiber area. **B,** Slow fiber area. Note significant atrophy of the medial gastrocnemius but no atrophy of the tibialis anterior. (Adapted from Roy *et al.*, 1987.)

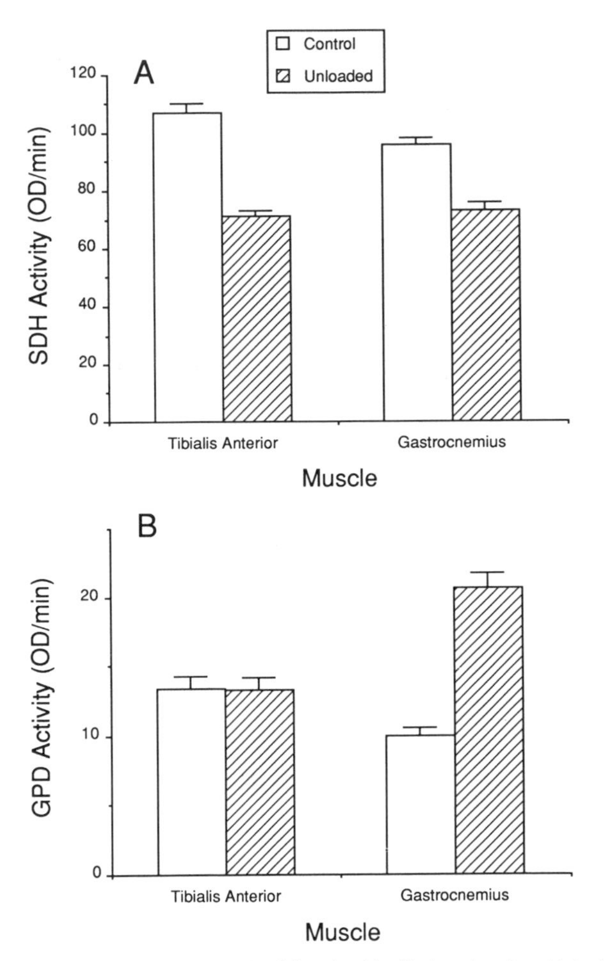

Figure 5.18. Muscle fiber enzyme activity following hindlimb unloading. Note the decrease in oxidative capacity (**A**) with an increase in glycolytic capacity (**B**). **GDP,** glycerol phosphate dehydrogenase. (Adapted from Roy *et al.,* 1987.)

MECHANISM OF MUSCLE FIBER ATROPHY

Introduction to Muscle Protein Turnover

We know already that the muscle fiber size is related to the number of myofibrils arranged in parallel within the fiber. We also know that myofibrils comprise about 70% of the total protein within the muscle cell. Thus it follows that muscle force is related to the total amount of myofibrillar proteins within the fiber.

It may not be obvious, but all proteins (indeed, all cellular components) eventually wear out and die. The rate at which these different components are replaced depends on their location and function. The control center of the cell (the nucleus) is responsible for synthesizing new muscle proteins and repairing the portions of the cell that have been damaged or have aged. If protein synthesis is completely blocked, the cell cannot repair itself and eventually dies (the basis of some chemotherapeutic drugs).

Protein synthesis and protein degradation are always occurring within the muscle cell. We believe that, in this way, cells retain the ability to adapt to a new environment. If proteins were permanent after synthesis, there would be no way to change them. Thus proteins are constantly "turning over" to yield to the current cellular demands. It is this study of protein turnover in atrophying muscle that has provided new insights into the atrophy mechanism.

Protein Turnover during Hindlimb Unloading

Studies of protein turnover in muscle were pioneered in the laboratories of Ken Baldwin and Frank Booth. These investigators measured rates of synthesis and degradation in muscles following various altered-use models and have painted a picture of muscle mass regulation following altered use. Specifically, hindlimb unloading has been studied by Don Thomason, working in both laboratories. Recall the data that demonstrated that both plantaris and soleus muscle mass decreased continually throughout the unloading period. For about the first 30 days of unloading, mass decreased at a slower and slower rate (Figure 5.16; Thomason *et al.,* 1987; and Thomason and Booth, 1990). After about 30 days mass stayed relatively constant. We know that muscle mass (or change in mass) simply represents the net balance between protein synthesis and degradation. Mass could decrease even if degradation rate decreased, as long as synthesis rate decreased more. Similarly, mass could increase even if synthesis rate decreased, as long as degradation rate decreased even more. What was the explanation for muscle mass changes following hindlimb unloading?

Changes in Synthesis and Degradation

Thomason *et al.,* (1989) showed that, very soon after hindlimb unloading (within a day), soleus muscles decreased their protein synthetic rate by about 50%, and this rate remained relatively constant for the remainder of the unloading period (up to 7 days have been measured). The protein synthesis rate decrease occurred in spite of the fact that soleus EMG activity continued to increase (another lesson that demonstrates that increased electrical activity does not translate into increased strength). Since we saw that muscle mass continued to decrease during this rate of constant protein synthesis, the data predicted that the protein degradation rate must have increased from the 4th to 14th day of unloading. Thereafter, degradation decreased to a value less than control, at which time degradation must have become relatively constant (Figure 5.19). All of the predictions are based on the fact that mass always represents a balance between synthesis and degradation. This hypothesis for the time course of protein degradation remains to be tested.

The take-home lesson from these experiments is that in hindlimb unloading, it takes about 30 days for a muscle to reach a new state of homeostasis even under constant unloading conditions. The phenomenon must be even more complicated under conditions when muscle state changes continually (*e.g.,* exercise, immobilization, continuous passive motion).

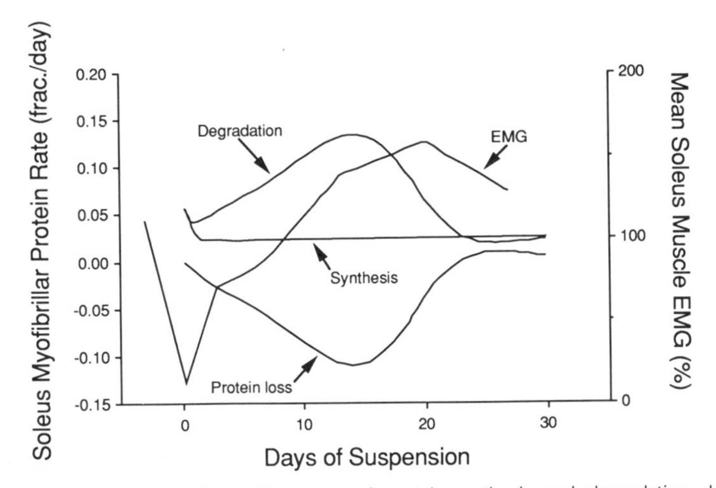

Figure 5.19. Time course of muscle mass and protein synthesis and degradation during rat hindlimb unloading. (Figure drawn after Thomason *et al.,* 1987.)

Cellular Control of Synthesis and Degradation after Unloading

What factors caused the synthetic and degradation rates observed? Clearly, electrical activity was not the cause of either the synthetic or the degradation rates (compare Figure 5.15 to 5.19). We can take a step back from measuring the rates of protein synthesis and look at the events that precede protein synthesis, namely, mRNA information.

Recall that the central dogma of molecule biology is that DNA is produced by replication. DNA single strands are copied into mRNA by transcription, and mRNA single strands are used to make protein by the process of translation. Thus, potentially, replication, transcription or translation can all be involved in changing the level of any cellular protein. Booth and his colleagues have looked at several different mRNAs (Babij and Booth, 1988a and 1988b). They found no change in mRNA coding for either α-actin (measured after 1 day of unloading) or the β-myosin heavy chain (measured after 7 days of unloading), in spite of the fact that the synthesis rates of both proteins had decreased precipitously during these time periods. After 7 days, α-actin mRNA concentration decreased by about 30%, but synthetic rate decreased by about 60%! Finally, administration of clenbuterol abolished the 30% decrease in α-actin mRNA but had no effect on protein synthetic rate!

They concluded that while hindlimb unloading resulted in downregulation of both transcription and translation, the dramatic synthesis rate decrease was primarily due to the downregulation of translation. Apparently, there was more than enough mRNA in the muscle cell, such that α-actin mRNA concentrations did not limit the synthetic rate. Recently, Booth reviewed other plasticity models and showed that regulation can occur (and does occur) at many different levels within the cell. The unifying principles that explain why regulation should occur at one point or another remain to be elucidated (Booth and Thomason, 1991).

ADAPTATION TO DENERVATION

We now consider a "completely different" type of decreased-use model—denervation (Figure 5.20). Denervation was used in early muscle plasticity studies as a model of disuse. However, as we have repeatedly seen, a denervated muscle is a "different" kind of muscle. For example, we saw that the denervated soleus muscle responds to electrical stimulation patterns much differently from how the normally innervated muscle responds. As such, we now consider denervation not so much as a typical model of decreased use but rather as a model illustrating the intimate relationship between muscles and nerves. We will also consider the unique aspects of

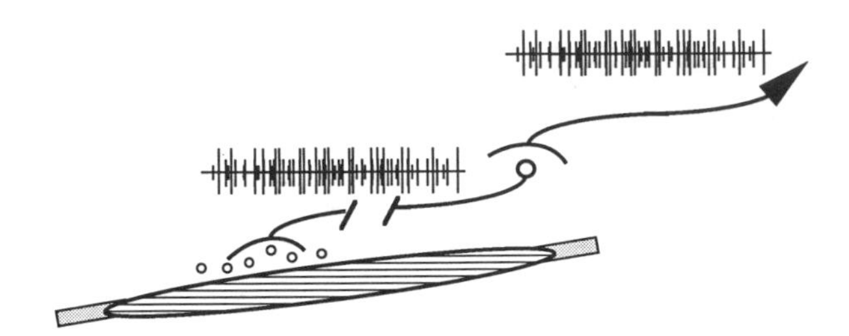

Figure 5.20. Schematic diagram of denervation model. Lower motor neuron is transected. (Compare to Figures 5.1 and 5.8.)

denervation, which have significant clinical implications. How does one treat a denervated muscle? Are the adaptive responses to tension and stretch similar? How do denervated muscle fibers become reinnervated? Do muscle properties return to normal following reinnervation?

Muscle Strength Changes following Denervation

The most obvious muscle change following denervation is muscle atrophy. This atrophy affects both fast and slow muscle fibers, resulting in decreased fiber diameter and decreased muscle force. Analogous to previously mentioned decreased-use models, denervated muscles increase their contractile speed due to fiber type conversion in the slow-to-fast direction. In spite of this conversion, the muscles still generate much lower forces than normal.

Muscle Fiber Atrophy following Denervation

While the mechanism of muscle fiber atrophy is not fully understood in any experimental model, recent experiments have implicated some intramuscular proteolytic enzymes in the atrophic process. An interesting denervation/reinnervation experiment was performed by Marie Badalmente, who applied the protease inhibitor leupeptin to primate muscles following median nerve denervation and surgical repair (Badalmente *et al.,* 1989). Muscle fibers normally contain enzymes within them that are capable of digesting cell constituents. In other words, the fibers have the capability to self-destruct if these proteolytic enzymes are activated. One family of enzymes is activated automatically when the intracellular calcium levels rise too high (the calcium-activated neutral proteases, CANPs). Since intracellular calcium levels are not normally high, increased free calcium within a muscle fiber signifies that muscle fiber integrity has been lost (the cell membrane has been broken) and subsequent regeneration should occur (see a further discussion of this

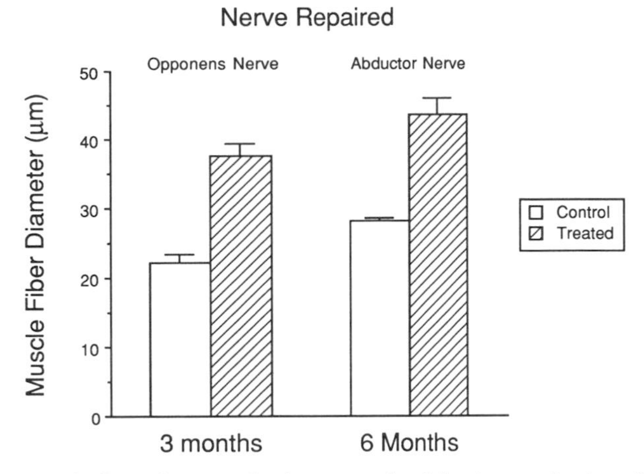

Figure 5.21. Muscle fiber diameter of primate muscles following application of leupeptin to denervated muscles. Treated muscles received application of leupeptin, the calcium protease inhibitor. Control muscles were simply denervated. (Data from Badalmente *et al.*, 1989.)

mechanism in Chapter 6). Badalmente and her colleagues injected monkeys with the CANPs inhibitor leupeptin in order to attenuate CANPs' effect. They first confirmed that CANPs were indeed suppressed by leupeptin, and they then showed that the muscle fiber diameter of leupeptin-treated abductor pollicis and opponens pollicis muscles was greater than that of the nontreated denervated muscles, suggesting less atrophy (Figure 5.21). Additionally, a greater number of axons survived when muscles were treated with leupeptin. These data may suggest that maintenance of muscle properties following denervation may have a beneficial effect on the reinnervating nerve. Alternatively, CANPs may have a direct axon-sparing role. Clearly, further work is required in this area.

Muscle Fiber Changes

Perhaps the most interesting muscle responses to denervation are the subtle changes around the fiber that signal to the outside world that the nerve and muscle are no longer in communication. For example, recall in muscle development that the myotube is literally covered with acetylcholine receptors (AChR) until the nerve "arrives." After innervation, the density and number of extrajunctional AChRs decrease dramatically, and the only remaining receptors are those at the neuromuscular junction (NMJ). Follow-

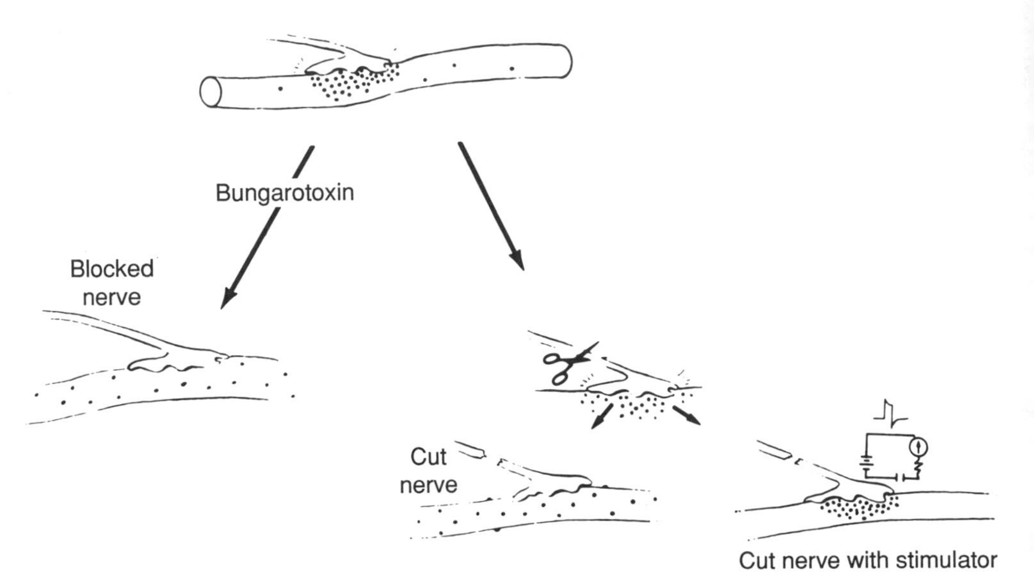

Figure 5.22. Relationship between the distribution of ACh receptors and the presence or absence of the nerve. Note that by blocking the neuromuscular junction with a toxin (in this case α-bungarotoxin), ACh receptors are expressed along the entire muscle fiber length. This is analogous to what happens if the motor nerve to a muscle fiber is cut (**lower right panel**). However, if muscle stimulation is superimposed upon the denervated fiber, such ACh receptor expression does not occur. This may imply that electrical stimulation of a denervated muscle fiber inhibits reinnervation.

ing denervation, we observe a reversal of the development process in that a proliferation of extrajunctional AChRs occurs (Figure 5.22). Some view this response as a sort of "signal" that causes nerve sprouts and a preparation to form a new neuromuscular junction. The fact that denervation mimics part of the development of the neuromuscular junction is one reason why developmental biologists use denervation as an experimental model (see experiments described in Chapter 1).

What other signals are available to guide the incoming nerves? Recently, an extracellular matrix molecule known as neural cell adhesion molecule (NCAM) was implicated in the denervation-reinnervation process. Recall once again that in development, NCAM is expressed on the surface of the primitive myotubes to perhaps guide incoming nerves to the muscle fiber. In denervation, NCAM is again expressed by the muscle fiber (Figure 5.23). In fact, the NCAM molecule appears to be expressed *any time* in which the muscle fiber is "receptive" to incoming nervous innervation. Let us digress for a moment to discuss some of the experimental demonstrations of muscle fiber innervation by foreign nerves.

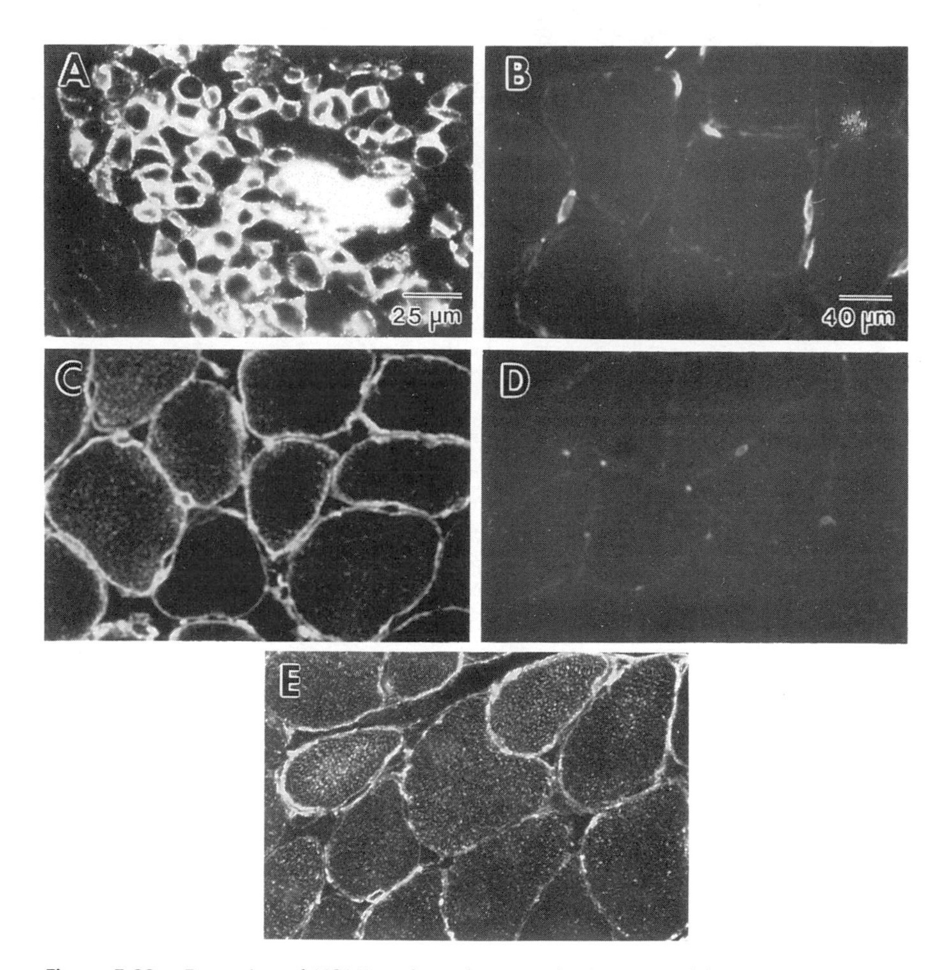

Figure 5.23. Expression of NCAM under various muscle conditions. (**A**) Small fibers during development. (**B**) Normal muscle. (**C**) Denervation. (**D**) Denervation with stimulation. (**E**) Neuromuscu- lar blockade with α-bungarotoxin. The bright "halo" around the fibers represents the antibody to NCAM. Thus in conditions **A, C,** and **E,** NCAM is present. (Micrographs kindly provided by Jonathan Covault, University of Connecticut.)

Muscle Fiber "Receptivity" to Nerves

A nerve can be surgically implanted into a normal skeletal muscle and will often continue to grow (Figure 5.24). It will continue to grow and sprout small neuronal processes, but these processes will not form new synapses with the muscle fiber. Why not? Apparently, the muscle fiber signals the nerve that it is normally innervated, and thus it remains refractory to further innervation. If,

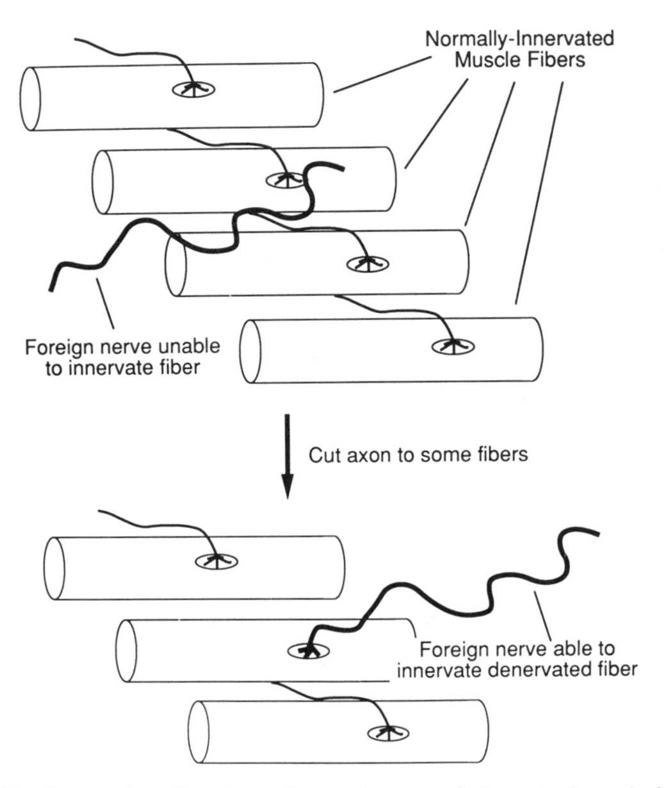

Figure 5.24. Synapse formation does not occur in a normally innervated muscle fiber (upper panel). Note that a foreign nerve (*thick wiggly line*) growing in a muscle composed of innervated fibers does not form a synapse. However, if the motor nerve to a muscle fiber is cut (lower panel), a new synapse will form.

however, the muscle's normal nerve is cut, the nerve that was surgically implanted into the muscle will now form functional synapses (Figure 5.24). In fact, if the muscle is only partially denervated (only a fraction of the nerve is cut), the implanted nerve will innervate only those fibers whose nerve branches were cut! Quite an effective signaling method is working here.

Experimentally, other treatments can cause muscle fibers to change their receptivity to innervation by foreign nerves. As you might predict, these are treatments that change the activity level of the muscle fibers. For example, a muscle can be paralyzed with α-bungarotoxin (a toxin that binds postsynaptically to nerves and blocks the ACh receptor), and the fibers will become receptive to innervation. Conversely, if a denervated muscle is electrically activated, it will not permit innervation by the incoming nerve in spite of the fact that it has no normal nerve of its own (Figure 5.24). Why not?

Joshua Sanes and Jonathan Covault have shown that all of these conditions correlate with the muscle fiber's expression of NCAM (Figure 5.23). Thus under conditions of high activity (normally innervated muscle or electrically stimulated, denervated muscle), NCAM is not expressed and synapse formation does not occur. However, when muscle fiber activity is low (in denervation, pharmacologic paralysis, or development), NCAM is expressed, and nerves synapse with the fiber (Sanes and Covault, 1985).

Formation of New Muscle Synapses

Sanes and his colleagues, as well as others, have implanted nerves into denervated muscles to determine where along the fiber length the synapse is formed. If nerves are implanted close to the original NMJ, the synapse forms at exactly the original location. If, however, a site remote from the original NMJ is chosen, a new, ectopic synapse will form. In an effort to determine what was special about the NMJ, Glicksman and Sanes (1985) experimentally killed muscle fibers by mechanically crushing them. Normally, the muscle fibers would regenerate (Chapter 6) to form a new muscle. However, these investigators prevented muscle regeneration by repeated exposure to intense x-rays, leaving only the muscle-free ghosts—essentially only the basal lamina filled with muscle fiber debris. Recall that the basal lamina is the structure ensheathing muscle fibers during development. It is outside of the sarcolemma and is thus an extracellular structure. Interestingly, reinnervation of these muscle-free ghosts proceeded in a relatively normal fashion. Relatively normal NMJs were formed in precisely the location of the original NMJ, and the nerve presynaptic terminals were "loaded" with normal-appearing ACh vesicles. These data provide strong evidence that at least in the early stages of synapse formation, the basal lamina provides a sufficient molecular signal to the nerve for synaptogenesis. What about our old friend NCAM? This experiment reinforces an important lesson: Just because one event is associated with another event does not mean that it causes it. Clearly NCAM expression is *associated* with reinnervation, but it alone does not *cause* it.

MUSCLE FIBER SPECIFICITY TO REINNERVATION

We have discussed the conditions surrounding reinnervation of a muscle fiber. Let us take this discussion one step further: How does a nerve select which muscle fiber *type* to innervate? This is a hot research area, and definitive answers are not yet available. However, a number of experiments have provided clues leading us to believe that nerves generally innervate the nearest receptive fiber, regardless of type. But the story is not quite that simple.

Motor Unit Properties following Cross-Reinnervation

To study the specificity of reinnervation, a nerve that normally innervates slow fibers is given the opportunity to innervate slow or fast fibers (or *vice versa*). This can be done by surgically cross-reinnervating the triceps surae muscles with the antagonistic motor axons from the common peroneal nerve. In this case, flexor neurons are forced to innervate extensor muscles. Such an experiment was performed by Tessa Gordon, Dick Stein, and their colleagues using the cat medial gastrocnemius muscle (Figure 5.25). After performing the surgical reanastomosis of the proximal flexor nerve with the distal extensor nerve stump (and *vice versa*), they waited 1–2 years and again measured motor unit properties (as discussed in Chapter 2, page 90) (Gordon *et al.*, 1988).

Motor Unit Tension following Reinnervation

These investigators had already documented the relationship between motor unit tension and axon size as a function of unit type (Figure 5.26**A**). As expected, a wide range of normal MG motor unit tensions was observed, ranging from only 10–20 mN to over 1000 mN (1 mN equals 1.0×10^{-3} N, or 9.8 g). Also, as expected based on the size principle, the normal MG units that generated the smallest tensions had the smallest axon action potentials and, thus, the smallest axon size. Units developing the highest tensions had the largest axon sizes. Finally, low-tension units were identified as S units, the intermediate-tension units as FR units, and the high-tension units as FF units.

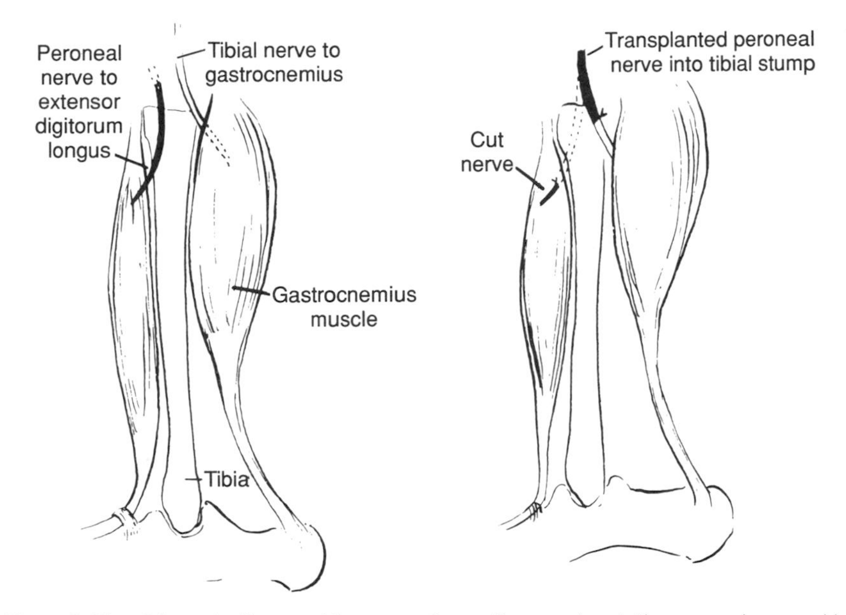

Figure 5.25. Schematic diagram of the cross-reinnervation experiment. The peroneal nerve, which normally innervates the anterior compartment muscles, is surgically reattached to the distal stump of the tibial nerve, which normally innervates the plantar flexors.

Following reinnervation a number of changes occurred. Before you read the results, try to predict what *could* have happened. Remember that the MG normally contains all three unit types, which generate over two to three times the average force of comparable units in the flexor muscles that are normally supplied by the peroneal nerve. What tensions do you think were generated by the reinnervated MG? It turns out that the reinnervated MG motor unit tensions were roughly the same as the normal MG motor units, suggesting that each peroneal nerve axon innervated more fibers in the MG than it had in the dorsiflexors. This might have been because fewer axons grew out of the peroneal nerve to innervate the denervated MG fibers. Another explanation might have been that the peroneal nerve axons had the intrinsic capability of innervating more fibers and simply seized the opportunity. Axon action potential was still directly correlated with motor unit tension, implying that the "size principle" still held for the reinnervated muscle. In fact, the regression relationship between motor unit tension and axon action potential was almost identical before and after reinnervation (Figure 5.26**B**).

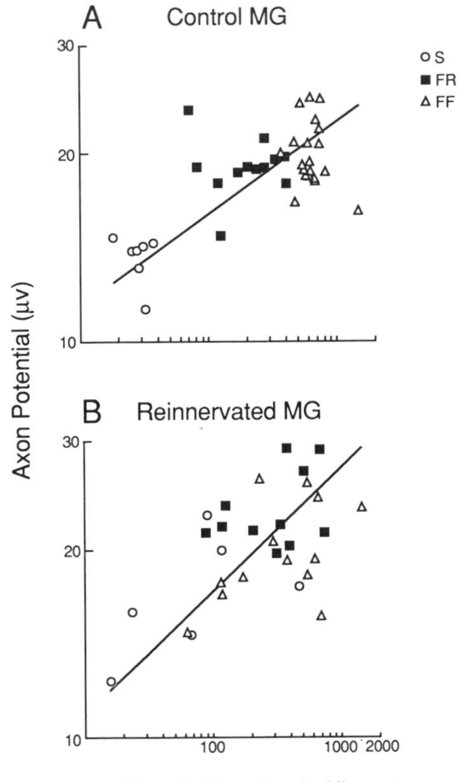

Figure 5.26. Relationship between motor unit tension and action potential amplitude for (**A**) normal and (**B**) cross-reinnervated cat medial gastrocnemius. Each symbol represents a different motor unit type. Normally, S units are associated with the smallest axons, represented by low axon potentials. However, following reinnervation, this relationship is dramatically altered. (Adapted from Gordon *et al.*, 1988.)

Motor Unit Specificity following Reinnervation

What about the specific muscle fiber types innervated by the reinnervating axons? Did axons of the appropriate size find the appropriate fiber types? The answer was clearly negative (Figure 5.27). Whereas in the normal MG, the low-tension units were of the S type, following reinnervation motor unit tension was not related to motor unit type (Figure 5.26**B**). Thus some S units

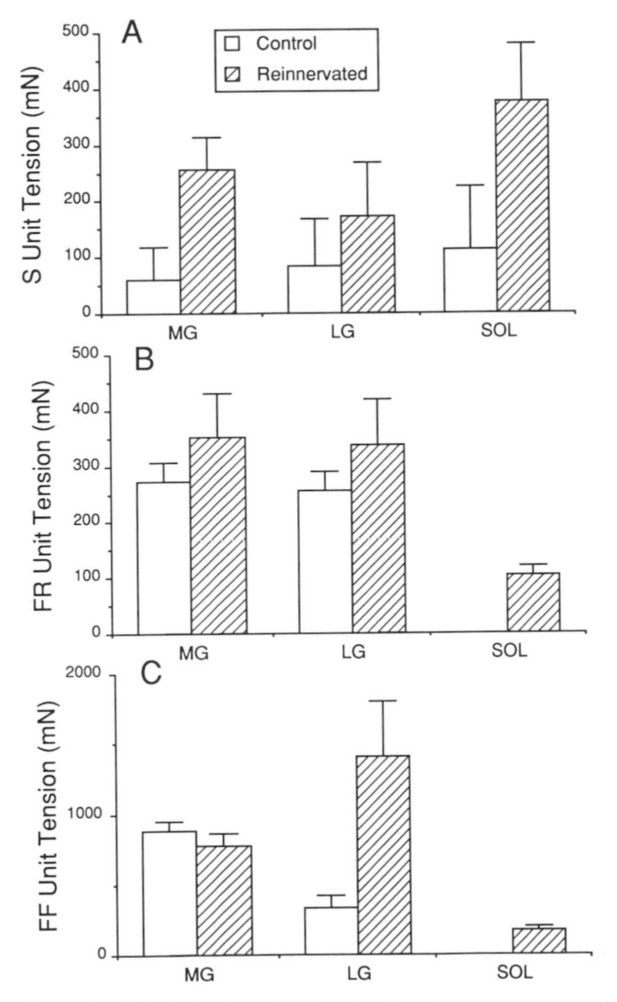

Figure 5.27. Bar graph of tetanic tension of different motor unit types from control (*open bars*) or cross-reinnervated (*hatched bars*) cat skeletal muscles. (Adapted from Gordon *et al.*, 1988.)

developed high tensions, and some FF units developed very low tensions. Reinnervation was not perfectly fiber type specific.

Other Consequences of Reinnervation

A few notable results were observed with the soleus muscle. The normal cat soleus is composed of 100% S units, 100% SO fibers. However, following reinnervation, the motor unit percentage was 39% S units, 22% FR units, and 39% FF units. The fiber type percentage was not explicitly measured, but it was noted that there were absolutely no type FG muscle fibers in the reinnervated soleus. No FG fibers, but many FF units? Right. The data demonstrate two things: First, it was not possible to completely transform the soleus muscle fibers into each and every fiber type, and, second, motor unit classification and muscle fiber type classification are not necessarily interchangeable. Recall that normal muscle FF units are composed of FG fibers, but this is obviously not necessarily the case since FF units were identified in the absence of any FG fibers. This result highlights the complex nature of motor unit performance and suggests caution in going from fiber type distribution to speculating on performance (Chapter 4, page 199).

Using the indirect computational methods of Burke (1981) (see Chapter 2 for discussion), Gordon and her colleagues calculated that the innervation ratio of the S and FR units increased while the innervation ratio of the FF units decreased (Figure 5.28). They also measured the fiber areas of the various muscle fiber types and found that, generally, the SO and FOG fibers increased in size, while the FG fibers decreased in size. Can you explain how these

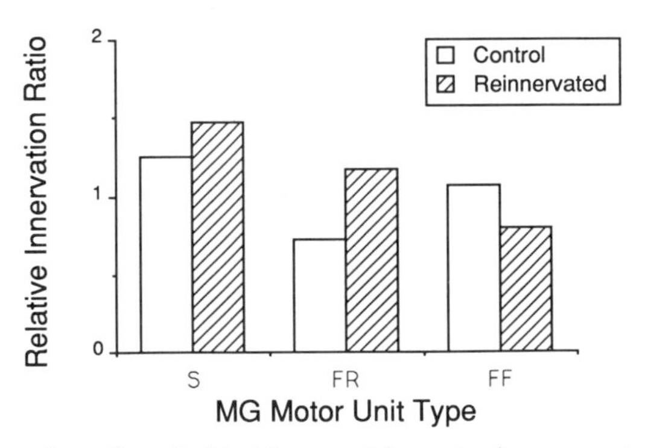

Figure 5.28. Innervation ratios calculated for normal (*open bars*) or cross-reinnervated (*hatched bars*) cat skelelal muscles. (Adapted from Gordon *et al.*, 1988.)

changes could result in the motor unit tension values measured? Consider each unit type separately: S unit innervation ratio increased by 18% and SO fiber size increased by 47%, but S unit tension increased disproportionately by 23%. The only plausible explanation for this result is that the specific tension of the S and units increased (since tension increased more than that expected simply based on an increased innervation ratio and the measured fiber size). Analogously, FR unit innervation ratio increased by 60% and FOG fiber size increased by 35%, but FR unit tension increased only by 25%. Again, a reasonable explanation is that specific tension of the FR units decreased (since the tension increased less than that expected based on the increase in innervation ratio and the measured fiber size). Can you explain the FF units result? (Hint: Specific tension was predicted to increase but for reasons different from those of the S units. This is a good exercise to see if you understand the contributing factors to motor unit tension.) Since the MG motor unit specific tensions were "correct" following reinnervation (in the sense that they were ordered correctly although their absolute values were different) but innervation ratio and fiber area were not, the conclusion was that specific tension was the only parameter uniquely determined by the innervating axon. The main limitation of this interpretation is that calculation of innervation ratios and specific tensions using this model are very indirect. Specific tension is always the term that is "left over" in order to balance the motor unit tension equation. Future experiments are required to clarify these issues.

To summarize, denervation muscle represents a "different beast" since loss of neuromuscular communication leads to dramatic alterations in muscle structure, function, and adaptive ability. In this exciting research area, we are beginning to understand the factors that regulate nerve-muscle interaction. Stay tuned for future developments.

Evidence for Specificity of Reinnervation

The previous denervation-reinnervation discussion might lead one to the conclusion that reinnervation is completely random. However, this is far from the truth. Many studies have demonstrated that many axons growing out from a denervated nerve stump are able to navigate to the correct distal stump and reinnervate the appropriate muscle. This was shown in a most dramatic way earlier this century by Cajál, who cut the motor nerve to a fast muscle and then constructed a Y-chamber in order to give this reinnervating nerve a choice of either the original muscle, or another tissue (Figure 5.29). If reinnervation were completely random, an equal number of axons would have chosen each "branch" of the Y (Cajál, 1928). However, most of the reinnervating axons

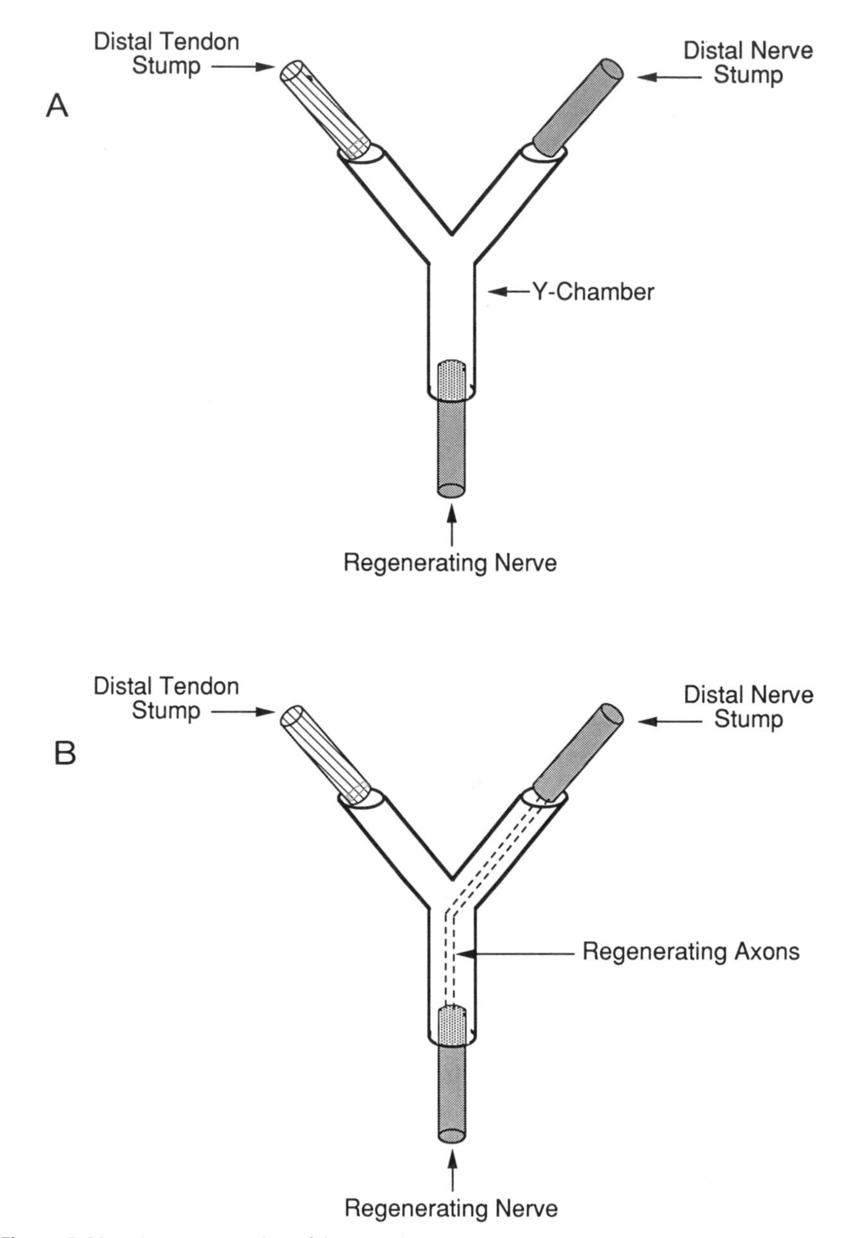

Figure 5.29. **A,** Representation of the experiment in which a severed nerve was given the option of growing through a Y-tube into either the original nerve, or a tendon. **B,** In all cases, the nerve selectively chose to grow toward the distal nerve stump. This experiment suggests some type of attraction of nerve for incoming axons.

chose to innervate the correct distal stump, providing evidence that some type of signal permitted specific reinnervation. More recent experiments have implicated various chemicals such as growth factors and specific extracellular proteins as the chemical signals for providing navigational aid to incoming axons.

A more recent demonstration of reinnervation specificity was provided by Rende *et al.* (1991), who used the "double labeling" technique to trace the connections between axons and motoneuron cell bodies in the spinal cord (Figure 5.30). The experimental model was the rat TA. The majority of the motoneuron cell bodies to the TA are located at spinal cord level L3. Rende *et*

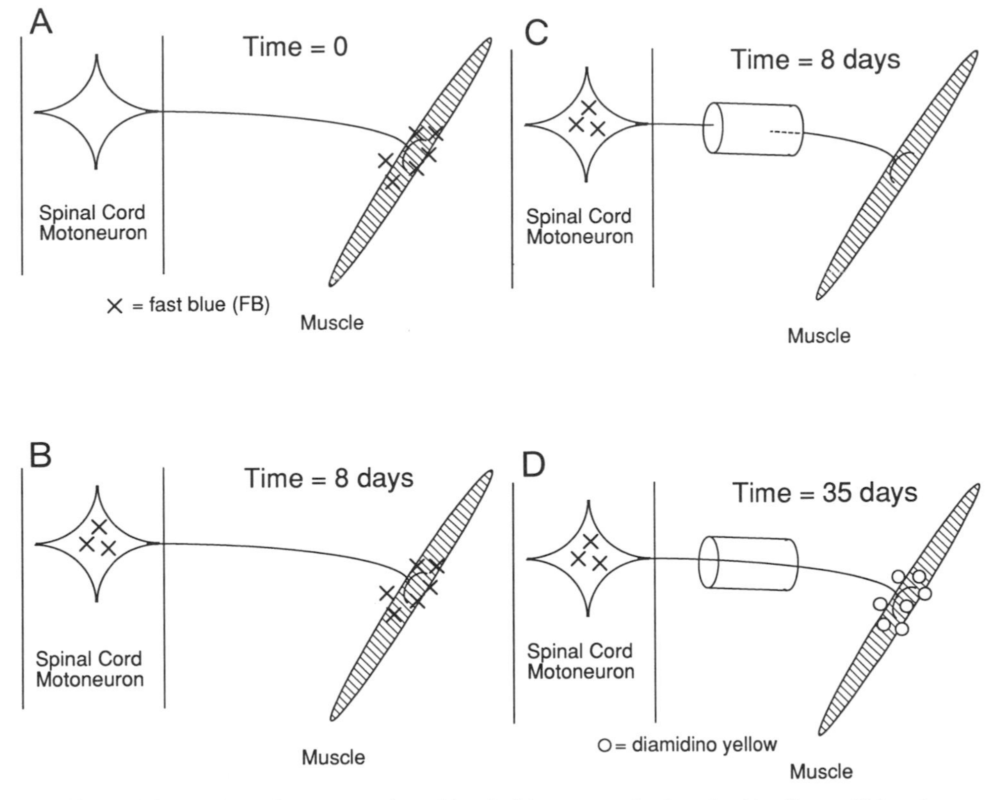

Figure 5.30. Schematic representation of the double tracer method used to identify specificity of connections between motoneurons in the spinal cord and peripheral muscle targets. **A,** Original connections between nerves innervating TA labeled using fast blue (FB) over an 8-day period. **B,** After 8 days, FB labels cell body. **C,** Nerve cut and allowed to grow back through a 10-mm gap in a plastic cube. **D,** After TA reinnervation, connections between TA and spinal cord labeled using diamidine yellow (DY).

al. used two separate tracer dyes to follow the connections between the TA and spinal cord motoneurons before and after denervation-reinnervation. They started by injecting the blue tracer known as "fast blue" into the TA muscle. Fast blue (FB) is taken up at the neuromuscular junction and moved via axonal transport into the motoneuron cell bodies (Figure 5.30**A**). After 8 days (enough time for the axons to transport FB to the motoneuron cell bodies) they cut the sciatic nerve (which contains the motor nerves to the TA, other dorsiflexors, and plantar flexors, as well as many sensory nerves) and placed both cut ends of the nerve in a small plastic tube, leaving a long (10-mm) gap between ends (Figure 5.30**B**). This configuration allowed the outgrowing axons to make a choice between several nerve fascicles, some correct, some incorrect. After 30–35 days (enough time for many outgrowing axons to arrive at the muscles), they re-injected the TA muscles with a different tracer known as "diamidino yellow dihydrochloride," or more simply, DY. DY acts similarly to FB in that it is transported back to the motoneurons where it can be observed microscopically (Figure 5.30**C**). Before reading on, try to predict the appearance of motoneuron cell bodies that were correctly innervated. Which tracers would be present? How would motoneurons appear that had originally projected to the TA but were not correctly innervated?

The results, shown schematically in Figure 5.31, provided convincing evidence that 30% of the reinnervating axons found the TA muscle! Motoneurons showing both the FB and DY tracer were those that had originally been labeled with FB and then subsequently labeled with DY after correct reinnervation. However, many other motoneurons at the L3 level showed only FB, which suggested either that they were originally labeled and failed to reinnervate anything or that they were originally labeled and reinnervated the wrong muscle (Figure 5.31). Current research in this area will use this type of model to investigate the signals (chemical, mechanical, hormonal) that control specificity of reinnervation. Ultimately, improved reinnervation specificity will improve functional recovery of patients following nerve injury. It looks like our main hope is not so much in expert microsurgical technique (note 30% specific reinnervation when *no* surgery was performed) but rather in identifying the appropriate biologic signals (Lundborg, 1988).

SUMMARY

Skeletal muscle response to decreased use is characterized by muscle fiber atrophy, decreased muscle force generating capacity, and a slow-to-fast muscle fiber type conversion (if the disuse is extreme enough). Generally, muscle oxidative capacity and endurance do not change. This essentially represents a reversal of the increased use response. Glycolytic response

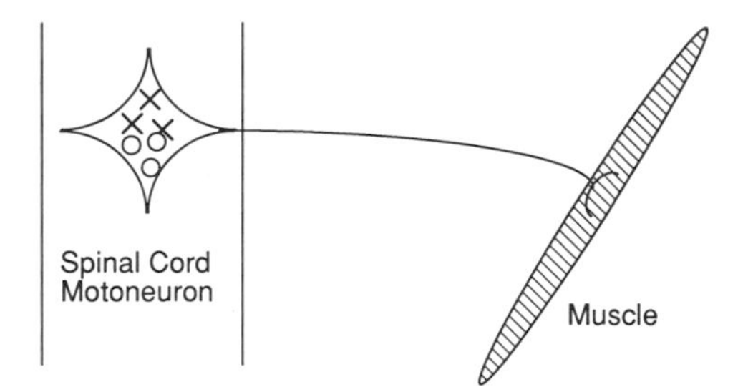

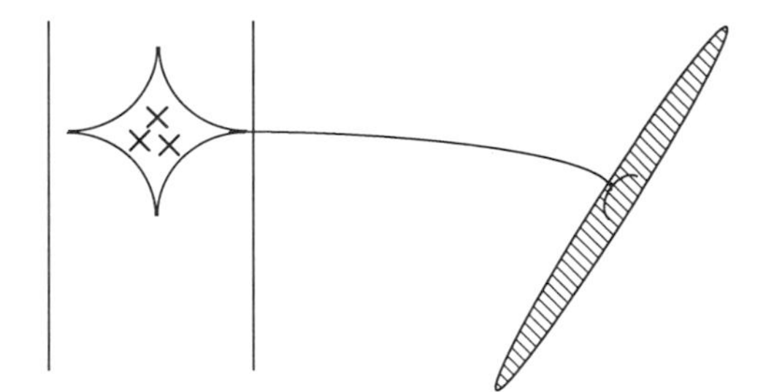

Figure 5.31. Schematic representation of innervation specificity following denervation of the rat sciatic nerve and permitting regrowth across a 10-mm gap within a plastic tube. Symbols refer to the tracers described in the text and Figure 5.30. Note that if the reinnervating axon is directed toward one of the originally innervated fibers, both tracers are in the cell body. However, if axons are directed toward the wrong target (where no diamidine yellow was applied) or if the axon failed to regenerate, only fast blue is found in the cell body.

seems to match the fiber type: Increased percentage of fast fibers is accompanied by increased glycolytic capacity. The magnitude of the atrophic and adaptive response is related to the *change* in use experienced by the muscle. Thus the often-used antigravity muscles atrophy to a greater extent than their less often used antagonists. Regulation of muscle mass represents the balance between protein synthesis and degradation. In the hindlimb unloading model, muscle loss was initiated at the translational level. Pretranslational and posttranslational regulation have been observed in unloading and other models of decreased use. When the nerve to a muscle is cut, the muscle fiber becomes a qualitatively different entity. We observe a reversal of some of the developmental processes that serve as signals to nerves for reinnervation to occur. The reinnervation process itself does not appear to be fiber type specific, although some signals are conveyed between the reinnervating stump and its target. At this point we have discussed muscle adaptation to the relatively "normal" perturbations. However, in some cases the level of use is so intense or extreme that the muscle itself becomes injured. In still other cases muscle malformation during development imparts specific properties to the muscle. In the next chapter, we consider these muscle responses to exercise-induced injury, surgical trauma, and muscle disease.

REFERENCES

Alford, E.K., Roy, R.R., Hodgson, J.A., and Edgerton, V.R. (1987). Electromyography of rat soleus, medial gastrocnemius, and tibialis anterior during hindlimb suspension. Exp. Neurol. 96:635–649.

Babij, P, and Booth, F.W. (1988a) Actin and cytochrome c mRNAs in atrophied adult rat skeletal muscle. Am J. Physiol. 254:C651–C656.

Babij, P., and Booth, F.W. (1988b). Clenbuterol prevents or inhibits loss of specific mRNAs in atrophying rat skeletal muscle. Am. J. Physiol. 254:C657–C660.

Badalmente, M.A., Hurst, L.C., and Stracher, A. (1989). Neuromuscular recovery using calcium protease inhibition after median nerve repair in primates. Proc. Natl. Acad. Sci. 86:5983–5987.

Booth, F.W., and Thomason, D.B. (1991). Molecular and cellular adaptation of muscle in response to exercise: perspectives of various models. Physiol. Rev. 71:541–585.

Burke, R.E. (1981). Motor units: anatomy, physiology, and functional organization. In: Brookhart, J.M., Mountcastle, V.B., Brooks, V.B., Geiger, S.R., eds. Handbook of Physiology. Bethesda, MD: American Physiological Society, 345–422.

Cajál, R. (1928). Degeneration and regeneration of the nervous system. London: Oxford University Press.

Edgerton, V.R., Barnard, R.J., Peter, J.B., Maier, A., and Simpson, D.R. (1975). Properties of immobilized hind-limb muscles of the galago senegalensis. Exp. Neurol. 46:115–131.

Fournier, M., Roy, R.R., Perham, H., Simard, C.P., and Edgerton, V.R. (1983). Is limb immobilization a model of muscle disuse? Exp. Neurol. 80:147–156.

Gordon, T., Thomas, C.K., Stein, R.B., and Erdebil, S. (1988). Comparison of physiological and histochemical properties of motor units after cross-reinnervation of antagonistic muscles in the cat hindlimb. J. Neurophysiol. 60:365–378.

Grimby, G., Broberg, C., Krotkiewska, I., and Krotkiewski, M. (1976). Muscle fiber composition in patients with traumatic cord lesion. Scand. J. Rehabil. Med. 8:37–42.

Haggmark, T., Eriksson, E., and Jansson, E. (1986). Muscle fiber type changes in human skeletal muscle after injuries and immobilization. Orthopaedics 9:181–185.

Hauschka, E, Roy, R., and Edgerton, R. (1987). Size and metabolic properties of single muscle fibers in rat soleus after hindlimb suspension. J. Appl. Physiol. 62:2338–2347.

Jiang, R., Roy, R., and Edgerton, R. (1990). Expression of a fast fiber enzyme profile in the cat soleus after spinalization. Muscle Nerve. 13:1037–1049.

Lieber, R.L, Fridén, J.O., Hargens, A.R., Danzig, L.A., and Gershuni, D.H. (1988). Differential response of the dog quadriceps muscle to external skeletal fixation of the knee. Muscle Nerve. 11:193–201.

Lieber, R.L., Fridén, J.O., Hargens, A.R., and Feringa, E.R. (1986a). Long-term effects of spinal cord transection on fast and slow rat skeletal muscle. II. Morphometric properties. Exp. Neurol. 91:435–448.

Lieber, R.L., Johansson, C.B., Vahlsing, H.L., Hargens, A.R., and Feringa, E.R. (1986b). Long-term effects of spinal cord transection on fast and slow rat skeletal muscle. I: Contractile properties. Exp. Neurol. 91:423–434.

Lieber, R.L., McKee-Woodburn, T., Fridén, J., and Gershuni, D.H. (1989). Recovery of the dog quadriceps after ten weeks of immobilization followed by four weeks of remobilization. J. Orthop. Res. 7:408–412.

Lundborg, G. (1988). Nerve injury and repair. New York: Churchill Livingstone.

Morey, E.R. (1979). Spaceflight and bone turnover correlation with a new rat model of weightlessness. Bioscience. 29:168–172.

Rende, M., Granato, A., Monaco, M.L., Zelano, G., and Toesca, A. (1991). Accuracy of reinnervation by peripheral nerve axons regenerating across a 10-mm gap withing an impermeable chamber. Exp. Neurol. 111:332–339.

Roy, R., Bello, M., Bouissou, P., and Edgerton, R. (1987). Size and metabolic properties of fibers in rat fast-twitch muscles after hindlimb suspension. J. Appl. Physiol. 62:2348–2357.

Sanes, J.R., and Covault, J. (1985). Axon guidance during reinnervation of skeletal muscle. Trends Neurosci. 8:523–528.

Simard, C.P., Spector, S.A., and Edgerton, V.R. (1982). Contractile properties of rat hindlimb muscles immobilized at different lengths. Exp. Neurol. 77:467–482.

Spector, S.A., Simard, C.P., Fourier, M., Sternlicht, E., and Edgerton, V.R. (1982). Architectural alterations of rat hindlimbs skeletal muscles immobilized at different lengths. Exp. Neurol. 76: 94–110.

Thomason, D.B., Biggs, R.B., and Booth, F.W. (1989). Protein metabolism and B-myosin heavy-chain mRNA in unweighted soleus muscle. Am. J. Physiol. 257:R300–R305.

Thomason, D.B., and Booth, F.W. (1990). Atrophy of the soleus muscle by hindlimb unweighting. J. Appl. Physiol. 68:1–12.

Thomason, D.B., Herrick, R.E., Surdyka, D., and Baldwin, K.M. (1987). Time course of soleus muscle myosin expression during hindlimb suspension and recovery. J. Appl. Physiol. 63:130–137.

Weibel, E.R. (1980). Stereological methods. Vol. 1: Practical methods for biological morphometry. New York: Academic Press.

Chapter

6

Skeletal Muscle Response to Injury

OVERVIEW

Numerous examples support the statement that skeletal muscle is one of the most adaptable tissues in the body. Whether the level of use increases or decreases, muscle responds accordingly. However, at times the amount of use, metabolic load, or level of stress on a muscle fiber is so great that the fiber actually suffers damage. When this happens, all or part of the muscle cell degenerates and is subsequently replaced. The replacement process is termed regeneration. Skeletal muscle mounts a vigorous regenerative response following injury. Such a response has important implications for both the normal developmental process and the potential use of regenerating muscle in the treatment of diseased muscle. The processes of degeneration and regeneration are the subject of this final chapter.

INTRODUCTION

In addition to skeletal muscle's ability to adapt, it also possesses a great ability to regenerate. For years, skeletal muscle's capacity to regenerate was doubted. Some believe that this was because of the distinguished Oxford anatomist Wilfred E. Le Gros Clark, who performed studies of gunshot wound healing around World War II and claimed that while muscle *repaired* itself following injury, regeneration of *new* muscle cells was not possible. There are still histology and physiology textbooks today that patently state that muscle cells do not regenerate. However, muscle cells certainly do regenerate.

Causes of Muscle Regeneration

Numerous experimental and pathologic conditions induce muscle cell degeneration and regeneration. Experimentally, muscles can be crushed, minced, transplanted, exercised, or grafted in order to observe this whole process. Clinically, muscles experience degeneration following blunt trauma, surgical manipulation, ischemia, and in direct response to the myotoxic effect of local anesthetics. All of these "treatments" have as a common theme the disruption of the muscle cell. The process of degeneration and regeneration proceeds much as you might expect based on your knowledge of muscle. In fact, much of the regenerative process recapitulates the developmental process that was discussed in Chapter 1. We will now consider the cellular basis of the muscle regenerative response by first describing the "typical" degeneration-regeneration cycle followed by a discussion of various experimental models that have provided insights into the mechanism of muscle regeneration.

MORPHOLOGY OF THE DEGENERATION-REGENERATION CYCLE

The regeneration sequence conceptually resembles the process of repairing a damaged car: The existing damaged components are removed, and the replacement components are made and then inserted into their final location. This cycle in muscle, described in detail by Bruce Carlson, proceeds in much the same way: Damaged cellular components are digested, satellite cells proliferate to form new muscle fiber building material, and satellite cells fuse to form new myotubes and muscle fibers, thus recovering cellular function (Carlson, 1973).

Digestion of Damaged Cellular Components

Following physical damage to the muscle cell, the degeneration process begins with digestion of damaged cellular components by endogenous protease enzymes from muscle cell lysosomes and by exogenous proteases released from infiltrating mononuclear macrophages. These macrophages serve as scavengers that "pick up" or phagocytose cellular debris (Figure 6.1**A**). The source of these infiltrating cells is the microcirculation surrounding the fibers, so that if circulation is absent for any reason (*e.g.,* the necrosis is due to ischemia or the circulation has been traumatized) then the repair process is delayed. In mammalian muscle, cellular infiltration begins within hours of the initial injury and is well developed within a few days.

During this digestion phase of degeneration, most cellular components are affected. A beautiful chronicle of cellular responses to degeneration has been published by Carpenter and Karpati (1984). Myofibrils lose their regularity and begin to appear disorganized in the region of the Z-disk. Mitochondria appear more rounded and lose their regular distribution within the cell. Actin and myosin filaments begin to lose their regularity and may even intertwine with the cytoplasmic processes of the macrophage. Glycogen particles, which are plentiful in normal (especially fast) muscle fibers, disappear, and the cell no longer stains positively with the periodic acid-Schiff (PAS) stain or for the enzymes used in glycogenolysis (*e.g.,* phosphorylase). Later, the muscle cytoplasm appears as a jumbled mesh of poorly staining filaments. Degeneration therefore represents the "dismantling phase" of the damaged cell.

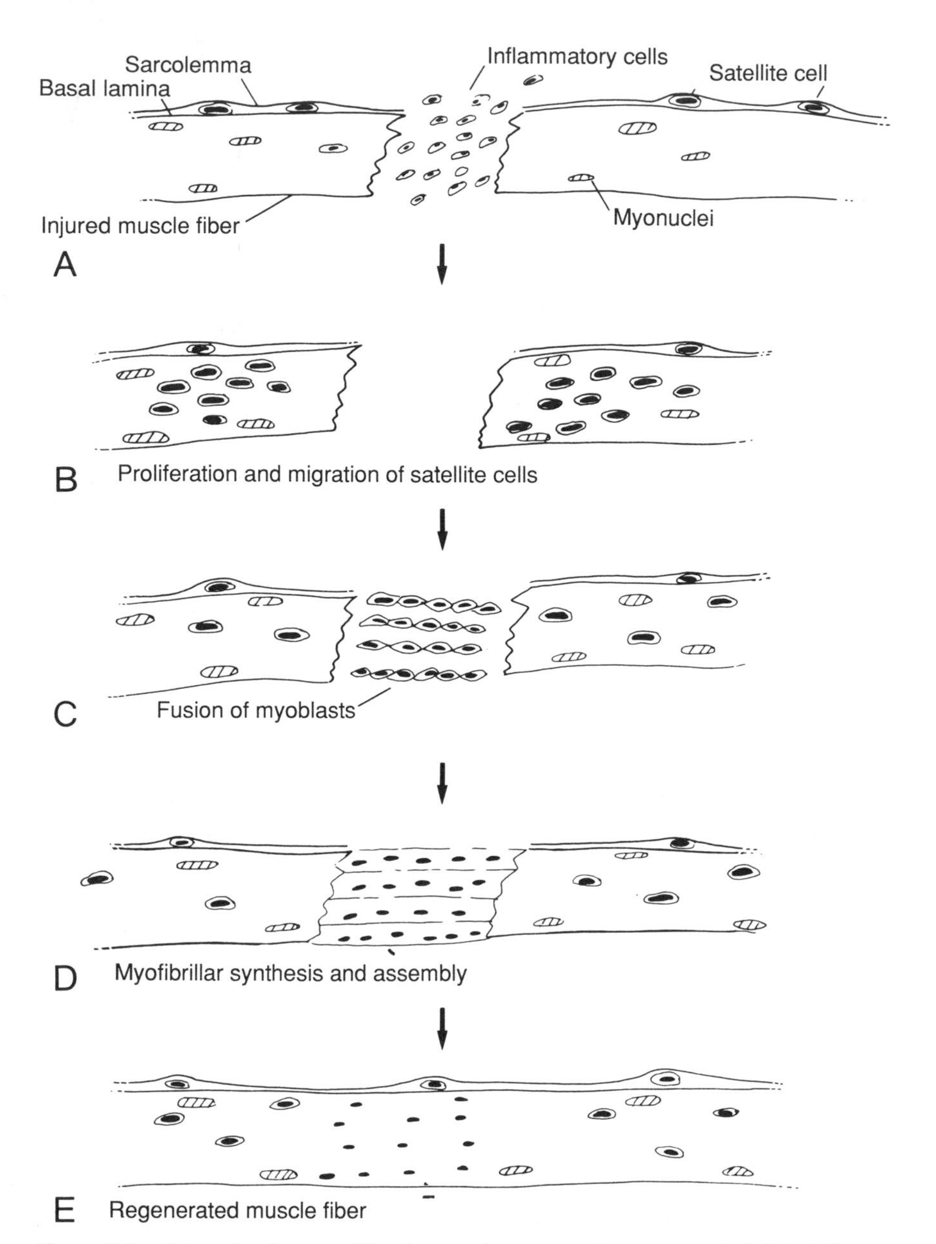

Figure 6.1. Schematic diagram of the degeneration-regeneration process. **A,** Cellular infiltration and inflammation to digest damaged cellular components. **B,** Proliferation of satellite cells. **C,** Fusion of satellite cells into myotubes to form new myofibrils. **D,** Synthesis of myofibrillar proteins to "fill" new fiber. **E,** Regenerated muscle fiber.

Proliferation of Satellite Cells

The next stage in the regeneration cycle is the proliferation of skeletal muscle satellite cells (Figure 6.1**B**). Recall from Chapter 1 that satellite cells arrange themselves peripherally between the basal lamina and sarcolemma during development during the time period when primary and secondary myotubes are forming. We haven't discussed satellite cells in detail so far in this text, but now they occupy center stage. Satellite cells are present along the periphery of muscle fibers at less frequent intervals than myonuclei. They *resemble* myonuclei in many ways but have a more developed cytoplasm and as such can be identified to represent from 1% to 10% of the total number of nuclei in the fiber periphery. Because there are relatively few satellite cells, it is not always easy to observe then. In fact, the discovery of the satellite cells was relatively recent (Mauro, 1961) and was received by the scientific community with some skepticism. However, we now know that these satellite cells represent unfused myoblasts that associate with myotubes during development but do not fuse with them. As such, many prefer the term "presumptive myoblasts" to "satellite cell" in order to specifically refer to their origin and function.

In spite of the then lack of understanding of satellite cell activity by the scientific community, it was well known that when a muscle fiber was injured, cellular proliferation occurred within the cell. The question was, What is the origin of the cells? In the late sixties, one school of thought claimed that the source of the proliferating cells was the nuclei of the damaged muscle fiber. They believed that following injury, the mature muscle fiber fragmented (along with a nucleus) into mononuclear myoblast-like cells that then recapitulated the developmental process. Others claimed that the satellite cell itself entered a dramatic mitotic phase and began to proliferate in response to some unknown stimulus (see description of this controversy in Mauro *et al.,* 1970).

The answer to this dilemma was offered in 1976 by Mike Snow in radioactive tracer experiments (Snow, 1976). Snow performed two series of experiments. In the first series, he caused developing muscle nuclei to take up radioactive thymidine (a component of the DNA molecule). The thymidine base was rendered radioactive by replacing one of the normal (^{2}H) hydrogen atoms with radioactive hydrogen, known as tritium (^{3}H). Pregnant rats were injected with ^{3}H-thymidine at several intervals during their pregnancy (Figure 6.2). The offspring were allowed to mature for 5–7 weeks. During growth, myonuclei became radioactively labeled whereas any label present in satellite cells was diluted past the point of detection. Then their muscles were minced and forced to degenerate and then regenerate. Snow then measured the

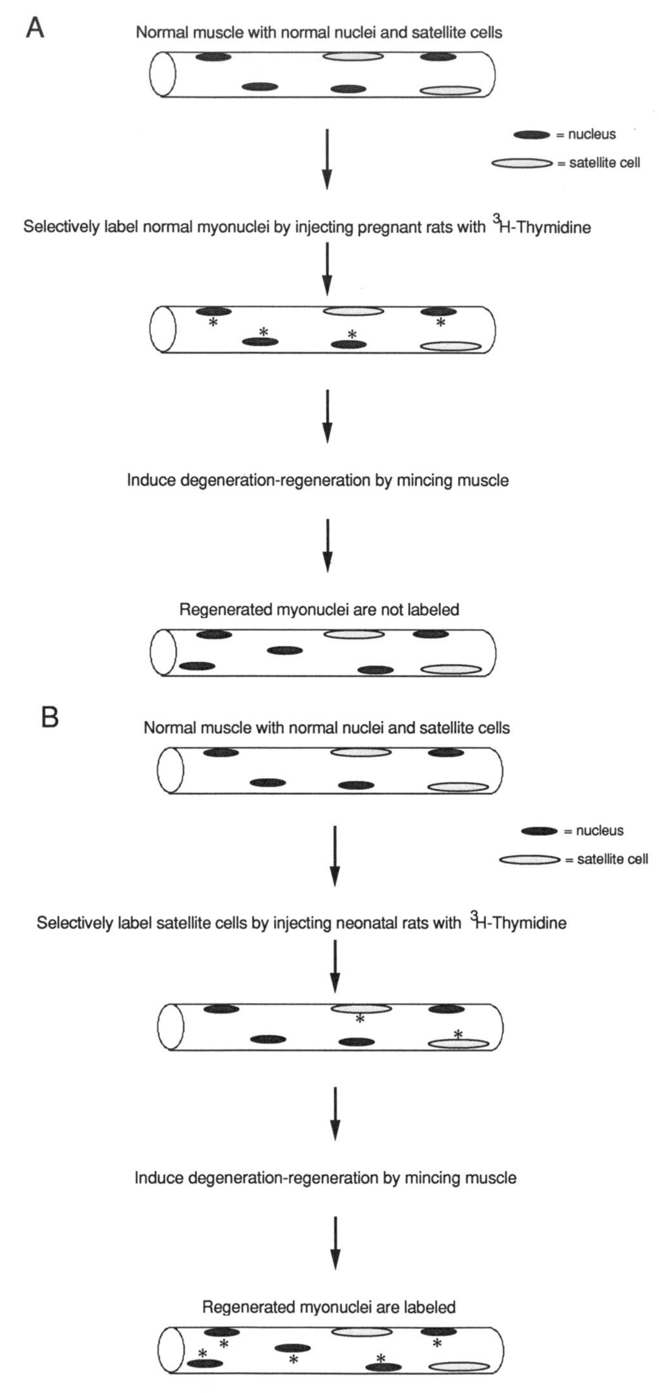

Figure 6.2. Experimental demonstration of the source of regenerating myonuclei. **A,** Normal myonuclei were radioactively labeled. Then, upon regeneration, cells in the new fiber were measured for radioactivity. Regenerated myonuclei were not labeled. **B,** Satellite cells were radioactively labeled. Then, upon regeneration, cells in the new fiber were measured for radioactivity. Regenerated myonuclei were labeled. These experiments showed that satellite cells are the source of myonuclei in regenerating fibers. (Data from Snow, 1976.)

radioactivity of the cells involved in the regeneration process. What do you predict would happen? Virtually *none* of the regenerating cells were radioactive, implying that the nuclei of the regenerated fibers came from something other than the normal myonuclei. To further investigate this phenomenon, Snow selectively labeled satellite cells by injecting neonatal rats with ^{3}H-thymidine. He again forced these muscles to degenerate and regenerate and found that the regenerating nuclei were indeed radioactive. These experiments established that the cellular source of myonuclei during regeneration was the satellite cells. In fact, satellite cell proliferation represents one of the most impressive biologic examples of high synthetic activity.

Signals for Satellite Cell Proliferation

What causes this impressive induction of satellite cell mitotic activity? What is the signal that induces satellite cell proliferation?

Recently, Richard Bischoff has performed a series of interesting experiments that have identified some signals controlling the behavior of satellite cells (Bischoff, 1986). In one experiment, Bischoff developed a method for "growing" isolated muscle fibers in culture, outside the animal, in order to test various "factors" that might cause satellite cell proliferation. If muscles were crushed and the crushed muscle extract placed on the cultured muscle fiber, satellite cells proliferated (Figure 6.3). However, neither "intact muscle extract" (extract from normal muscle) nor extract from other crushed tissues caused satellite cell proliferation. Bischoff concluded that satellite cells were under "positive control." That is, some factor(s) from crushed muscles caused satellite cells to proliferate when, normally, they would remain quiescent. This experiment suggested the appealing scenario that when a muscle fiber was damaged and released this mitogenic factor, that satellite cell proliferation would be induced, resulting in muscle repair.

Since the satellite cell is essentially "sandwiched" in between the muscle fiber sarcolemma and the basal lamina, Bischoff attempted to determine the relative influence of sarcolemma and basal lamina on the satellite cell proliferative response (Bischoff, 1990). The experimental system again involved isolation of single rat muscle fibers and their satellite cells. These cells were placed in a specially designed cell culture apparatus and allowed to proliferate under a variety of conditions. In the first condition, muscle fibers were killed by application of the local anesthetic Marcaine (Figure 6.4). This left satellite cells along with their basal lamina to proliferate. In another treatment, the fibers were again killed with Marcaine, but the satellite cells were removed from the basal lamina by spinning them off using a centrifuge.

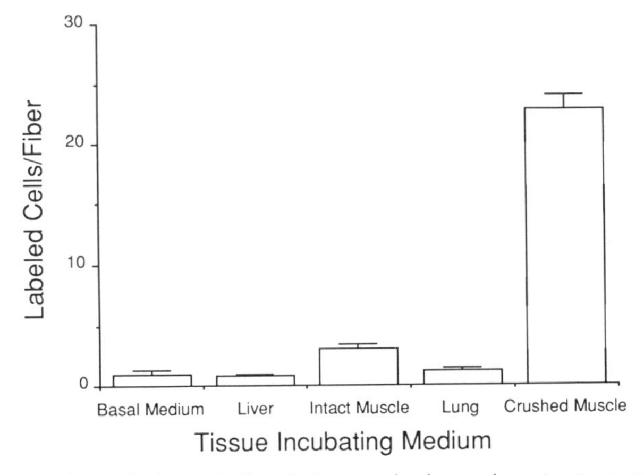

Figure 6.3. Experimental determination that a crushed muscle extract causes satellite cell proliferation. Muscle fibers with their associated satellite cells were incubated with various tissue extracts. Only with the extract from crushed muscle did satellite cells proliferate. This experiment demonstrates that normal muscle contains a mitogen. (Data from Bischoff, 1986.)

Finally, control fibers with both the intact muscle sarcolemma and the fiber basal lamina were allowed to proliferate.

Satellite cell proliferation was induced by incubating the culture systems in "muscle extract," the soluble portion obtained when muscle fibers were minced. Muscle extract induced cell proliferation (as we saw above), but the *magnitude* of the proliferation was greater if the satellite cells were *not* in contact with the sarcolemma. Bischoff concluded that the sarcolemma exerted a negative control on satellite cells to prevent proliferation. In this way it was possible to explain why one portion of a damaged fiber might mount a regenerative response (since its sarcolemma was damaged and thus not in contact with the satellite cell) which another portion might remain quiescent (since the satellite cell was still in contact with the sarcolemma).

Thus satellite cells are under both positive and negative control: positive control by some mitogenic factor within the muscle and negative control because of sarcolemmal contact. The molecular details of each of these control processes remain to be elucidated.

Fusion of Satellite Cells into Myotubes

Only a few days following injury, proliferation of satellite cells is noticeable, and within 3–4 days, their presence is unmistakable (Figure 6.1**B**). Now the regeneration process resembles, in many ways, the normal developmental

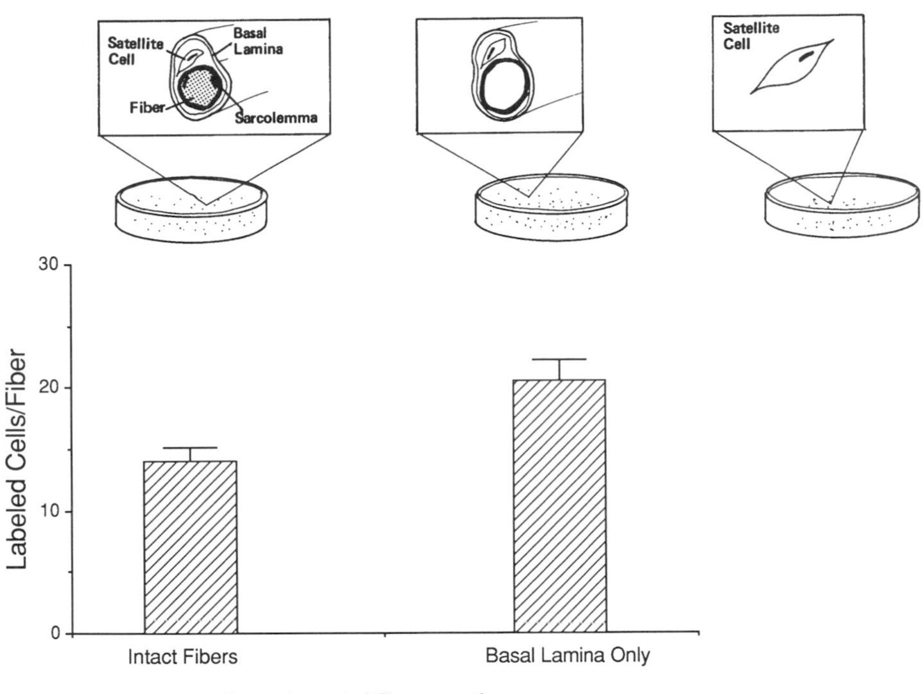

Figure 6.4. Schematic representation of experiments used to determine the effect of basal lamina and sarcolemma on satellite cell proliferation. Three experimental conditions were studied: the intact fiber with satellite cells in contact with both the sarcolemma and basal lamina (*top left*), fiber "ghosts" in which fibers were killed, leaving the satellite cell in contact with the basal lamina (*top middle*), and satellite cells dissociated from both the sarcolemma and basal lamina (*top right*). Satellite cells proliferated to a greater extent when the muscle fiber was removed. (Drawn from Bischoff, 1990.)

process. Satellite cells begin to align themselves along the basal lamina and fuse into myotubes (Figure 6.1**C**). The presence of the basal lamina has been considered to be of critical importance as a substrate for satellite cell proliferation and fusion. For example, it is known that satellite cells when placed in cell culture proliferate and fuse better on a collagen substrate (the primary component of the basal lamina) compared to a plastic substrate. It is known that the basal lamina is not a completely passive structure during regeneration, in that it expresses various extracellular matrix components along a certain time course. Thus its characteristics change continually throughout the regeneration process, and it is unlikely that it maintains a central controlling influence (Gulati *et al.,* 1983). In fact, a recent report demonstrated satellite cell proliferation and fusion under conditions where the basal lamina was actually absent! It may be, therefore, that other factors are actually responsible for the alignment of myoblasts as they fuse into myotubes. For example, a myoblast cell culture system was developed by Herman Vandenberg (Vandenberg, 1982). He showed that myoblasts grown in a simple, nonmoving culture system proliferated with random orientation and remain relatively immature (Figure 6.5). However, if the cells were cyclically lengthened and shortened, the cells became well aligned, took up greater amounts of amino acids, and synthesized more proteins. These data suggested that in addition to certain chemical factors, mechanical stress or at least strain had a pronounced effect on the myoblastic maturation process.

Synthesis of Contractile Proteins

As the myotubes mature and continue to differentiate, they synthesize new myofibrillar proteins and, as in development, deposit them in the outer subsarcolemmal region of the regenerating fiber. Thus regenerating fibers, like developing fibers, when stained for myofibrillar proteins look like doughnuts with light central cores (Figure 6.1**D**). As fibers continue to fill with contractile protein, muscle nuclei are usually pushed to the periphery, and the fiber takes on the appearance of a mature skeletal muscle fiber. Sometimes

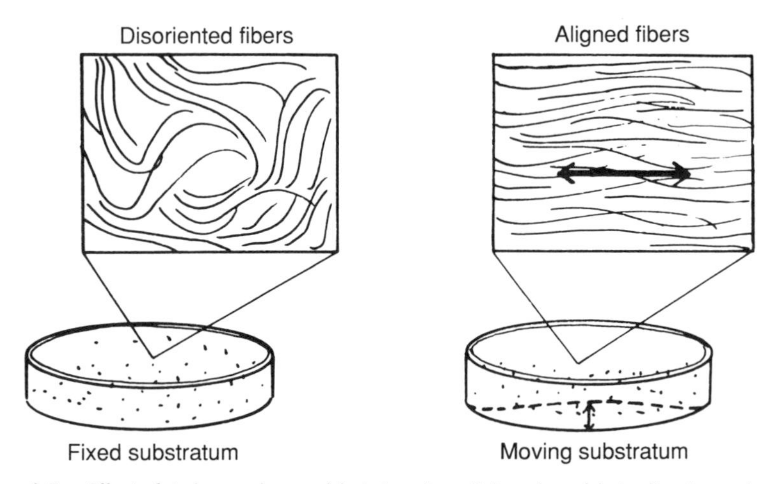

Figure 6.5. Effect of strain on avian myoblasts in culture. Cultured myoblasts align themselves and mature faster when mechanically strained *in vitro* (*arrow* represents strain direction). (Drawn from Vandenburg 1982.)

regenerated fibers fail to move the nucleus to the periphery, and a central nucleus remains. This central nucleus serves as a pathologic statement that the fiber has undergone degeneration and regeneration. The entire regeneration process requires about 6 months in most higher mammals, although regeneration in rat muscle is quite a bit faster, requiring only about 2 months! It should be obvious from this discussion that we do not fully understand the control and coordination of muscle events during regeneration. Many of these questions are also relevant to the developmental process and are being actively pursued.

MUSCLE PROPERTIES FOLLOWING REGENERATION

After regeneration occurs do muscle properties return to their original values? Do all properties recover to the same extent? We have seen dramatic demonstrations of muscle plasticity under various conditions. How plastic is regenerating muscle? Answers to these questions have important clinical implications where performance is affected following injury or trauma.

Recovery of Muscle Properties in Cat Muscle

To address the question of muscle recovery following regeneration, John Faulkner and his colleagues (1980) induced cat extensor digitorum longus muscle regeneration by surgically removing the muscle, transecting the nerve and neurovascular supply, and replanting the entire muscle (Figure 6.6). Muscles were allowed to recover for 40–400 days, and contractile properties were measured at specific intervals. After a few weeks, muscles developed measurable tension, and maximum tetanic tension continued to increase for about 8 months, at which point it stabilized to about 50% of the level reached in control muscles (Figure 6.7). This relative recovery was increased if the nerve was left intact. In contrast to maximum tension (P_O), Faulkner *et al.* found that many contractile properties demonstrated complete recovery after graft stabilization (stabilization was defined as the time when fibers in the central muscle core were the same size as fibers in the periphery). For example, time-to-peak tension, half-relaxation time, maximum contraction velocity (V_{max}), fiber length and individual fiber cross-sectional area returned to control levels after 6 months. Apparently not all of the fibers regenerated, since although fiber area and length were normal, muscle mass and P_O decreased significantly, implying a decrease in fiber number. Alternately, perhaps the force generated per fiber decreased. Future studies will clarify these issues.

Metabolically, the regenerated muscle demonstrated a decrease in endurance following transplantation. Capillary density and succinate dehydrogen-

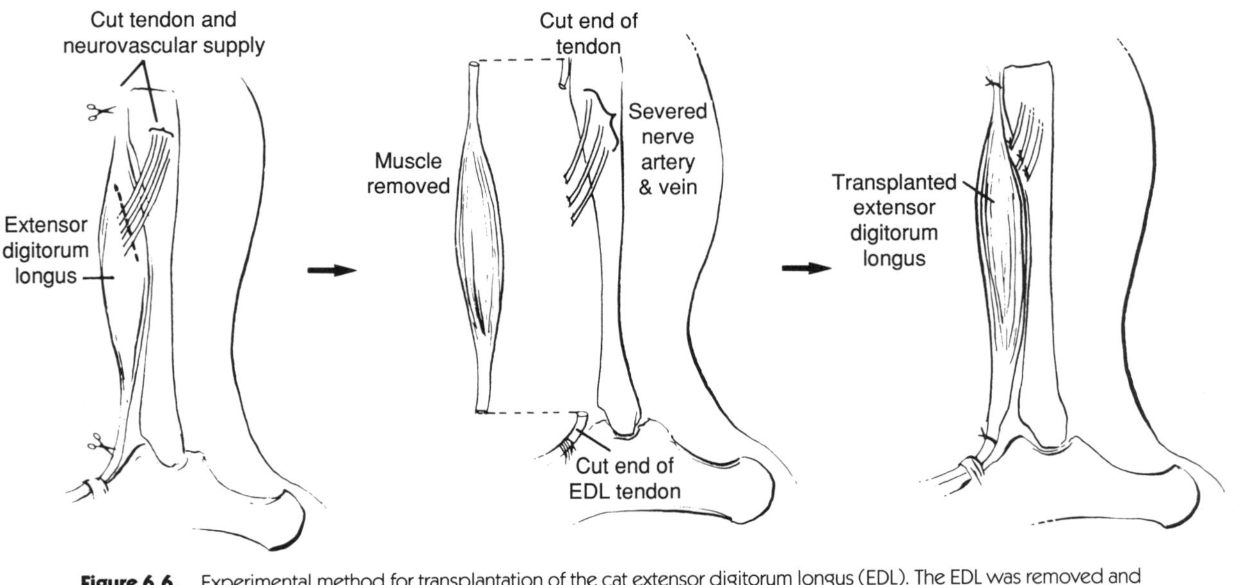

Figure 6.6. Experimental method for transplantation of the cat extensor digitorum longus (EDL). The EDL was removed and the neurovascular supply severed. It was then replaced in the original site where reinnervation and regeneration could occur. (Adapted from Faulkner *et al.*, 1980.)

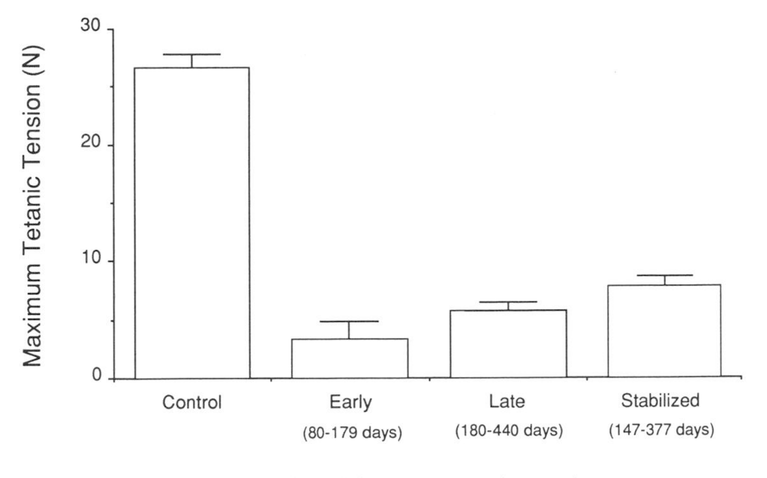

Figure 6.7. Time course of tetanic tension in the cat EDL following regeneration. (Data from Faulkner *et al.*, 1980.)

ase activity had recovered to only about 50% of control levels, rendering the muscle dramatically more fatigable. Fortunately, since we have learned that muscle endurance is very plastic in response to increased muscle use (Chapter 4), it is possible that training of these muscles might significantly improve fatigability. This is an area of ongoing investigation, and early experimental results are promising! (Incidentally, muscle endurance in regenerated rat muscle is much greater than that observed in cat muscle.)

Plasticity of Regenerating Muscle Fibers

Faulkner and his colleagues (Donovan and Faulkner, 1987) also investigated a regenerating muscle's adaptive capacity by comparing surgical transplantation (where the muscle was removed without its neurovascular supply and placed in a new site) to surgical transposition (where the muscle was removed with its vascular supply to a new site and innervated by the new nerve; Figure 6.8). In *transplantation,* the muscle fibers degenerated due to ischemia, and in addition, the investigators injected the local anesthetic bupivacaine in order to obtain a uniform regenerating population of fibers. In surgical *transposition,* the fact that the vascular supply was left intact ensured fiber survival. Implantation of the new motor nerve would thus be similar to a cross-reinnervation experiment.

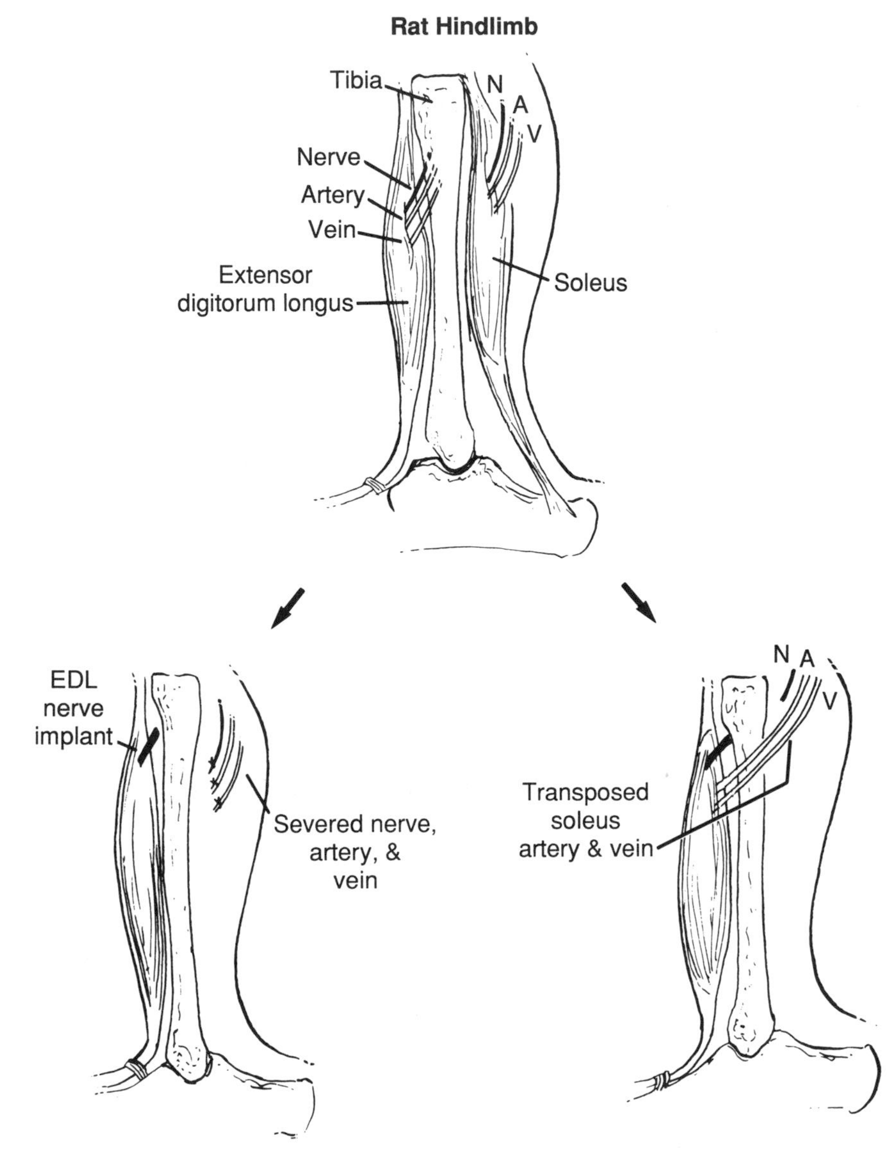

Figure 6.8. Schematic diagram of muscle transplantation (*left panel*) and transposition (*right panel*) of rat soleus muscles into the EDL site. (Adapted from Donovan and Faulkner, 1987.)

Donovan and Faulkner compared the adaptation of rat soleus and extensor digitorum longus (EDL) muscles following either transposition or transplantation. In all cases, *transposition* resulted in muscle fibers' recovering the greatest strength (Figure 6.9**A**). For example, the transposed EDL P_O was 830 mN compared to the transplanted EDL P_O of only 590 mN. Similarly, the transposed soleus P_O was 710 mN compared to the transplanted soleus P_O of 660 mN. However, what was perhaps more interesting was the change in contractile speed and fiber type distribution following the treatments. In all cases, the *transplanted* muscle became more like the muscle that had been in the recipient site. Thus the transplanted soleus (SOL) time-to-peak twitch (TPT) tension was 29 msec, very close to the control EDL TPT of 24 and very different from either the transposed SOL TPT of 67 msec or the control SOL TPT of 83 msec (Figure 6.9**B**). In the same way, the transplanted EDL contained only 33% fast fibers, closer to the control soleus fast fiber percentage of 15% than either the transposed EDL percentage (86%) or the control EDL percentage (97%; Figure 6.9**C**). Thus in all cases, the *transplanted* muscles, consisting of regenerating fibers, were the most adaptable fibers present.

These experiments raised the intriguing possibility that since regenerating fibers were highly adaptable, a specific strategy to enable muscle fiber adaptation might be to enter the regeneration cycle. While this is certainly not the typical mechanism of muscle adaptation, the possibility exists that it might operate secondary to surgical reattachment of skeletal muscle or muscle fiber adaptation to intense exercise that is accompanied by fiber breakage (see below).

CLINICAL APPLICATION TO MUSCULAR DYSTROPHY

An improved understanding of skeletal muscle regeneration promises to provide improved functional recovery following surgical repair and grafting. However, nowhere is the promise of muscle regeneration so bright as in the application of modern techniques to the repair of diseased muscle tissue.

Duchenne Muscular Dystrophy

The most common type of genetic dystrophy that affects children is Duchenne muscular dystrophy (DMD) named for its descriptor. DMD is a devastating myopathy which is sex-linked, striking young boys (about 1 in 3500) early in the first years of life. By the age of about 13 most DMD patients must use a wheelchair for mobility, and weakness of the intercostal muscles often leads to death by age 20. A recent exciting discovery has shown that a specific protein, name "dystrophin," is present in normal muscle but absent from

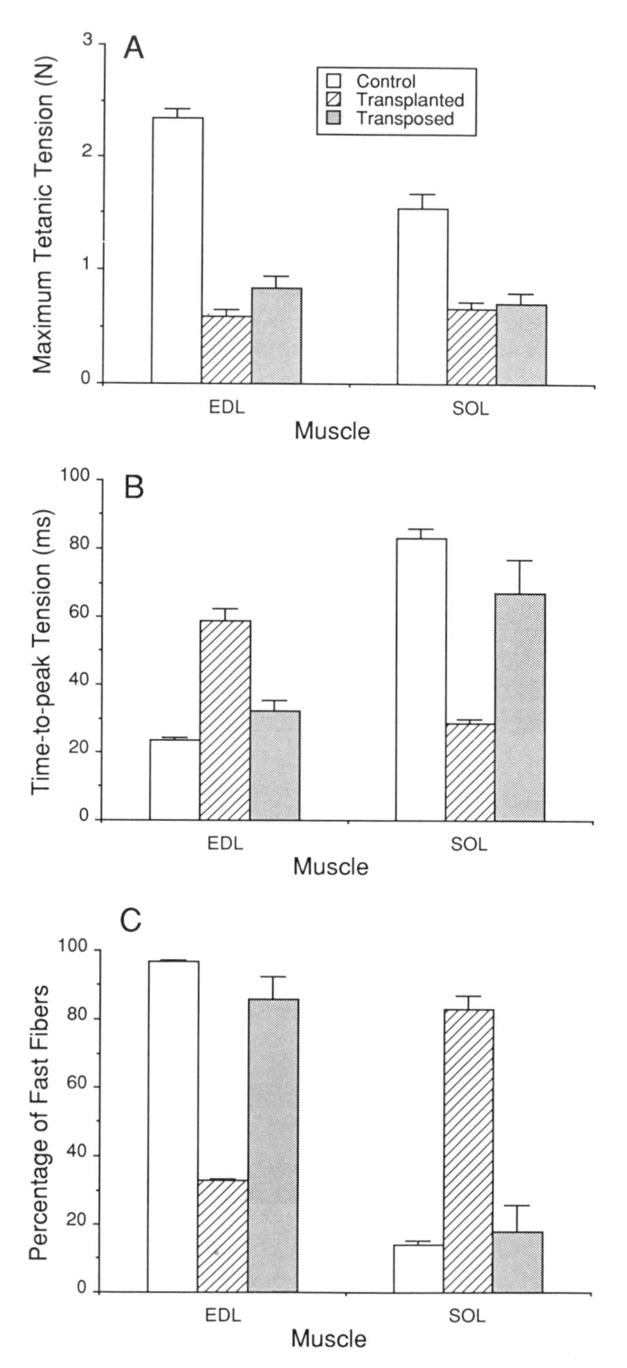

Figure 6.9. **A,** Maximum tetanic tension of muscles following transplantation and transposition. **B,** Time-to-peak tension of muscles following transplantation and transposition. **C,** Fast fiber percentage of muscles following transplantation and transposition. (Data from Donovan and Faulkner, 1987.)

dystrophic muscle. This protein is associated with the sarcolemma (Ohlendieck *et al.,* 1991) and thus might explain why the dystrophic sarcolemma is "leaky."

Recently, an animal model of DMD was discovered that mimicked some (but, importantly, not all) of the morphologic effects of DMD. The so-called MDX mouse muscles show segmental clusters of necrotic fibers at the age of 10–12 days. The muscles then degenerate and regenerate over the next few months, with regenerating fibers achieving near-normal diameters but with centrally placed nuclei (see above). The centrally placed nuclei thus provide a morphologic label for degenerated and regenerated fibers.

Therapeutic Implantation of Myoblast Cells in MDX Mice

A potentially thrilling set of experiments has suggested that it might be possible to "fix" affected muscle from MDX mice by implanting normal myoblastic cells into the DMD muscle during early development. This work has been performed in different laboratories around the world. One set of illustrative experiments was performed by George Karpati, Sterling Carpenter, and colleagues in which nondystrophic myogenic cells that had been labeled with ^{3}H-thymidine (as described earlier) were implanted into the quadriceps of young (3- to 15-day-old) MDX mice (Karpati *et al.,* 1989; Figure 6.10). At this young age, MDX mice have not yet begun to show muscle necrosis. MDX mice were allowed to grow, and the developed quadriceps muscles were screened for the presence of radioactive myonuclei. Karpati *et al.* found heavily labeled myonuclei in the *host* fibers, implying that implanted normal nuclei had fused with the MDX myoblasts and formed what are called "mosaic" or "chimeric" muscle fibers—fibers with nuclei of different origin. Subsequent work from this and other laboratories has also demonstrated that following formation of mosaic fibers, many of the normal degenerative features of dystrophic muscle are not seen—the fibers are "fixed" by the addition of normal myoblasts.

A similar experiment was performed by Partridge *et al.* (1989) in which chimeric muscle fibers were formed as described above. Importantly, using a relatively low ratio of normal donor to MDX nuclei, these investigators showed that the fibers converted from dystrophin-negative to dystrophin-positive! Thus the normal cells induced the expression of the dystrophin protein. Subsequent experimentation has shown that dystrophin is expressed even in muscle fiber regions where the original MDX nuclei are present (Huard *et al.,* 1991).

Obviously, the therapeutic potential of these treatments is immense, and we will surely be hearing of human clinical trials occurring in the near future.

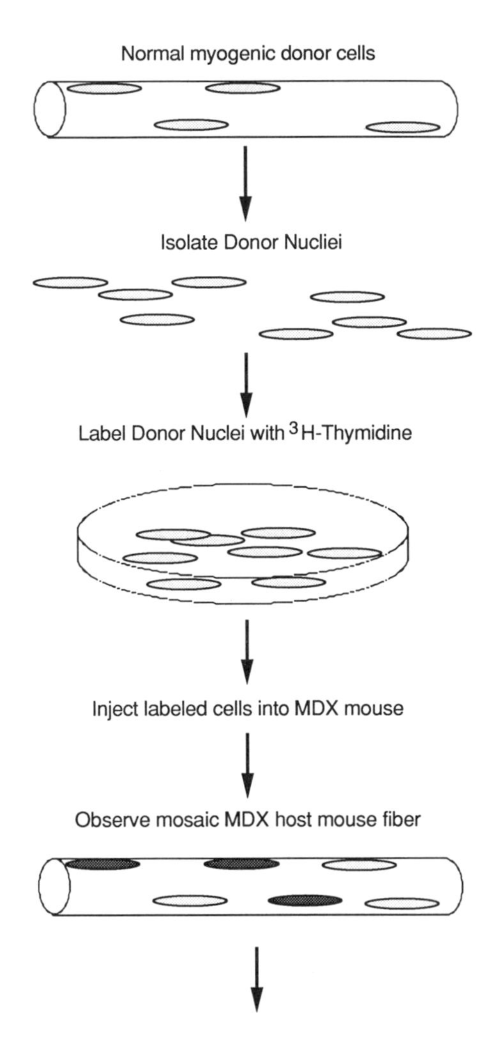

Figure 6.10. Experimental manipulation of mouse MDX cells by fusion with normal satellite cells. Incorporation of normal myogenic donor cells into MDX cells results in expression of dystrophin and normal cellular morphology. (Adapted from Karpati *et al.*, 1989.)

MUSCLE RESPONSE TO EXERCISE-INDUCED INJURY

I have repeatedly stated that skeletal muscle actions associated with lengthening (eccentric) contractions are also associated with high muscle forces. Numerous investigators have demonstrated that when eccentric exercise is performed, muscle damage and muscle soreness result.

Mechanics of Eccentric Contractions

Our most basic understanding of the eccentric contraction phenomenon is based on the discontinuity of the force-velocity relationship for shortening and lengthening. We showed in Chapter 2, for example, that when a muscle shortens at about 1% V_{max}, tetanic tension drops to about 95% P_O. However, when a muscle is forced to lengthen at 1% V_{max}, tension rises precipitously to over 125% P_O (Figure 6.11)! In Chapter 2, we explained this behavior based

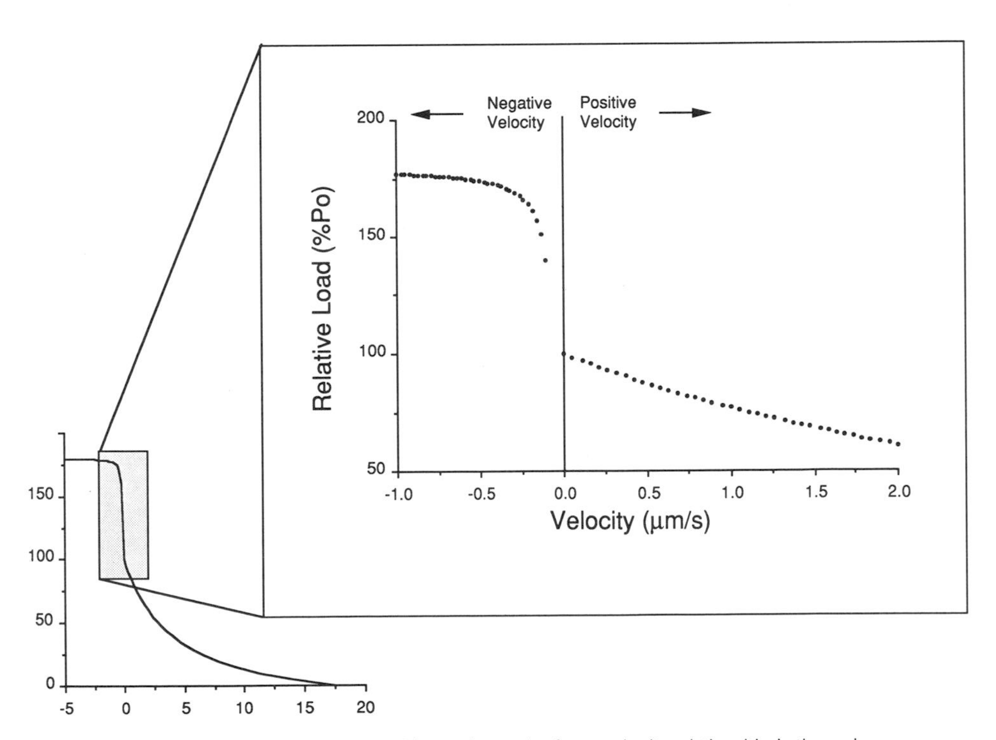

Figure 6.11. Enlargement of the positive and negative force-velocity relationship in the region near zero velocity. Note the force increase per unit negative velocity is over six times the force decrease per unit positive velocity.

on the asymmetrical mechanical properties of the cross-bridge. However, it is becoming increasingly clear that a number of unexplained mechanical phenomena are associated with eccentric contractions (Harry *et al.,* 1989, and Morgan, 1989). The details of these phenomena are beyond the scope of this textbook. I raise the fact to point out that in spite of the fact that eccentric contractions are a normal part of the gait cycle (Chapter 3, page 152) and are experienced by most muscles in the body, relatively little is known of the physiology of eccentric contractions. Many investigators agree that eccentric contractions, if performed at a high intensity, can cause injury, but as the physiologist Herman deVree (1983) pointed out, "it is difficult to imagine why a structure would be injured when performing the very act for which it was designed!" In fact, we have seen that most physiologic experiments have been performed with maximal activation under steady-state conditions (isometric or isotonic contractions) when, in reality, these conditions are rarely achieved. "Eccentric contraction physiology" is an exciting area of research promising new vistas in therapeutic and exercise treatment.

A common eccentric contraction experienced during the gait cycle is the active lengthening of the quadriceps musculature as the foot strikes the ground (Figure 6.12). As foot strike occurs and the momentum of the body carries it forward and down, the quadriceps lengthen and the knee flexes. This is especially pronounced if deceleration is required, such as in downhill running or walking down steps. How many of us have felt sore quadriceps after a long downhill hike? Indeed, downhill running in animals and humans remains a favorite physiologic model to induce muscle injury.

Human Models of Eccentric Contraction

One finding on which many agree is that following an intense bout of eccentric exercise, soreness is maximum about 24–48 hours later. This phenomenon has been termed delayed onset muscle soreness (DOMS). There is no question that DOMS is uniquely related to eccentric contraction and not to exercise itself. Unfortunately, it is difficult to quantify soreness. Another objective parameter that characterizes human response to eccentric exercise is the circulating level of creatine kinase.

As we saw in Chapter 2, creatine kinase (CK) is an enzyme found in striated muscle that is involved in conversion of ADP to ATP according to the reaction

$$\text{Creatine Phosphate} + \text{ADP} \rightarrow \text{Creatine} + \text{ATP} \qquad (6.1)$$

(In fact, ATP is so rapidly regenerated from ADP by CK that ATP levels remain almost unchanged during muscle contraction.) CK is found inside muscle

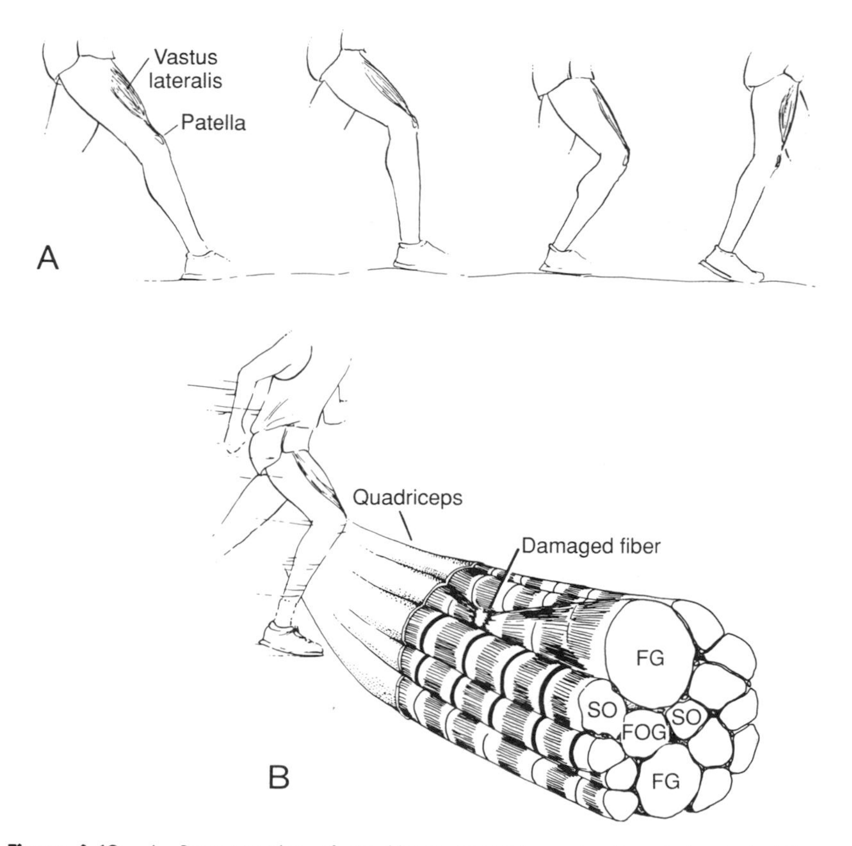

Figure 6.12. **A,** Representation of quadriceps eccentric contraction during foot strike and subsequent toe-off. As the momentum of the body forces knee flexion, vastus lateralis is forced to actively lengthen. **B,** Schematic illustration of damage to specific muscle fibers following eccentric contraction. (Compare to Figure 6.20.)

fibers and, under normal conditions, remains there. However, when exercise is extremely intense and a cell breaks, CK is released into the bloodstream, where it can be detected. (In fact, different isozymes of CK exist in heart and skeletal muscles, so that serum cardiac CK levels can be used to diagnose myocardial infarction). CK therefore serves as an indirect measure of myofiber integrity.

Serum CK Levels Following Eccentric Contraction

Bill Evans measured serum CK levels following intense eccentric exercise in young college students (Evans *et al.,* 1985). He presented two major experimental findings. First, he showed that CK levels were elevated a few days following the exercise bout, peaked 5 days following the exercise bout, and remained elevated for several days thereafter (Figure 6.13). Think about what this implies regarding muscle fiber breakdown. Clearly, the data suggest that muscle fibers do not simply "break" in response to exercise and release their contents. The CK data suggest that the muscle fibers experience some type of injury that initiates a cascade of events culminating in fiber breakage. The cascade may continue for several days, as illustrated by the protracted elevated CK levels.

Evans *et al.* also importantly demonstrated that if subjects had been trained prior to eccentric exercise that the magnitude and duration of the elevated CK levels were greatly attenuated (Figure 6.13). These trained subjects also had elevated CK levels prior to the experimental exercise bout, suggesting that they were usually experiencing greater muscle fiber turnover. The study therefore presented two important results: (*a*) Muscle damage due to eccentric contraction is delayed and prolonged, and (*b*) previous training has a "protective" effect on further muscle damage. Obviously, the data imply that training is beneficial if high exertion levels are anticipated.

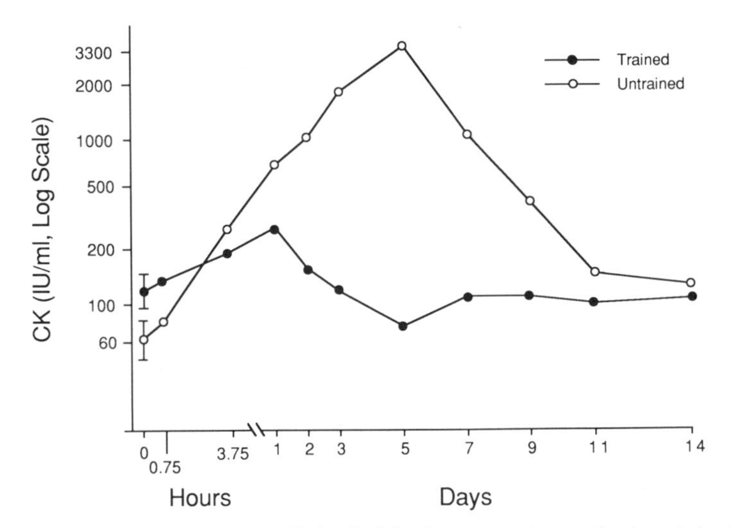

Figure 6.13. Time course of serum CK levels following eccentric exercise in untrained (*open circles*) and trained (*filled circles*) subjects. Note the delay between the exercise bout and subsequent peak serum enzyme levels. (Redrawn from Evans *et al.*, 1985.)

Muscle Ultrastructure following Eccentric Contraction

What morphologic changes occur in skeletal muscle following such intense eccentric contractions? Jan Fridén, working in Björn Ekblom's laboratory quantified the extent and type of muscle injury that occurred in humans following "model" eccentric contractions (Fridén *et al.,* 1981 and 1983). Subjects were asked to pedal *against* a motor-driven ergometer that was moving in the opposite direction to their applied force (Figure 6.14). Subjects generated extremely high power levels for 30 minutes at 80%–100% of their Vo_2max! Immediately and 3 and 6 days postexercise, small biopsies were taken from their vastus lateralis muscles. The most striking changes observed in the muscles were the disorganization of the myofibrillar material, especially at the Z-disk (Figure 6.15). The nature of the disruption was relatively focal, often extending only a few sarcomeres. This myofibrillar disruption was accompanied by breakage of the myofibrillar cytoskeleton as evidenced by significant redistribution of proteins associated with the cytoskeleton such as vimentin, laminin, and desmin. Fridén *et al.* demonstrated that the ultrastructural disruption was only observed after eccentric contraction and, again, the magnitude of the disruption was greatly attenuated if the subjects had been previously trained.

When isokinetic strength was measured from subjects following eccentric exercise, a small but significant decrease in isokinetic torque was observed at high angular velocities even 6 days following the exercise bout. These investigators cautiously interpreted their data as indicating preferential damage to the fast muscle fibers.

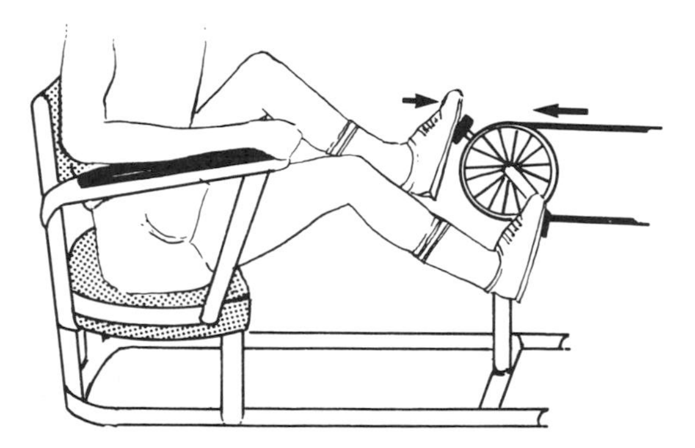

Figure 6.14. Method for inducing eccentric muscle contractions in humans. (Redrawn from Fridén *et al.,* 1983.)

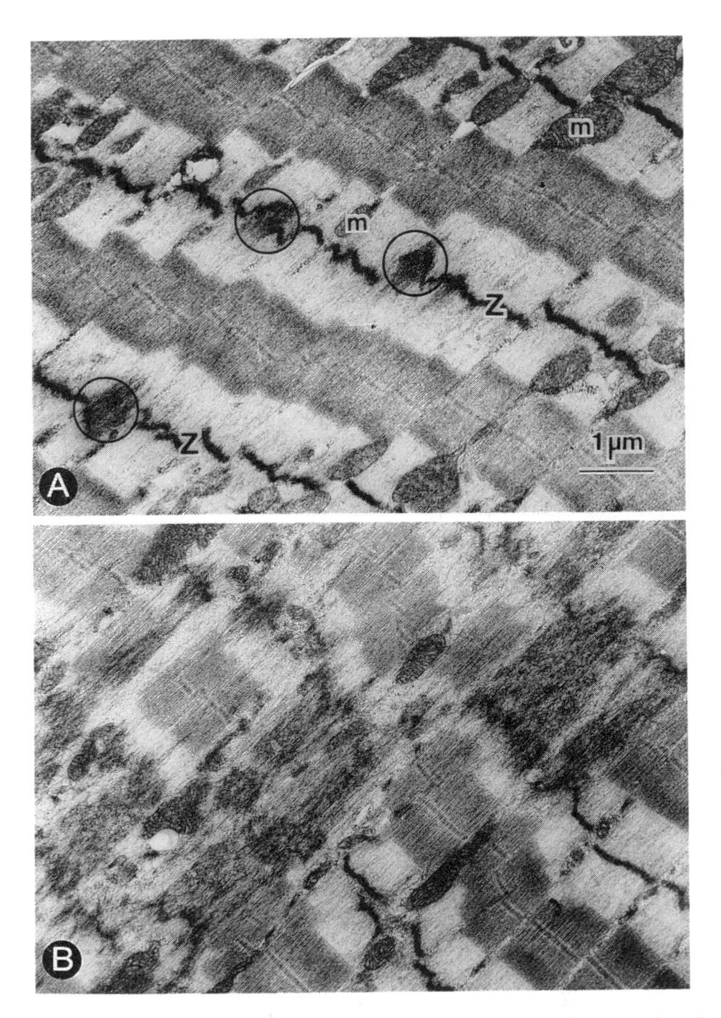

Figure 6.15. Ultrastructural changes in human muscle following eccentric contraction. *Z* = Z-disk. *m* = mitochondrion. (Micrograph courtesy of Dr. Jan Fridén.)

Many questions regarding eccentric contraction–induced exercise remain. What are the cellular signals that initiate the damage process? What can be done to prevent the muscle damage? What can be done to facilitate recovery following muscle damage? How often should repeat exercise bouts be experienced in order to maximally strengthen the muscle but so as not to cause excessive damage?

As we have seen throughout this text, significant advances in understanding

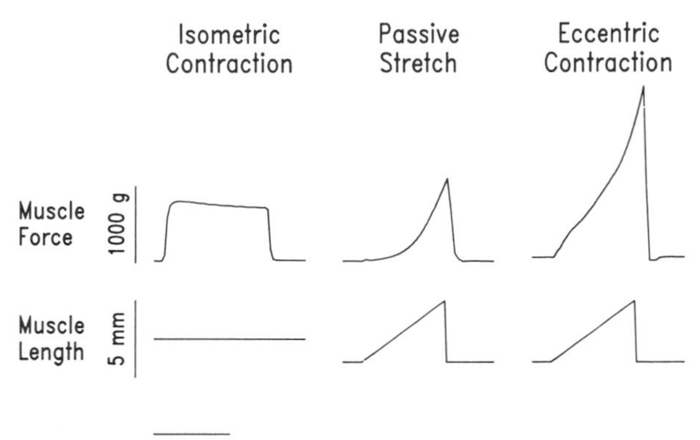

Figure 6.16. Tension comparison under different conditions for the rabbit tibialis anterior muscle. Contractile records from isometric contraction (IC) (*left panel*), passive stretch (PS) (*middle panel*), and eccentric contraction (EC) (*right panel*). The upper trace depicts muscle force; the lower trace depicts muscle length. In the IC panel, muscle length was held constant while the muscle was stimulated for 400 msec at 40 Hz, resulting in a tension equivalent to about 50% P_O. In panel PS, the unstimulated muscle was stretched 25% of the muscle fiber length in 400 msec (corresponding to a strain rate of 63% per second). In panel EC, the stimulation pattern from IC and the length change from PS were superimposed to yield the eccentric contraction shown. Note that tension was greatest for EC followed by PS and IC. Muscle stimulation coincided with lengthening. (From Lieber RL, Fridén JO, McKee-Woodburn TG. Muscle damage induced by eccentric contractions of twenty-five percent strain. J Appl Physiol 1991;70:2498–2507.)

are only made using well-designed and carefully planned experiments. When mechanism questions are addressed, excellent animal models are required to provide unequivocal answers.

Recall that in isolated skeletal muscle, a muscle that is eccentrically activated generates more tension than a muscle that is isometrically activated (Figure 6.16). Muscle can also generate relatively high tensions even if it is only passively stretched. These isolated contractile events can serve as models for eccentric contraction–induced injury studies.

Eccentric Exercise of Isolated Muscles

John Faulkner and his colleagues were the first to induce eccentric contractions directly in isolated animal muscles to investigate muscle function and the cellular response to injury (McKully and Faulkner, 1985). They attached the distal portion of the mouse EDL muscle to a specially designed motor that could forcibly lengthen the muscle by a controlled amount. In this way, they exercised animals eccentrically, isometrically, and isotonically, and compared the muscular response.

Force Changes following Eccentric Exercise

McKully and Faulkner found that following a single 30-minute bout of "exercise," P_O decreased the most when the exercise was eccentric compared to isometric or isotonic (Figure 6.17). The decrease in P_O after eccentric contraction was greatest 5 days postexercise and recovered to control levels after about 30 days. During this time period, significant cellular infiltration was observed—again only in the eccentrically exercised muscles. Thus while all groups decreased in P_O, only the eccentric group mounted the degeneration-regeneration response.

We repeated the type of experiment performed by Faulkner's group in order to investigate the fiber type-specific effects of eccentric contraction (Lieber and Fridén 1988, and Lieber *et al.,* 1991). Again, three groups of experimental animals were used: eccentric contractions (EC), isometric contractions (IC), and passive stretch (PS; Figure 6.16). This ensured that stretch itself did not damage the muscle. Muscles were "exercised" in one of the three modes for 30 minutes, and muscle contractile properties were again

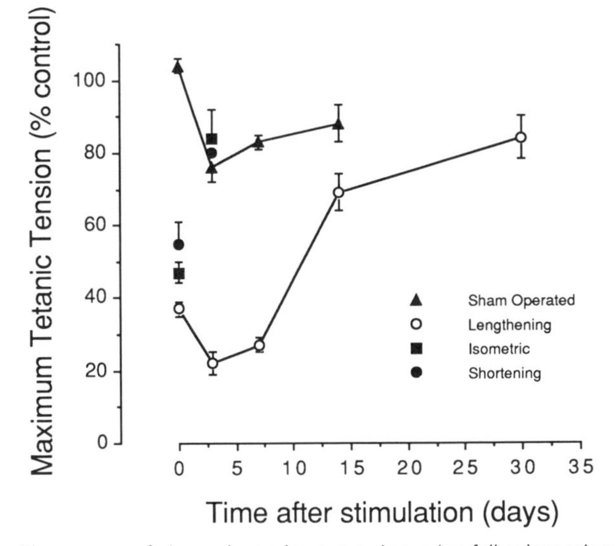

Figure 6.17. Time course of change in maximum tetanic tension following various experimental treatments. (From McCully and Faulkner, 1985.)

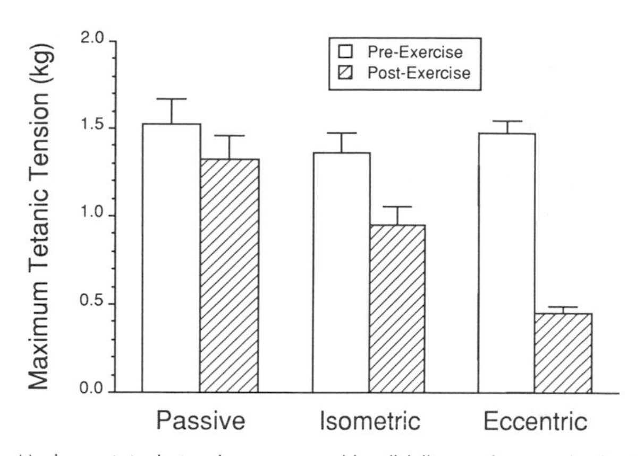

Figure 6.18. Maximum tetanic tension generated by tibialis anterior muscles before (*open bars*) and after (*hatched bars*) one of the three experimental treatments. P_O decreased following all treatments. However, the magnitude of the decrease was greater for EC, followed by IC, followed by PS. (Data from Lieber *et al.*, 1991).

measured. We will discuss contractile and morphologic results serially in order to build a case for a putative damage mechanism.

We found, as expected, that after 30 minutes of EC, IC, or PS exercise, P_O decreased to the greatest extent in the EC group compared to either the IC or PS group (Figure 6.18). In addition, the tension decrease was accompanied by a muscle "slowing" as evidenced by a significant decrease in twitch and tetanic rate of change in tension (dP/dt).

Time Course of Tension Change during Exercise

For all three treatments, peak tension decreased monotonically with time (Figure 6.19). No abrupt drop in tension was observed for EC treatment, which would have suggested that the damage had occurred as a discrete, coordinated "event" or tear along the length and width of the muscle.

In order to determine whether the time course of "injury" could be estimated during the exercise treatment period itself, we defined an added component of tension due to EC (Figure 6.19, dotted line). The rationale was that EC represents a sort of summation of PS and IC induced simultaneously. This added tension, denoted P_{ADD}, was calculated as the difference between the peak EC tension level (P_{EC}) and the sum of the peak IC tension (P_{IC}) and peak PS tension (P_{PS})—that is, $P_{ADD} = P_{EC} - (P_{IC} + P_{PS})$. Notice that early in the treatment period, P_{ADD} exceeded 500 g, but decayed relatively rapidly and became zero after about 7 minutes. After this, it actually became slightly

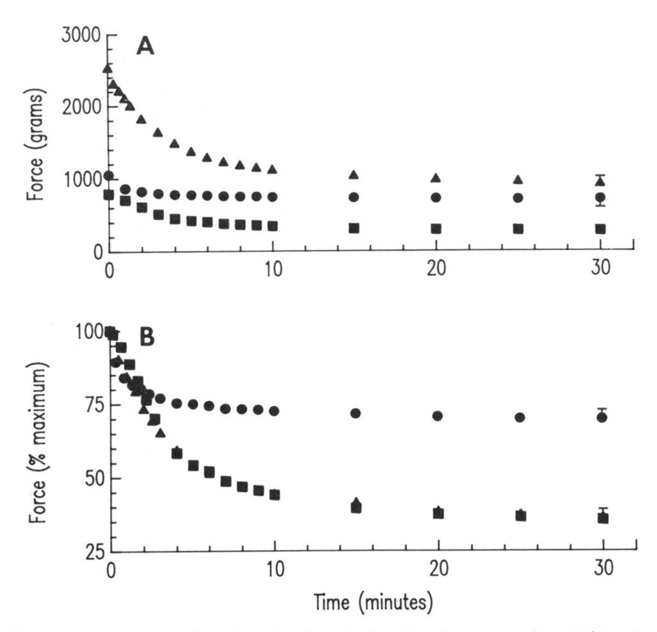

Figure 6.19. **A,** Time course of tension decline during the three experimental treatments. In each case, data were acquired on-line during 30 minutes of either EC (*triangles*), IC (*squares*), or PS (*circles*). Note that EC resulted in the greatest tension. The dotted trace represents the parameter P_{ADD}, the addition tension that is calculated as the difference between the EC tension and the sum of the IC and PS tensions. See text for explanation of the significance of P_{ADD}. Mean ± SEM is shown only for the last symbol for clarity. (Data from Lieber *et al.*, 1991.)

negative. Thus early in the treatment period the muscle experienced 500 g more tension than it would have experienced as the simple algebraic sum of IC and PS. The significance of this finding is addressed below when we discuss a potential damage mechanism.

Muscle Morphological Changes Following Eccentric Exercise

In an effort to understand the basis for the contractile results, we looked at the morphology of the muscles at both the light microscopic and electron microscopic levels. While the morphology of samples from the PS and IC groups appeared normal, the most obvious result was that the eccentrically exercised muscles exhibited increased portions of abnormal fibers when viewed in cross-section. These fibers appeared rounded, more lightly stained by hematoxylin and eosin, and approximately four times the normal size (Figure 6.20). Only fibers from muscles in the EC group demonstrated this

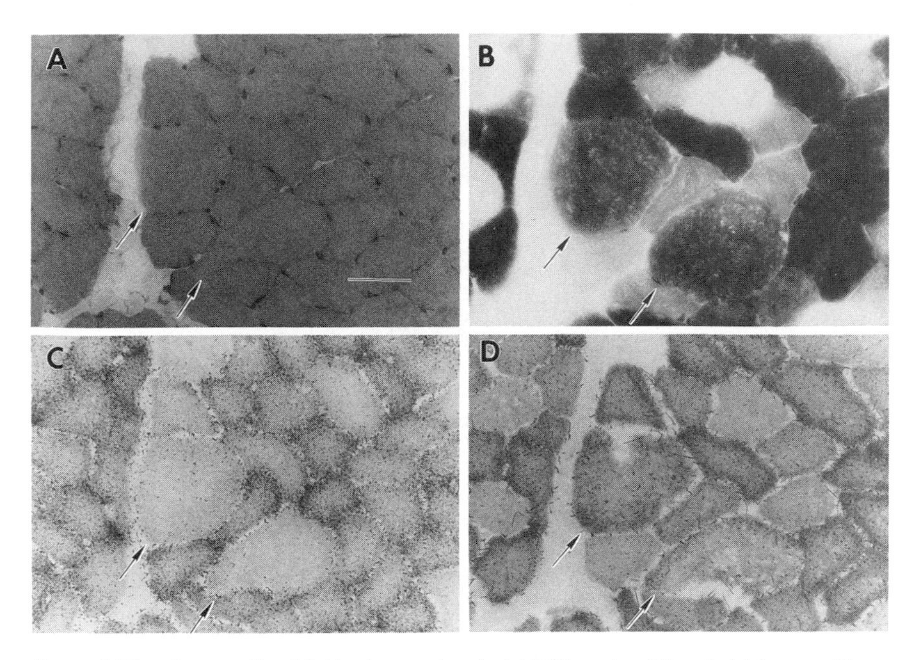

Figure 6.20. Cross-sectional light micrographs of rabbit TA under different staining conditions. Arrows point to enlarged fibers, which are of the FG fiber type. Calibration bar = 50 μm. **A,** Hematoxylin and eosin. **B,** myofibrillar ATPase following preincubation at pH = 9.4. **C,** Succinate dehydrogenase. **D,** α-glycerophosphate dehydrogenase. (Data from Lieber and Fridén, 1988.)

abnormal appearance, and they were always depleted of glycogen, confirming that they had been activated.

What fiber type were these enlarged fibers? Fiber type was determined by staining serial sections, as described in Chapter 2, for myofibrillar adenosine triphosphatase (ATPase) activity, succinate dehydrogenase (SDH) activity, and α-glycerophosphate dehydrogenase (αGP) activity. We found that all enlarged fibers were exclusively of the fast glycolytic (FG) fiber type (Figure 6.20)! What do you think this means? Our impression was that it provided insights into the damage mechanism, as described below.

While no ultrastructural abnormalities were observed in any of the muscles from the IC or PS groups, a significant portion of the fibers in the EC group displayed various degrees of disorganization of the sarcomeric band pattern such as Fridén *et al.* (1983) had observed in human eccentric exercise. Streaming and smearing of the Z-disk material, focal loss of Z-disks, and extension of Z-disks into adjacent A-bands were commonly seen (Figure 6.21). The Z-disk smearing was always located in specific locations within the same

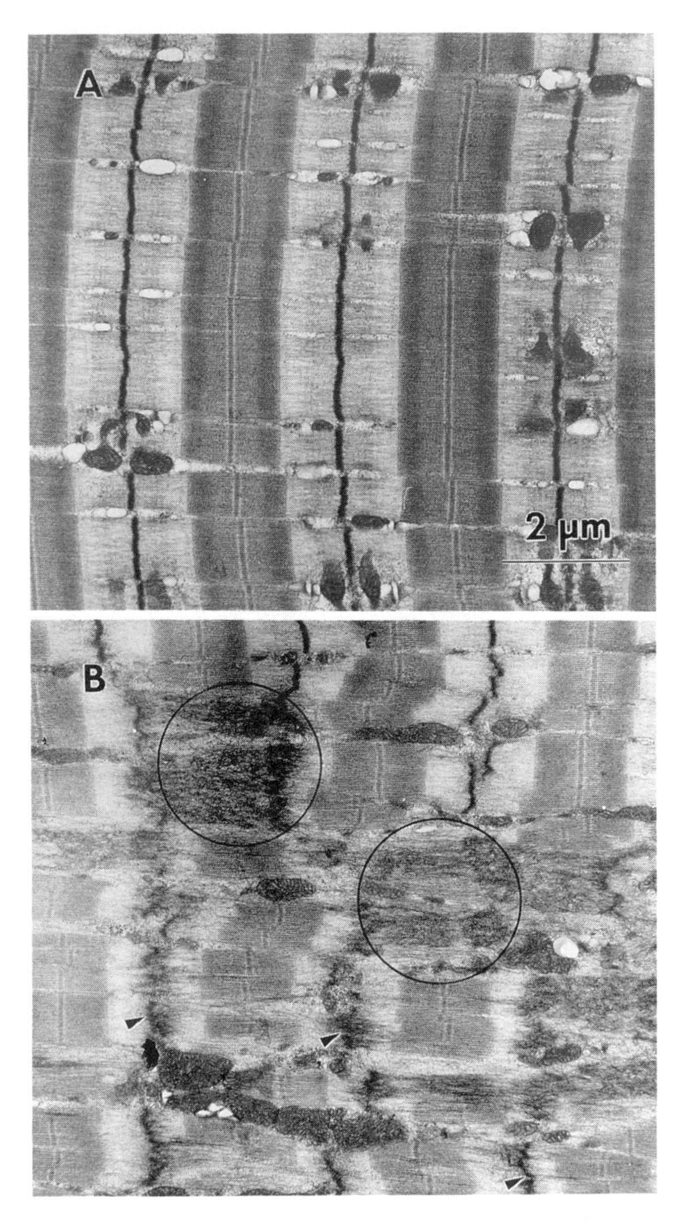

Figure 6.21. Longitudinal electron micrographs of rabbit TA. **A,** Sample from IC group, which shows normal striation pattern and Z-disks perpendicular to the long myofibrillar axis. **B,** Sample from EC group, which demonstrates various disrupted regions. Note streaming and smearing of the Z-disk material (*arrowheads*) and extension of the Z-disks into adjacent A-bands (*circled areas*). (From Lieber RL, Fridén JO, McKee-Woodburn TG. Muscle damage induced by eccentric contractions of twenty-five percent strain. J Appl Physiol 1991;70:2498–2507.)

fiber and never extended across the whole fiber (and infrequently over more than three sarcomeres). In these areas, the thick filaments were out of register, overlapping one another and/or the I-bands. Fine structural deviations appeared randomly distributed across and along the fibers.

Putative Mechanism of Eccentric Contraction-Induced Damage

The fact that only FG fibers demonstrated histologic abnormalities suggested that fiber oxidative capacity is important in determining the extent of fiber damage that occurs immediately following EC.

If muscle fiber oxidative capacity is a determining factor in fiber damage, we could hypothesize a damage scheme that occurs as follows:

1. Early in the exercise period (say within the first 10 minutes) FG fibers fatigue.
2. Based on their inability to regenerate ATP, they enter a rigor or high-stiffness state.
3. Subsequent stretch of stiff fibers mechanically disrupts the fibers, resulting in the observed cytoskeletal and myofibrillar damage.

This hypothesis is appealing for several reasons. First, it explains the well-known "protective" effect of endurance training on EC-induced damage, mentioned above. Endurance training is known to result in an increased muscle oxidative capacity, and, therefore, FG to fast oxidative glycolytic (FOG) fiber subtype conversion. Because FOG fibers do not fatigue and enter rigor as readily as the FG fibers, EC-induced damage would be expected to be lower following endurance training. This concept may be important in training regimens. An appeal of this hypothesis is that it makes a testable prediction. We predict that the EC-induced damage will occur early in the treatment period—within the first few minutes. Interestingly, this was precisely the time during which P_{ADD} was significant (Figure 6.19). After about 7 minutes, P_{ADD} dropped to zero, implying that the damage was done. We would also predict that a muscle that has fatigued (say due to 20 minutes of IC) could be damaged significantly if the stretch came late in the treatment period. Damage would not necessarily require a large number of stretches. Perhaps this is also why many muscular injuries occur after muscles have been previously fatigued. Whether these hypotheses hold true remains to be tested. It may be possible to use external force or torque measurements to determine the time period during which muscles are vulnerable (or not vulnerable) to eccentric exercise–induced muscle damage.

A second damage mechanism that could depend on fiber oxidative capacity relates to the other cellular processes that rely on oxidative metabolism. We could again hypothesize a damage scheme that occurs as follows:

1. Early in the exercise period FG fibers fatigue.
2. Based on their ability to regenerate ATP, mitochondria lose their calcium buffering capacity.
3. Increased intracellular calcium results in activation of the calcium-activated neutral proteases (which we saw were active in denervation), lysosomal proteases, and other cellular processes that are calcium-mediated.

Interestingly, Bob Armstrong's laboratory has recently provided direct evidence of increased mitochondrial calcium concentration in rats trained by prolonged downhill running. They also showed that agents that bound calcium attenuated the magnitude of the injury (Duan *et al.,* 1990). Armstrong has also prevented a review that emphasizes the critical role of intracellular calcium regulation in mediating muscle injury (Armstrong, 1984 and 1990).

Inflammation in Muscle Damage

Our previous muscle injury discussion focused on events that occurred *inside* the cell. However, recent studies suggest that inflammation of muscle tissue from circulating cells *outside* the muscle may significantly affect muscle function. These studies suggest that, while mechanical injury may cause the initial injury, further injury may result from subsequent tissue inflammation. This hypothesis implies that the inflammatory process, which we have seen can repair tissue, may also cause further damage.

This hypothesis was tested directly by Joe Cannon working in Bill Evans' laboratory (Cannon *et al.* 1990, 1991). These investigators measured plasma CK levels in the same way as that described for Figure 6.13. However, they performed these experiments on "young" (<30 years old) and "old" (>55 years old) subjects. To study inflammation, they also treated both groups with large doses of vitamin E, a substance that is known to stabilize cell membranes and protect them against oxidative damage such as occurs during inflammation. The results were surprising, to say the least! Cannon and colleagues observed the typical transient CK response in young individuals, whether or not they had received vitamin E treatment. They also observed an increase in blood levels of inflammatory cells in all young subjects. However, the elderly subjects demonstrated a much different response. The group that did *not* receive vitamin E had a *smaller* CK transient response, similar to that described previously for training (*filled circles,* Figure 6.13). That is, the older, unprotected muscle demonstrated a smaller CK response. Since older muscles are known to be more vulnerable to exercise-induced injury, the simplest interpretation of these data is that CK levels alone are not an exact measure of muscle damage. This was confirmed by inspection of the CK data from the older subjects who did receive vitamin E. Their CK response looked more like that seen for young, untrained subjects (*open circles,* Figure 6.13). What do we conclude about CK levels and muscle injury? The definitive answer is not yet available, but we must acknowledge that, while increased CK levels are *associated* with muscle injury, they are not necessarily the direct result of the injury. Perhaps increased CK levels represent muscle remodeling which must occur following injury.

In terms of circulating inflammatory cells, the older subjects again demonstrated an unexpected response: the group that did *not* receive vitamin E demonstrated an attenuated level of circulating neutrophils which was "increased to normal" by pretreatment with vitamin E. Experiments such as these will have profound effects on our strategies for treating skeletal muscle injuries. It goes without saying that a rational basis for the prevention of muscle inflammation via the use of anti-inflammatory agents is not yet available.

SUMMARY

In this chapter, we saw the dramatic, dynamic, and impressive response of skeletal muscle fibers to traumatic injury. Skeletal muscle tissue regenerated vigorously following injury. The main player in the regenerative response is the satellite cell, which, under cellular control, proliferates, fuses, and synthesizes proteins to restore fiber function to normal. Such regeneration may someday be exploited in the therapeutic application of myogenic cells to myopathic tissue. Muscle injury during exercise is selectively associated with eccentric contractions. While high tensions are clearly required for muscle damage, the best evidence to date suggests that a disruption of the muscle cell's ability to buffer calcium provides the cellular signal to degenerate and regenerate a new fiber.

REFERENCES

Armstrong, R.B. (1984). Mechanism of exercise-induced delayed onset muscular soreness: a brief review. Med. Sci. Sports Exerc. 16:529–538.

Armstrong, R.B. (1990) Initial events in exercise-induced muscular injury. Med. Sci. Sports Exerc. 22:429–435.

Bischoff, R. (1986). A satellite cell mitogen from crushed adult muscle. Dev. Biol. 115:140–147.

Bischoff, R. (1990). Interaction between satellite cells and skeletal muscle fibers. Development 109:943–952.

Cannon, J.G., Orencole, S.F., Fielding, R.A., Meydani, M., Meydani, S.N., Fiatarone, M.A., Blumberg, J.B., and Evans, W.J. (1990). Acute phase response in exercise: interaction of age and vitamin E on neutrophils and muscle enzyme release. Am. J. Physiol. 259:R1214–R1219.

Cannon, J.G., Meydani, S.N., Fielding, R.A., Fiatarone, M.A., Meydani, M., Farhangmehr, M., Orencole, S.F., Blumberg, J.B., and Evans, W.J. (1991). Acute phase response in exercise. II. Associations between vitamin E, cytokines, and muscle proteolysis. Am. J. Physiol. 260:R1235–R1240.

Carlson, B. (1973). The regeneration of skeletal muscle: a review. Am. J. Anat. 137:119–150.

Carpenter, S., and Karpati, G. (1984). Pathology of Skeletal Muscle. New York: Churchill Livingstone.

Donovan, C.M., and Faulkner, J.A. (1987). Plasticity of skeletal muscle: regenerating fibers adapt more rapidly than surviving fibers. J. Appl. Physiol. 62:2507–2511.

Duan, C., Delp, M.D., Hayes, D.A., Delp, P.D, and Armstrong, R.B. (1990). Rat skeletal muscle mitochondrial [Ca 2^{+}] and injury from downhill walking. J. Appl. Physiol 68:1241–1251.

Evans, W.J., Meredith, C.N., Cannon, J.G., Dinarello, C.A, Frontera, W.R., Hughes, V.A., Jones, B.H., and Knuttgen, H.G. (1985). Metabolic changes following eccentric exercise in trained and untrained men. J. Appl. Physiol. 61:1864–1868.

Faulkner, J.A., Niemeyer, J.H., Maxwell, C., and White, T.P. (1980). Contractile properties of transplanted extensor digitorum longus muscles of cats. Am. J. Physiol. 238:C120–C126.

Fridén, J., Sjöström, M., and Ekblom, B. (1981). A morphological study delayed muscle soreness. Experientia 37:506–507.

Fridén, J., Sjöström, M., and Ekblom, B., (1983). Myofibrillar damage following intense eccentric exercise in man. Int. J. Sports Med. 4:170–176.

Gulati, A.K., Reddi, A.H., and Zalewski, A.A. (1983). Changes in the basement membrane zone components during skeletal muscle fiber degeneration and regeneration. J. Cell. Biol. 97:957–962.

Harry, J.D., Ward, A.W., Heglund, N.C., Morgan, D.L., and McMahon, T.A. (1989). Crossbridge cycling theories cannot explain high speed lengthening behavior in frog muscle. Biophys. J. 57: 201–208.

Huard, J., Labrecque, C., Dansereau, G., Robitaille, L., and Tremblay, J.P. (1991). Dystrophin expression in myotubes formed by the fusion of normal and dystrophic myoblasts. Muscle Nerve. 14:178–182.

Karpati, G., Pouliot, Y., Stirling, C., and Holland, P. (1989). Implantation of nondystrophic allogenic myoblasts into dystrophic muscles of mdx mice produces "mosaic" fibers of normal microscopic phenotype. In: Kedes, L.H., and Stockdale, F.E., eds. Cellular and Molecular Biology of Muscle Development. New York: Alan R. Liss, pp. 973–985.

Lieber, R.L., and Fridén, J. (1988). Selective damage of fast glycolytic muscle fibers with eccentric contraction of the rabbit tibialis anterior. Acta Physiol. Scand. 133:587–588.

Lieber, R.L., Fridén, J.O., and McKee-Woodburn, T.G. (1991). Muscle damage induced by eccentric contractions of twenty-five percent strain. J. Appl. Physiol. 70:2498–2507.

Mauro, A. (1961). Satellite cells of skeletal muscle fibers. J. Biophys. Biochem. Cytol. 9:493–495.

Mauro, A., Shabiff, S.A., and Millorat, A.T., eds. (1970). Regeneration of striated muscle and myogenesis. Amsterdam: Excerpta Medica.

McCully, K.K., and Faulkner, J.A. (1985). Injury to skeletal muscle fibers of mice following lengthening contractions. J. Appl. Physiol. 59:119–126.

Morgan, D.L. (1989). New insights into the behavior of muscle during active lengthening. Biophys. J. 57:209–221.

Ohlendieck, K., Ervasti, J.M., Snook, J.B., and Campbell, K.P. (1991). Dystrophin-glycoprotein complex is highly enriched in isolated skeletal muscle sarcolemma. J. Cell. Biol. 112:135–148.

Partridge, T.A., Morgan, J.E., Coulton, G.R., Hoffman, E.P., and Kunkel, L.M. (1989). Conversion of mdx myofibres from dystrophin-negative to -positive by injection of normal myoblasts. Nature. 337:176–179.

Snow, M.H. (1976). Myogenic cell formation in regenerating rat skeletal muscle injured by mincing. Anat. Rec. 188:201–217.

Vandenburg, H.H. (1982). Dynamic mechanical orientation of skeletal myofibers in vitro. Dev. Biol. 93:438–443.

Index